# THE BIOLOGY OF LIFE SPAN

# THE BIOLOGY OF LIFE SPAN

## A Quantitative Approach

**Leonid A. Gavrilov**
*A.N. Belozerskii Laboratory of Molecular Biology and Bioorganic Chemistry, Moscow State University, USSR*

and

**Natalia S. Gavrilova**
*All-Union Institute for Medical Information, Moscow, USSR*

Revised and Updated
English Edition

*Edited by*
*Academician V.P. Skulachev*

*Translated from the Russian by*
*John and Liliya Payne*

harwood academic publishers
chur • london • paris • new york • melbourne

**Harwood Academic Publishers**

Post Office Box 197
London WC2E 9PX
United Kingdom

58, rue Lhomond
75005 Paris
France

Post Office Box 786
Cooper Station
New York, New York 10276
United States of America

Private Bag 8
Camberwell, Victoria 3124
Australia

Originally published in Russian in 1986 as Биология продолжительности жизни : количественные аспекты (Biologiia prodolzhitelnosti zhizni : kalichestvenniye aspekty) by Nauka, Moscow


**Library of Congress Cataloging-in-Publication Data**

Gavrilov, L. A. (Leonid Anatol'evich)
[Biologiia prodolzhitel'nosti zhizni. English]
The biology of life span : a quantitative approach / by Leonid A. Gavrilov and Natalia S. Gavrilova ; translated from the Russian by John and Liliya Payne.
p.cm.
Translation of: Biologiia prodolzhitel'nosti zhizni.
Includes bibliographical references and index.
ISBN 3-7186-4983-7 (hardcover).
1. Longevity--Physiological aspects. I. Gavrilova, N. S. (Natal'ia Sergeevna)
QP85.G3313 1991
612.6'8--dc20 90-26125
CIP

## CONTENTS

# PREFACE

The aim of the present book is to bring together the most important facts and ideas connected with the observation that organisms have a limited life span. An interest in this problem is just as natural to everyone as the desire for life itself. For this reason, the book is intended to be read by a very wide circle of readers. However it is specifically designed for those who, having satisfied their initial curiosity, wish to participate in the advancement of this scientific discipline which is of great importance to humanity. For such readers, the book is intended to serve both as an introduction to the field of their future research and as a source of essential data. Finally, investigators in a variety of disciplines (biologists, gerontologists, physicians, demographers, mathematicians etc.) may be interested in the book because of its broad interdisciplinary approach to the problem of life span, and be stimulated to participate in a scientific debate on the wide range of issues which it raises.

While working on this book, we have tried to show that many of the fundamental problems of the biology of life span can be successfully resolved on the basis of factual information which is already available at the present time, and with the use of very modest techniques. What is more, in the course of our own investigations we have been surprised to discover that some of the main issues which are still hotly debated could have been resolved some thirty years ago! All that was required was to take a somewhat wider view of the problem, and to undertake an analysis of the demographic tables on human life span. Excessive specialisation and reductionism led to this highly important information not being properly evaluated by biologists. The result was that anthropological investigations suffered inordinately from a scarcity of factual data, and were consequently bedevilled with prejudices, unfounded speculations and the repetition of banal truths. In this book, we have attempted to correct the situation by showing how a quantitative analysis of traditional demographic life tables can lead to non-trivial results which make biological sense. Carrying out this research has convinced us that it is not a simple task to find the necessary life tables: we were often forced to apply to the statistical bureaux of various individual countries. In order to simplify the search for other future investigators, we have therefore compiled and published a bibliography with a selection of complete human life tables (Gavrilova et al., 1983). It was also necessary to collate the widely dispersed information on the life span of other organisms. This material was scattered throughout hundreds of books and articles, many of which seem to be unknown even to the specialists in the field. The present book therefore includes an extensive bibliography of publications which contain the most complete and reliable life tables.

The contents of the book are presented in an order which follows the logic of contemporary research in the field. This research basically takes an inductive path. Firstly, the initial data on life span are summarised in the form of life tables. Then on the basis of these tables regularities are discovered and working hypotheses are formulated. Finally, inductive generalisation reaches its highest point in the construction of a mathematical model which provides a quantitative explanation for the observed regularities. It is at this stage that the foundations for a theory of life span can be created, and at the same point that a number of problems remain unsolved. Whilst working on the book, we have tried not only to provide a comprehensive review of the problem, but also to subject it to logical and systematic scrutiny. It is therefore best to start reading the book at the beginning, not omitting the elementary mathematics which is essential to an understanding of the subsequent chapters.

We will be grateful to our readers for any questions, observations and suggestions connected with the book. These should be sent to the following address: Dr. Leonid A.Gavrilov, Ph.D., A.N.Belozersky Laboratory of Molecular Biology and Bioorganic Chemistry, Moscow State University, Moscow, 119899 USSR.

Moscow, December 1990

L.A.Gavrilov

N.S.Gavrilova

## ACKNOWLEDGEMENTS

We are indebted to approximately one hundred colleagues from many countries who have supplied us with life span data and their recent publications, as well as helpful comments and criticisms of our studies.

We especially want to acknowledge our gratitude to the following colleagues who supplied us with very important life tables: Mr. Åke Nilsson, senior vital statistician in the Population Unit, and Mr. Nils Welander in the Statistical Information Unit, Sweden; Mr. Jens-Kristian Borgan and Mr. Gerd Skoe Lettenstrom from the Central Statistical Bureau in Norway; Mr. Risto Kolari, senior statistician from the Central Statistical Office of Finland; Mr. Ole Kaysen Petersen from the Central Statistical Office of Denmark; Dr. Eugenio Fornas from Spain; Mr. Yoel Floresheim from the Demography Division of the Central Statistical Bureau in Israel and Dr. Uri Goldbourt, Head of the Epidemiology and Biostatistics Section in the Chaim Sheba Medical Centre, Israel; Dr. Shigezaku Hishinuma, President of the Japanese Institute of Gerontology; Dr. Herman Le Compte, President of the International Association on the Artificial Prolongation of the Human Specific Lifespan, Belgium; Dr. Michel A. Hofman from the Netherlands Institute for Brain Research; Dr. A. Lery from the Institut National de la Statistique et des Études Économiques, France, and Dr. Graeme Caughley, senior principal research scientist from the Division of Wildlife and Rangelands Research, Australia.

We are grateful to Professor Nathan Keyfitz from the International Institute for Applied Systems Analysis in Austria for stimulating discussions. We thank Dr. Ansley J. Coale from Princeton University, Dr. Roy L. Walford from the University of California, Professor Max A. Woodbury and Professor Kenneth G. Manton from Duke University and Professor Jean Bourgeois-Pichat, France, for their interest, advice and constructive criticism of our papers.

We are also grateful to numerous colleagues in the USSR, particular Professor V.N.Maksimov and Dr. V.N.Nosov from Moscow State University, Dr. V.A.Borisov from the Institute for Sociological Studies and Z.B.Mazurkevich, for the pleasure of shared knowledge, expertise and enthusiasm.

The publication of this book was made possible by the strong support of Professor Vladimir P. Skulachev, USSR, and great technical assistance from Victoria G. Semyonova, USSR.

# CHAPTER 1. INTRODUCTION

## 1.1. Why is it necessary to study life span?

The biology of life span is the science of the mechanisms which determine the life span of organisms. Amongst its most interesting problems it is possible to cite the problem of the heritability and variability of lifetimes (Gavrilov, 1980; Jacquard, 1982), the problem of sex differentials in lifetimes (Hazzard, 1986), and the problem of the changing life span of organisms in the process of evolution (Malinovsky, 1983; Cutler, 1984a). Thus one of the tasks of the biology of life span is to explain the causes of individual differences in lifetimes, as well as the causes of interpopulation and interspecies differences. The practical interest of these investigations is to open up the possibility of predicting and controlling the lifetimes of organisms, and most importantly, to discover ways of extending the lives of human beings (Gavrilov, 1990).

The biology of life span was formed as an independent scientific discipline at the beginning of the century, thanks to the classic investigations of the well-known American scientist Raymond Pearl (Pearl, 1922; 1931; Pearl and Milner, 1935; 1936; 1941; Pearl et al., 1927; Pearl and Parker, 1921; 1922a; 1922b; 1922c; 1922d; 1924a; 1924b; Pearl et al., 1923; Pearl and Pearl, 1934). The term "biology of life span" itself seems to have occurred for the first time in a work by one of Raymond Pearl's students, the Russian professor Vladimir V.Alpatov (Alpatov and Gordeenko, 1932).

Because the study of the mechanisms which determine the length of life is closely linked with investigations into the processes of aging, the biology of life span owed its subsequent development to the biology of aging and gerontology. It is sufficient to note that such universally known gerontologists as Alex Comfort, Bernard L.Strehler and George Sacher also made an invaluable contribution to the study of life span (Comfort, 1956; 1957a; 1957b; 1958; 1963; Strehler, 1961; Sacher, 1977). It is also indicative that, in the USSR, the biology of life span subsection (chairman - Dr.L.A.Gavrilov, Ph.D.) of the Moscow Society of Naturalists was founded in 1981 on the basis of the gerontology section of the same Society. At the present time, the ideas and methods of the biology of life span are so widely used in gerontology that some gerontologists go so far as to regard the biology of life span as one of the branches of their own discipline.

This view does not seem to us, however, to be fully justified. It is a fact that the biology of life span, in distinction to gerontology, is just as much interested in the mechanisms which determine life spans in wild animal populations (which are of great importance for ecology and the theory of evolution), and life spans in extreme situations (which are of special interest also for toxicology and radiobiology). What is more, the biology of life span, in distinction to gerontology, attaches great significance to the biological mechanisms which permit a certain proportion of organisms to perish at an early stage in their development, long before the first signs of aging appear. Thus although the biology of life span is close to the biology of aging, it nevertheless has its own specific goals and cannot be reduced to this latter science. What is more, as will be demonstrated below, the biology of life span has its own historical roots which connect it with demography and population biology. The result is that the biology of life span has developed its own characteristic style of research: the use of precise quantitative methods, a probabilistic approach to natural phenomena, and a desire to explain the mechanisms of particular processes by their external manifestations in the population under investigation. Thus from the point of view of its methodology, the biology of life span seems to be close to such disciplines as quantitative genetics or the kinetics of biological processes.

Since the end of the 1970s, there has been a noticeable rise in interest towards the problems of the biology of life span. One particular manifestation of this is the sharp increase in the number of research groups which are working in the field, and a corresponding increase in the number of publications (Anisimov et al., 1988; Blokhinov, 1982; Gavrilov et al., 1983; Cutler, 1984; Economos, 1980a; 1980b; 1983; 1985).

This phenomenon would appear to reflect a general tendency in modern biology: the tendency to study not only the spatial, but also the temporal organisation of living organisms. It is a fact that the basic achievements of biology were until very recently connected with an account of the spatial organisation of living matter - from molecules and cells to the organism and the biosphere.

Against the background of these achievements, the gap in our knowledge of the temporal organisation of living matter became more and more obvious. But how is it possible to understand what life is without explaining why it is limited in time and what these limits are determined by?! This has to be a fundamental problem of natural science, one which will play a key role in our whole scientific world-view! It was inevitable that the temporal organisation of living matter would come to occupy a worthy place along with the other fundamental scientific problems, and this is indeed now taking place. The transition to the study of temporal organisation in modern biology has led not only to a revival in the biology of life span, but also to the accelerated investigation of biological growth, biological clocks and biological rhythms, and to the intensive development of disciplines such as biokinetics and chronobiology.

This inevitable expansion of research into the biology of life span is also connected with the fact that experiments in life extension have now stopped being unrealistic plans and have become a whole new field of scientific research (Emanuel, 1982; Nikitin, 1984; Frolkis and Muradian, 1988). For example, a special research program called "The Extension of Life", the first in the history of world science, has been created in the USSR and is now in full progress (Chebotarev, 1979; 1986; 1987). This multi-disciplinary program, supported by the USSR Ministry of Health and combining the efforts of dozens of teams from the country's medical and academic research institutes, aims "to experimentally develop new approaches to life extension, and to test the ones which are possible on human beings" (Chebotarev, 1979, p.8). Thus what was once considered science fiction is now treated as an executable scientific project. The completion of this project would open up a new era in the history of mankind - the mastery not only of space (including cosmic space), but also of time. It is difficult to overestimate the historical significance of this project and the role of the biology of life span in its fulfilment.

Of course, a really important project will inevitably provoke arguments, doubts and fears about its possible negative consequences in the event that it is successful. In this regard, the life extension project is no exception. For example, the fear is sometimes expressed that the complex ecological situation on our planet will deteriorate even more rapidly. In the opinion of some opponents, an increase in life span will accelerate the undesirable growth of the world population, and an increase in the proportion of older people who are incapable of work and need constant help will compound this "demographic explosion". Some opponents also believe that the succession of generations which is essential for social and biological progress will slow down, and may even degenerate.

Since a discussion of the moral, ethical, socio-economic and political aspects of life extension lies outside the scope of this book, we will not consider these aspects in any detail, instead recommending the reader to consult the specialist literature on the topic. With this goal in mind, we can recommend a book by the distinguished Polish demographer Edward Rosset, The Span of Human Life, one of whose chapters is entitled "What is the purpose of the struggle for the extension of human life?" (Rosset, 1979). Since the main focus of attention in Edward Rosset's book is a discussion of the humanistic and moral aspects of life extension, it can well be supplemented with a book by the American authors Joel Kurtzman and Phillip Gordon, No More Dying: The Conquest of Aging and Extension of Human Life (Kurzman and Gordon, 1976). This deals with some other aspects of the problem (see especially the chapter entitled "Is it really necessary?").

At the same time, it seems desirable to answer some of the objections which arise from a misunderstanding of the contemporary demographic situation. The authors of the present book have often been faced with the opinion that "there are more than enough people already, so what is the point of an increased life span?". However if we consult the demographic statistics, it turns out that in the developed countries, where the possibility of a radical prolongation of life might be expected to arise first, there is no undesirable population growth. On the contrary, many states have been forced to expend significant sums on increasing the birth-rate in order to avoid a drop in population. Amongst these countries we can cite France, Poland, Czechoslovakia, Romania, Hungary and Bulgaria (Bruk, 1981). A negative natural population growth is now observable in Austria, Denmark, West Germany, Hungary, Luxemburg and Sweden (World Health Statistics Annual, 1987). In East Germany the number of deaths has begun to exceed the number of births since 1972, and from 1950 to 1978 the population fell by 9% (Bruk, 1981). At the present time, the list of such examples might be extended.

An analogous phenomenon can also be observed in some regions of the USSR. As a result of the death-rate overtaking the birth-rate, a process of natural population decrease has begun in the Kalinin, Pskov, Novgorod, Tula, Yaroslavl, Kostroma and other districts and regions of the country (Bednyi, 1979; Goskomstat SSSR, 1989).

Thus the notion that the developed countries are overpopulated is fallacious and connected with excessive concentrations of population in particular areas (megalopolises), i.e. with the problem of the optimal distribution of population.

Another objection is that an increase in life span has already led to a significant growth in the proportion of chronically ill and economically inactive pensioners, therefore life prolongation will only result in more acute economic problems. It is difficult however to agree with this view. It is true that we have observed an aging of the population in the developed countries (a growth in the proportion of elderly people in the community), but up to very recently this has not been brought about by increased longevity. Instead, it has been caused exclusively by a reduction in the birth-rate! This fact was reliably established by demographers a long time ago in collaboration with the UN Population Commission (Sauvy, 1954). What is more, however paradoxical it might sound, the increase in life span in the 20th century has on the contrary prevented the aging of the population (this is connected with the fact that the growth in life span has basically occurred because of a reduction in mortality at early ages) (Sauvy, 1954).

Of course, a further increase in life span will lead to an increase in the proportion of formally old people. If however this prolongation of life is achieved by slowing down the process of aging (which is what is intended), the proportion of ill and helpless people might even decrease.

To conclude this discussion of the problem of life prolongation, we would like now to express our own opinion. It seems to us that the research into life prolongation must be pursued, but at the same time the possible consequences of a significant increase in human life span should be investigated. The question whether it is *necessary* to prolong life will simply disappear when it becomes clear how this *can* be achieved. Indeed, if those who object in principle to life prolongation are forced to support their position by their own example, and the remaining population lives much longer than they do, then discussions about this question will quickly die down because of the small numbers of the opposition! In fact, rejecting the means of prolonging life will at that time seem as absurd as it would seem now to commit passive suicide by rejecting modern and effective methods of treating infection. It is inevitable that the means for prolonging life will be developed and practised, since the struggle for the prolongation of life is the struggle for the lives of human beings. The Western reader may be interested to learn that this view conforms even with the Marxist conception of the future of mankind, according to which the value of human life is self-evident and self-sufficient (Frolov, 1983).

Returning to the biology of life span, we would like to note that the aims of the investigation do not reduce to the problem of the prolongation of human life, however significant and compelling this might be. We have already noted the importance of the biology of life span in investigating the fundamental problems of biology (explaining the temporal organisation of living matter). Apart from this, research into the biology of life span may already be of use in solving a wide range of everyday practical problems.

For example, in the study and forecasting of life span in demography, there have long been attempts to find a biological standard of longevity against which the direction and magnitude of the influence of different social factors might be measured (Urlanis, 1978). A similar problem is faced by health care, since in evaluating its effectiveness the scope for a reduction in human mortality must be determined, and the results achieved in different regions must be analysed in the light of the ecological and genetic differences in the populations under comparison. Ecology, toxicology, pharmacology and radiobiology are at a stage when there is an acute need for effective methods of evaluating the effect of various factors on lifetimes, especially when these effects are delayed. In medical demography, it is important to determine the influence of environmental factors on human mortality and human life span against a complex background of socio-economic differences in the populations under comparison. This information is necessary for the colonisation of new regions and the optimalisation of population distribution. Finally, life span is an economically important characteristic in agriculture, and problems of genetics and selection related to this characteristic are of great practical importance (Milne, 1985). It would be possible to extend this list with a number of similar problems, but the examples given seem to prove satisfactorily that the biology of life span is a noble and necessary science.

## 1.2. Short history of the subject

In order to gain a better understanding of the characteristics of the biology of life span as a science, and in order to become aware of the style and nature of its approach, it is useful to know how and by whom it was created. A short excursus into the history of the discipline is therefore of more than purely historical interest. Since for our purposes we do not need to delve too deeply into past centuries, we shall begin our historical exposition with the first systematic compilations of facts concerning lifetimes, i.e. with the 17th century.

Life span investigators have every reason to be proud of their discipline's history. The beginnings of the field are straddled by such great scientists as Christian Huygens (1629-1695), Gottfried Wilhelm Leibniz (1646-1716), Edmund Halley (1656-1742), Leonard Euler (1707-1783) and Pierre Simon Laplace (1749-1827). The contribution of these men, surveyed in detail in a book by E.Rosset (Rosset, 1979), was basically connected with the compilation of statistical information concerning human mortality in the form of life tables.

The first such table was constructed in 1662 by the Englishman John Graunt (1620-1674) for the inhabitants of London. On the basis of Graunt's work, the famous Dutch physicist Christian Huygens was one of the first to compute the mean human life span and suggest the use of such tables in calculating the probability of surviving to a given age. Unfortunately, the table itself is now of purely historical interest, since Graunt did not have reliable statistical information. Information of this kind (for the city of Breslau, now Wroslaw) was later presented to the Royal Society in London by the well-known German mathematician Wilhelm Leibniz (see Rosset, 1979). On the basis of this material, the eminent English astronomer Edmund Halley, after whom the well-known comet is named, constructed the first reliable life table. His method of calculating tables for stationary populations, known as the Halley method, was used in the construction of all tables right up to the end of the 19th century. In 1760, the Halley method was supplemented by the famous Swiss mathematician Leonhard Euler, who published a work entitled General Investigations into Mortality and the Multiplication of the Human Race.

Life tables also interested the well-known French naturalist G.L.L.Buffon, who tried to link life span with an organism's period of growth. According to Buffon, the biological life span of human beings and of other species is 6-7 times greater than their growth periods. In the case of human beings, this corresponds to 90-100 years.

In 1812, the great French scientist Pierre Laplace developed a probabilistic interpretation of life tables and proposed a direct method for their construction which is still widely used in the computation of life tables for laboratory animals. His pupil and disciple, the well-known Belgian scientist Adolphe Quetelet (1796-1874), became one of the founders of the modern method of constructing life tables.

This initial historical stage in the development of the subject can be described as the period of descriptive human mortality statistics. Scholars with wide interests and encyclopaedic knowledge took part in its development. They understood that nature is indivisible, and did not see an opposition between man and other living creatures, or between life span and other characteristics of the organism. For example, the above-mentioned Adolphe Quetelet was an astronomer, a mathematician and a statistician. As well as being a professor of mathematics, he was the head of the Belgian national statistical service and constructed demographic mortality tables (Rosset, 1979). This did not prevent him from writing a book entitled Anthropology and applying the statistical regularities he discovered both to human beings and to other living creatures.

Some modern demographers and biologists ought not to forget that they share a common history and many common problems. The desire to draw artificial boundaries in the study of man is evidence only of narrow-mindedness and can lead only to scientific stagnation. Fortunately, thanks to the development of a systems approach, a tendency has developed in recent years which is favourable to the integration of medico-biological and demographic research, and life span research in particular (Gavrilov et al., 1986; Gavrilov et al., 1982).*

In the 19th century, the accumulation of reliable statistical data and the development of sophisticated analytic methods created the preconditions for the first attempts to explain the quantitative laws of life span. In 1825, the English actuary (life insurance specialist) Benjamin Gompertz (1779-1865) published a work (Gompertz, 1825) which became the keystone of the biology of life span. Gompertz gave a theoretical foundation for, and used

---

*This integration of research has been greatly aided by the activity of the International Institute for Applied Systems Analysis (IIASA) in Austria. In particular, the Population Project developed in this Institute under the guidance of Professor Nathan Keyfitz might serve as a good foundation for the further integration of demographic and medico-biological investigations of life span.

concrete examples to demonstrate the idea that the force of mortality (the relative rate at which a population dies out) increases with age according to the geometric progression law. What is more, he noted that alongside this mortality there must also exist a chance mortality which does not depend on age. Gompertz's observation was taken into account in 1860 by another English actuary, William Makeham (1823-1891), who presented the age-related change in mortality as the sum of a constant (the Makeham term) and an exponential (the Gompertz function). This was the birth of the well-known Gompertz-Makeham equation, which has great importance for the biology of life span to this day (Makeham, 1860).

The problem was further developed in the works of the well-known English expert on mathematical statistics, one of the founders of biometry, Prof.Karl Pearson (1857-1936). In 1901 Pearson founded the journal Biometrika, remaining its editor until he died. In the first issue of this journal Pearson published his article "On the inheritance of the duration of life and on the intensity of natural selection in man" (Beeton and Pearson, 1901). This work signifies the beginning of truly biological research into life span. Nevertheless, the remarkable American scientist Raimond Pearl must be considered the real founder of the biology of life span. In our opinion, the biology of life span was born as an independent discipline in 1922 with the appearance of Pearl's book The Biology of Death (Pearl, 1922). This book is in fact not concerned at all with problems of death (the dying process), but is totally dedicated to problems of the biology of life span. Pearl published a total of several dozen works on this problem, including a large series of articles entitled "Experimental studies on the duration of life" (Alpatov and Pearl, 1930; Pearl and Miner, 1935; 1941; Pearl et al., 1927; Pearl and Parker, 1921; 1922a; 1922b; 1922c; 1922d; 1924a; 1924b; Pearl et al., 1923). In his investigations, Pearl dealt with practically all the problems in the biology of life span, including the genetic, ecological, physiological and comparative-evolutionary aspects. Many of these works have preserved their importance and relevance to the present day, and methodologically almost all of Pearl's works can serve as a model for contemporary investigators. Finally, it must be noted that Pearl created a school for research into life span, one of whose co-workers was the Russian scientist, Professor V.V.Alpatov (1898-1979) (Alpatov and Pearl, 1929; Alpatov, 1930). As has already been noted, Alpatov was one of the first (in 1932) to use the term "biology of life span" and to establish the foundations for systematic development in this field (Alpatov and Pearl, 1929; Alpatov, 1930; Alpatov and Gordeenko, 1932).

Since the biology of life span was founded by scholars of the exact sciences (mathematicians, physicists, astronomers, statisticians), it is a discipline which from the start, unlike most of the other areas of biology, began to take the shape of an exact science. In this regard it is related to such subdivisions of biology as biometry, quantitative genetics and biokinetics. Whereas in many areas of biology exceptional attention is paid to *experimental* techniques in the search for clear qualitative effects, in the biology of life span primary importance is attached to the method of quantitative *observation* with subsequent mathematical analysis of the resulting measurements.

This approach, which is less "aggressive" towards the natural environment, may seem at first sight to be too indirect, formal and unconvincing. However its fruitfulness has been demonstrated many times in the history of the natural sciences. For example, all the basic notions of genetics (the concept of genes and their mutations, the concept of allelic forms of genes and their pairwise association (the diploid concept), the conclusion that genes are linearly ordered and organised into cohesive groups, the determination of the relative distances between genes and the crossing-over phenomenon) were the result of this "formal" method: a statistical analysis of the splitting up of the characteristics of an integral organism in subsequent generations. These and many other examples prove that reason can be trusted no less than sense organs and instruments, and if the logic of our arguments leads us towards a particular conclusion, it is not worth rejecting this conclusion simply because it cannot be directly checked by seeing and touching. The desire to discover scientific truths "by pen and paper" was a characteristic feature of the founders of the biology of life span, and this tradition has been maintained to the present day.

The historical development of the biology of life span is closely interwoven with the historical development of statistics, demography and even the technical aspects of life insurance. This led to the discipline being formed as a statistical, population-based area of biology which easily borrows the methods and ideas of other sciences. In particular, the ideas and methods of the theory of reliability, which studies the "life span" of machinery, were incorporated with surprising ease and naturalness (Gavrilov et al., 1978). The populational nature of the biology of life span largely established the manner of its approach towards explaining the mechanisms which determine lifetimes. Whereas the biology of aging, like most of the other divisions of modern biology, is characterised by a "bottom-up" approach, the biology of life span finds a "top-down" approach more natural: an explanation of mechanisms as they are revealed at the level of populations of integral organisms. Of course, these two approaches are not exclusive. They supplement each other, as the whole experience of the development of the natural sciences shows.

To summarise this short historical survey, we can conclude that the biology of life span was created by world-famous scientists who laid a sound foundation for subsequent research.

## 1.3. The present state of the discipline

> "And the Lord said, Behold, the people is one, and they have all one language; and this they begin to do: and now nothing will be restrained from them, which they have imagined to do.
> Go to, let us go down, and there confound their language, that they may not understand one another's speech.
> So the Lord scattered them abroad from thence upon the face of all the earth..."
>
> Genesis 11, verses 6-8

Turning from the history of the biology of life span to the present state of the discipline, we observe with some surprise that despite the soundness and firmness of its foundations, nothing of comparable inspiration has as yet been built on them. What is more, on many issues we observe a standstill and even a regression. The life span question seems to be scattered between a multitude of disciplines:

(i) gerontology and the biology of aging (Comfort, 1964; 1979; Strehler, 1962; 1978; Walford, 1983; Lints, 1985; Mos and Hollander, 1987; Slob and Janse, 1988)

(ii) demography (Le Bras, 1976; Urlanis, 1978; Bourgeois-Pichat, 1979; Manton and Stallard, 1984; Waldron, 1986)

(iii) ecology (Caughley, 1976; 1977; Hutchinson, 1978;

Millar and Zammuto, 1983; Gibbons, 1987) and hydrobiology (Litton, 1987).

(iv) genetics (Tantawy and El-Helw, 1970; Luckinbill and Clare, 1985; 1986; Hogan, 1986), including medical genetics (Kaveggia, 1985)

(v) radiobiology (Storer, 1962; 1979; Lindop and Rotblat, 1961; Sacher, 1956; 1966; Thompson et al., 1985a; 1985b; Thomson and Grahn, 1988; Buldakov et al., 1987)

(vi) toxicology (Luchnik and Livchak, 1963; Van Leeuwen et al., 1985)

(vii) oncology (Hoel and Walburg, 1972; Smith et al., 1973; Conti et al., 1985; Portier et al., 1986)

(vii) zoology (Bever and Sprankel, 1986; Zammuto and Sherman, 1986), including ornithology (Beklova and Pikula, 1985; Paevsky, 1985; Tatner, 1986; Zammuto, 1986), ichthyology (Gläser, 1986), herpetology (Vinegar, 1975; Petranka, 1985; Plummer, 1985), entomology (Sivaprakasam et al., 1985; Young, 1985; Bartlett and Murray, 1986; Sathe, 1986) and zooparasitology (Joubert et al., 1986; Klein et al., 1987).

Specialists studying life span often fail to understand each other, sometimes do not try to understand each other, but most often simply do not suspect that analogous problems have long been studied in other fields. The situation that has arisen is very reminiscent of the Tower of Babel legend quoted in the epigraph.

In order to convirce oneself how great the disconnectedness of life span research can sometimes be, it is sufficient to compare the publications written for example by gerontologists (Strehler, 1962; 1978; Walford, 1983) and ecologists (Caughley, 1977; Hutchinson, 1978; Millar and Zammuto, 1983). The data these works use and the authors they cite turn out to be almost totally non-intersecting, despite the obvious coincidence of the issues that they raise and solve. The life span of plants (Solbrig, 1980) and the survival of seeds (Roberts, 1978) are studied in isolation from investigations into animals, although they could serve as good models for life span research. As a result, well-known mistakes are quite frequently repeated, and long-discovered methods and previously established generalisations are reinvented.

The isolation of life span research from reality appears in publications which contain a large amount of general discussion and economise on factual data. Especially disturbing and regrettable is the sad fact that many investigators have stopped publishing in full the life tables which they obtain. As Lints (1985, p. 98) justifiably remarks: "... contrary to Pearl's practice which was to publish complete life-tables of his experiments the actual practice of researchers is to simply report mean adult life span - eventually the median or 50% survivorship value - for each sex. This is unfortunate because such data do not allow detailed comparisons to be made between laboratories or among studies". As a result, many publications summarising years of work on the life spans of a large number of organisms are devalued by the incompleteness of the published information and consequently do not receive the attention and recognition they might deserve (Strehler, 1961; Redman et al., 1979; Conti et al., 1985; Portier et al., 1986; Arking, 1987; Mos and Hollander, 1987). Often, instead of publishing a complete life table, authors present data graphically in the form of survival curves which are totally useless for further scientific analysis. It is hardly therefore surprising that mathematical theories of life span are often not tested by their authors for a strict correspondence with factual data! In many cases the lack of accessible factual data leads to the end of the theoretical section signifying the end of the article. Theoretical research into life span consequently develops in isolation from experimental investigations, and both the theoreticians and the experimenters lose by this.

Nevertheless, the biology of life span has suffered above all from the disconnectedness of biological and demographic investigations. The clearest manifestation of this is the fact that the colossal archive of exact data on the human life span (hundreds of demographic life tables from a wide variety of the world's population groups) has been almost unused in human biology, despite the acute deficiency of factual information in the field. For a long time many biologists had no suspicion of the scale of this archive of accumulated medico-demographic data, and of how it might be used to investigate the biology of life span. Consequently some of the principal laws of the biology of the human life span, which might have been established as early as the 1950s, are only now beginning to be elucidated (Gavrilov and Gavrilova, 1979a; 1979b; Gavrilov et al., 1983; Gavrilov et al., 1985).

The temporary loss of contact with demography has had

the consequence that many researchers - biologists working on life span - lack the necessary minimum of demographic and statistical knowledge for the task. This was pointed out by the well-known Soviet demographer B.Ts.Urlanis in his most recent book The Evolution of Life Span (Urlanis, 1978). Dissatisfied with the way in which some gerontologists have tried to estimate the value of the biological species-specific life span, Urlanis himself made an attempt at its determination, remarking that the scholars in question "cannot be entrusted to solve the problem we have set ourselves, since some of them have only a dim notion of demographic categories" (Urlanis, 1978, p. 244).

B.Ts.Urlanis's remark is worth discussing at greater length. It is indeed the case that amongst a proportion of biologists, and especially amongst gerontologists, a rather strange conception of the species-specific life span has developed and established itself through extensive mutual citation. It is believed (without any proof) that for every biological species there is a life span limit, independent of the conditions of existence, beyond which survival is impossible even for a second. It is also maintained that the species-specific human life span limit "has remained unchanged across time, races and civilizations" (Economos, 1985). The exact magnitude of such a species-specific limit has not been established for any biological species, no method exists for measuring it, nor indeed is there any evidence that such a limit exists at all.

The theoretical unsoundness of this dogma has been frequently pointed out by demographers (Sauvy, 1961), by experts in probability theory (Feller, 1968), and by biologists who are qualified in statistical methods (Caughley, 1977). Nevertheless, when one of the authors of the present book was involved in a scientific debate on the issue through the editorial staff of a solid academic journal, he received the following response to his cautious remarks: "The issue of the existence of a species-specific limit to life has long been settled". Indeed, at the end of the debate, a summary of which has been published (Gavrilov, 1984c), our opponent was forced to admit that his ideas about the existence of a species-specific limit to human life did not correspond to any real data, but ... he then proposed the existence of three such limits! It soon became clear that even this strange hypothesis is not in agreement with factual data (Gavrilov et al., 1984). We shall return later in the book to a discussion of the problem of the species-specific life span (see ch.4).

In general it must be admitted that there still exists a significant information gap between specialists working in the life span field, and that by no means all works sucessfully combine valuable biological ideas with reliable factual data and effecient mathematical techniques. Nevertheless, we have every basis for optimism. The problem of life span was indeed scattered throughout many disciplines, but now the seeds that were sown have already started to sprout and promise a good harvest. If one and the same fact has been rediscovered several times, it can really be believed. The disconnectedness of the individual disciplines held back the progress of the biology of life span as a whole, but this led to the accumulation of masses of data which has not yet been analysed, a multitude of facts which need to be generalised, and a whole battery of ideas which await testing. It is now difficult to find a scientific field which offers so much promise to the diligent researcher.

The scope of life span research is exceptionally broad and continues to expand. In recent years, especially intensive work has been carried out on problems of life span genetics (Milne, 1985), as applied to humans (Diamond, 1987; Wainscoat et al., 1987), drosophila (Luckinbill and Clare, 1985; 1986; Kogan, 1986; Arking, 1987; Koivisto and Portin, 1987), and nematodes (Johnson, 1987; Friedman and Johnson, 1988; Johnson et al., 1988). Great attention has been paid to explaining the causes of interspecies differences in life span (Calder, 1985; Cutler, 1985; Adelman et al., 1988), and to the problem of the evolution of life span (Golikov, 1985; Service et al., 1988). There has been active discussion of the mechanisms behind the sex differential in life span (Waldron, 1985; 1987; Hazzard, 1986; 1987). There have been definite achievements and there is much promise in the study of the kinetics of survival (Gavrilov and Gavrilova, 1985; 1986; Gavrilov and Nosov, 1985; Gavrilov et al., 1985; 1986; Rosenberg and Juckett, 1987; Juckett and Rosenberg, 1988). Many works have appeared in recent years which are concerned with the mathematical modelling of mortality and life span (Manton et al., 1986; Piantanelli, 1986; Portier et al., 1986; Witten, 1986; Pohley, 1987; Guess and Witten, 1988), and with the critical analysis of such models (Gavrilov and Gavrilova, 1986; Gavrilov, 1987). Investigations concerned with life span extension have already been transformed into a new scientific discipline portrayed in separate monographs (Walford, 1983; 1986; Frolkis and Muradian, 1988). This is probably one of the most popular fields in the biology of life span (Kellog and Yost, 1986; Heicklen and Brown, 1987; Richie et al., 1987; Sawada and Carlson, 1987; Anisimov et al., 1988; Gavrilov, 1990).

Concluding this short survey of the present state of the biology of life span, we can say that works which generalise and interpret already accumulated factual material are becoming especially important at the present time. Works of this kind which contain radically new ideas will be the ones which lead to a scientific revolution in the field and contribute to its further development.

# CHAPTER 2. INDIVIDUAL DIFFERENCES IN LIFETIME

"... And he who goes very slowly may, always following a straight path, progress significantly further than he who runs and strays from this path"

René Descartes, "Discourse on Method"

## 2.1. Where should the study of life span begin?

If one analyses the contemporary scientific literature on life span, the exceptional variety of the existing approaches is particularly striking (see ch.1). Each approach is being developed independently of the others, and the impression is therefore created that the problem of life span is like a crossword puzzle: one can start to solve it with any randomly chosen question. In this conception, the study of life span can follow several parallel paths, and the choice of a particular path depends only on the individual investigator's tastes. Such a view is however plausible only at first sight. The problem of the biology of life span is much more like a knot which can only be untied at one end. To be successful, it is necessary to start the investigation by analysing individual differences in lifetime. This limitation is connected with the fact that the remaining approaches are based on assumptions whose correctness still needs to be proved. They are all either overtly or covertly concerned with the form of the lifetime distribution function, which describes individual differences in lifetime. For example, when differences in mean life span are studied for small samples and a comparison is made using the Student test, the assumption is being made that lifetimes have a normal distribution. If maximal life span values are used for the comparison, this is equivalent to the assertion that the lifetime distribution is cut off at the top, i.e. that there is an absolute limit to longevity. The methods of correlation, regression and variance analysis which are widely used in life span research likewise covertly contain a number of assumptions about the distribution of lifetimes amongst individuals. A list could be made of these and similar examples.

Some investigators believe that it is unnecessary to test such assumptions, and that their correctness is self-evident. However, as we shall demonstrate below, many of these "self-evident" assumptions (for example, the normal life span distribution law, and the notion of an absolute limit to longevity) are simply unsound when tested. Consequently, the conclusions that are based on them cannot be trusted. To start the study of life span, it is therefore essential to start at the very beginning - with an analysis of individual differences in lifetime. Naturally, in comparison with direct experiments on life extension, this research path might seem rather dull and long-winded. Nevertheless, there are grounds for believing that it will lead more quickly to success, since it is more in correspondence with the logic of scientific invesigation. We shall therefore follow Descartes' advice as presented in the epigraph, and we will not rush to give simple answers to complex problems.

## 2.2. Individual differences - the key issue in the biology of life span.

The life span of individuals of a single species, living under similar conditions, varies within very wide bounds. On the one hand, life can be cut off even before birth. According to recent data, this is the fate of 70-80% of fertilised human egg cells, and in the majority of cases the end comes at such an early stage that nobody usually notices it (Diamond, 1987). On the other hand, the lifetimes of particular individuals can significantly exceed the corresponding mean values for the whole population, and an absolute upper limit to longevity appears not to exist (we shall return to this question a little later). However even if we disregard all the cases of death before birth and all the facts of surprising longevity, it still seems to be the case that individual lifetime differences are quite large. For example, in the years 1974-1978 the mean life span of Swedish men was 72.2 years (Befolkningsförändringar, 1970-1981), and the probability of dying in the age interval 71-73 years was less than 6%. In some other countries this figure turns out to be even less. Of course, the high variability of the human lifetime might be explained by individual social and genetic differences, let alone by the climatic and biogeochemical characteristics of different regions in a country. But the same picture can be observed in populations of inbred laboratory animals. For example, in a population of 1415 female laboratory drosophila from line 107 (Pearl and Parker, 1924a) with a mean life span of 48 days, only 11% of all individuals died at age 43-49 days!

One of the characteristics of lifetime variability is its coefficient of variation (the relation of the standard deviation to the arithmetic mean, expressed as a percentage). Calculating these coefficients in the case of laboratory drosophila shows that the lifetime coefficient of variation is usually around 35%, and on occasion reaches 65% (Semyonova, 1983).

Thus the factual existence of a significant biological variability in lifetimes cannot be doubted, and this fact requires its own explanation and quantitative description. There is also another circumstance which brings this issue to the fore.

In distinction to most other quantitative characteristics (for example height or body temperature), an organism's life span can be measured only once. This trivial fact has far-reaching consequences. It turns out to be impossible in principle to study lifetimes at the level of the organism, since the individual dynamics of longevity cannot be investigated, and the errors of once-only observations cannot be determined. The only way therefore to study life span is through the statistical investigation of populations while genetic and environmental factors are varied. However such an approach cannot in principle provide anything more than information about the lifetime distribution of the population under investigation, and the effect of different factors on this distribution! Thus the problem of how to decipher data on lifetime distribution is the key to all further research.

An understanding of the law of life span distribution is useful even during the initial stages of investigation. When the results of survival experiments are represented as a series of figures for the life span of each organism, they are indeed too cumbersome to be effectively analysed. The need therefore arises for a more concise record of the results. This problem is partially solved by the construction of life tables (Dubina and Razumovich, 1975; Caughley, 1979).

Usually a life table is drawn up for a group of individuals born at the same time (cohorts), and the observed mortality is recorded until the death of the last member of the group. Such life tables are called cohort life tables and are widely used in the study of the life span of laboratory and wild animals (Leslie and Ranson, 1940; Miller and Thomas, 1958; Pearl and Parker, 1928a).

Let us review the method of constructing a cohort life table in more detail (Chiang, 1978). The numbers of the cohort are measured at equal time intervals of length $n$. The age of the cohort members will increase by a corresponding amount at each measurement. The basic parameters which enter into the cohort life table are: $l_x$ - the number surviving until age $x$; $d_x$ - the number dying in the interval $(x, x+n)$. The following relationship holds between $l_x$ and $d_x$:

$$l_x - l_{x+n} = d_x \tag{1}$$

or

$$l_{x+n} = l_x - d_x \tag{2}$$

The probability $q_x$ of dying during each age interval $(x, x+n)$ can be obtained by dividing $d_x$ by $l_x$:

$$q_x = d_x/l_x \tag{3}$$

The number $L_x$ of living members at age $x$ is usually calculated using the formula:

$$L_x = nl_{x+n} + (1 - \tfrac{1}{2})nd_x = n/2(l_x + l_{x+n}). \tag{4}$$

In order to calculate the expectation of life it is necessary first to compute the values of $T_x$ for each age, where $T_x$ represents the sum of the lifetimes of all cohort members aged $x$ and above:

$$T_x = L_x + \ldots + L_w \tag{5}$$

where $w$ is the beginning of the final age interval.

Finally, the expectation of life $e_x$ is calculated in the following manner:

$$e_x = T_x / l_x \tag{6}$$

where $x = 0, 1, \ldots, w$.

As an example of a cohort life table we can present a life table of *Drosophila melanogaster* imagoes (see Table 1). Fig.1 shows some typical dependencies of the life table functions on age.

**Table 1** Life table of male *Drosophila melanogaster* imagoes (calculated on the basis of data from Pearl and Parker, 1924a)

| Age interval (days) | Number living to age $x$ | Number dying in the interval $(x, x+n)$ | Probability of death in the interval $(x, x+n)$ | Mean number living in the interval $(x, x+n)$ | Total lifetime at age $x$ and above | Expectation of life at age $x$ |
|---|---|---|---|---|---|---|
| $(x, x+n)$ | $l_x$ | $d_x$ | $q_x$ | $L_x$ | $T_x$ | $e_x$ |
| 0-6 | 1407 | 18 | 0.01279 | 8388 | 63657 | 45.2 |
| 7-12 | 1389 | 12 | 0.00864 | 8298 | 55269 | 39.8 |
| 13-18 | 1377 | 26 | 0.01888 | 8184 | 46971 | 34.1 |
| 19-24 | 1351 | 85 | 0.06292 | 7851 | 38787 | 28.7 |
| 25-30 | 1266 | 100 | 0.07899 | 7296 | 30936 | 24.4 |
| 31-36 | 1166 | 130 | 0.11149 | 6606 | 23640 | 20.3 |
| 37-42 | 1036 | 188 | 0.18147 | 5652 | 17034 | 16.4 |
| 43-48 | 848 | 173 | 0.20401 | 4569 | 11382 | 13.4 |
| 49-54 | 675 | 206 | 0.30519 | 3432 | 6813 | 10.1 |
| 55-60 | 469 | 285 | 0.60768 | 1959 | 3381 | 7.2 |
| 61-66 | 184 | 83 | 0.45109 | 855 | 1422 | 7.7 |
| 67-72 | 101 | 62 | 0.61386 | 420 | 567 | 5.6 |
| 73-78 | 39 | 35 | 0.89744 | 129 | 147 | 3.8 |
| 79-84 | 4 | 3 | 0.75000 | 15 | 18 | 4.5 |
| 85+ | 1 | 1 | 1.00000 | 3 | 3 | 3.0 |

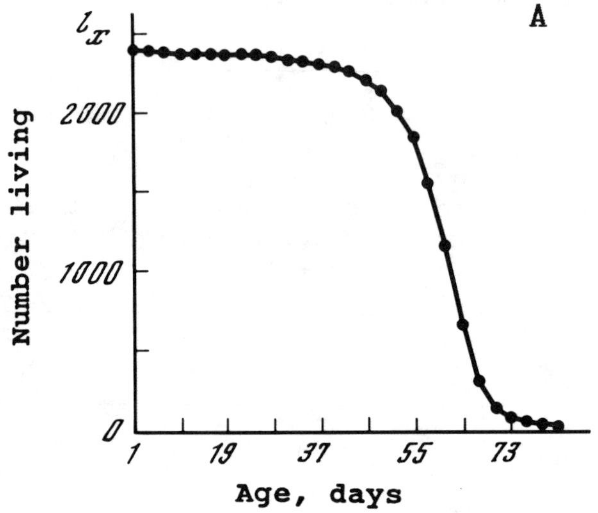


Survival curve ($l_x$)

Ordinate: number of living organisms

Abscissa: age, days

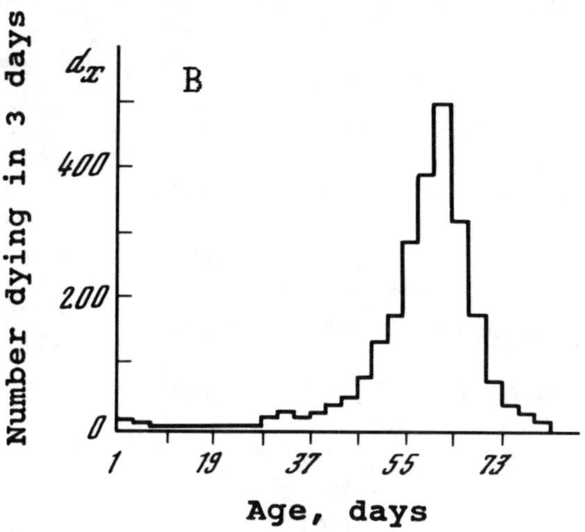


Number dying curve ($d_x$)

Ordinate: number of organisms dying in 3 days

Abscissa: age, days

**Figure 1** Graphs showing the dependence of life table functions on age.

Calculated and constructed on the basis of data on the survival of 2400 *Drosophila melanogaster* females published by Hall (1969).

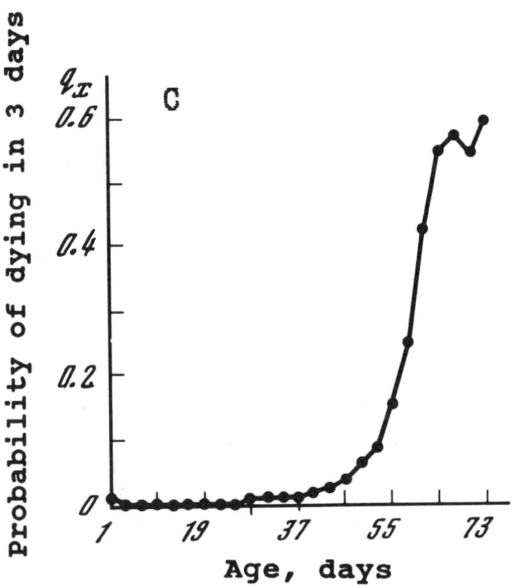


Probability of death ($q_x$)

Ordinate: probability of death (proportion of those dying) in the course of 3 days

Abscissa: age, days

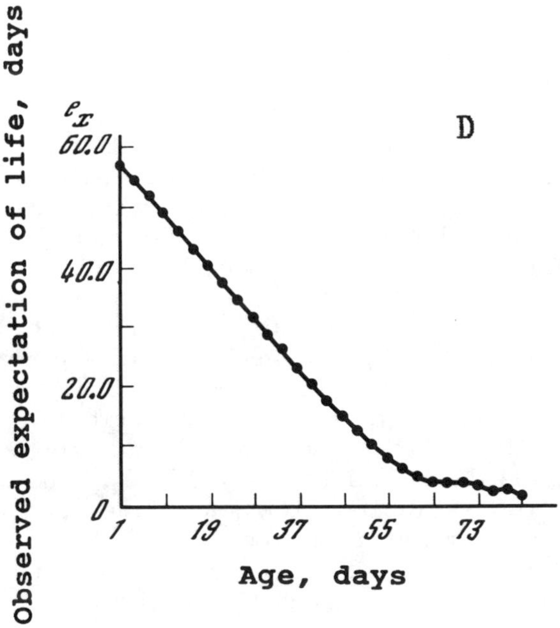


Life expectancy ($e_x$)

Ordinate: expectation of life, days

Abscissa: age, days

In order to construct life tables for human beings, the method just described would require an observation period of close to 100 years. Human life tables are therefore usually calculated in a different way, which will be described in ch.3.

At the present time, a large number of life tables have been constructed for laboratory and domesticated animals (Hickey, 1960), for wild animals (Caughley, 1966; Deevey, 1947), and even for plants (Solbrig, 1980). The compilation of these tables has required, and will continue to require a colossal amount of work, however their importance for the biology of life span is hard to overestimate. Essentially, these tables are the sole experimental source for life span research. However the published life tables for animals are scattered throughout a multitude of publications on a wide variety of topics: gerontology, oncology, ecology, radiobiology etc. By no means every work concerned with life span actually cites life tables, and to this day there is no exhaustive bibliography of all publications containing life tables for animals.

In view of this, bringing together the published life tables for animals will almost certainly assist future research in the biology of life span. Appendix 1 provides a list of the published life tables for animals with an indication of the conditions under which each experiment was carried out.

Nevertheless, the approach based on the use of life tables is far from being perfect. On the one hand, life tables (especially complete life tables for humans) are still too cumbersome. For example, a complete demographic life table which gives the values of each parameter for each year of life contains no less than 80 pairs of values and an overall quantity of information greater than 500 significant figures (Keyfitz, 1982).

On the other hand, when the tables are constructed, there is a certain loss of information connected with the grouping of data at discrete age intervals. This problem might be radically solved if the law of life span distribution were known. For example, if the distribution of organisms according to their life spans was normal, then any amount of data on survival, no matter how large, could be simply recorded using two variables: the mean life span and the standard deviation. Knowledge of the law of life span distribution would thus enable experimental results to be recorded in the most concise form with the minimal loss of information.

The next problem facing the experimenter is how to establish the significance of survival differentials as the

**Table 2** The inadequacy of the normal law as a description of the life span distribution of organisms

| Description of genotype | Initial population numbers | $\chi^2$ observed values | $\chi^2_{0.999}$ tabular critical values |
|---|---|---|---|
| *Drosophila melanogaster* | | | |
| Canton-S, females | 2400 | 1190.70 | 43.82 |
| Canton-S, females | 1200 | 413.60 | 49.70 |
| Canton-S, males | 1200 | 190.96 | 48.27 |
| Line 107, females | 1415 | 239.24 | 34.53 |
| Line 107, males | 1407 | 212.37 | 32.91 |
| $vg^{no}$b-pr/$vg^{no}$b-pr, males | 1061 | 190.75 | 26.13 |
| Canton-S, males | 2400 | 429.73 | 51.18 |
| vg/vg, males | 1048 | 165.24 | 26.13 |
| purple-speck, females | 1750 | 107.37 | 39.25 |
| vg/vg, females | 1026 | 174.37 | 22.46 |
| purple, females | 1006 | 110.83 | 36.12 |
| purple-speck, males | 1465 | 135.41 | 40.79 |
| $vg^{no}$b-pr/$vg^{no}$b-pr, females | 1099 | 198.48 | 27.88 |
| Quintuple, females | 988 | 210.67 | 58.30 |
| Quintuple, males | 1000 | 216.20 | 59.71 |
| Rats of the Wistar strain | | | |
| Wistar, males | 1430 | 116.24 | 37.70 |
| Wistar, females | 1045 | 114.69 | 40.80 |

experimental conditions are varied. In modern gerontology, this problem is usually "solved" by mechanically employing techniques of parametric statistics (for example, the Student test) which are based on the hypothesis of a normal life span distribution. However, as explicit investigations have shown (Gavrilov, 1980; Semyonova, 1983), the distribution of organisms according to their life spans deviates sharply from the normal distribution. The results of these investigations, which are based on a comparison of observed and theoretical distributions using the chi-square test, are presented in table 2.

To ensure that the results would have a high statistical significance, life tables were used which were constructed for large-sized populations (more than 1000 individuals). It should be noted that in each case the calculated values of the chi-square test exceed the corresponding tabular critical values. This means that for life span distributions it is possible to reject the normal distribution law, at the confidence level $p > 0.999$, as being incompatible with experimental data. It follows that any conclusions reached using the techniques of parametric statistics, especially with small samples, have only the appearance of scientific rigour and must be examined anew. Thus a knowledge of the genuine distribution law is essential both for establishing the significance of any differences when the number of observations is minimal, and for avoiding erroneous conclusions.

Finally, when the investigation reaches its last stage, the obtained results must be correctly interpreted. Unfortunately, a significant degree of laxity can be observed at present in this field, and a single set of principles for interpreting data on the survival of organisms does not exist. For example, it is often concluded that aging is slowing down. This conclusion is based on data connected with the increasing life span of organisms, and the substances which bring about this increase are called geroprotectors, i.e protectors against aging (Obukhova and Emanuel, 1984). However, as a number of authors have correctly observed (Lamb, 1977), by no means every factor which affects mortality and life span must also affect aging. Otherwise it would be necessary to count penicillin as a geroprotector, together with the mass of social and hygienic measures which led to a sharp increase in human longevity in the 20th century. This example demonstrates that what is needed at the present time is a comprehensive model for the survival of organisms, a model which would be sufficiently general to be accepted by the majority of gerontologists, but also sufficiently concrete to permit a unified interpretation of experimental results. It can easily be seen that the creation of such model presupposes a knowledge of at least the most general properties of the life span distribution of organisms.

## 2.3. The nature of variability in lifetimes

The very first question which is bound to arise in the study of individual lifetime differences is the following: what is the nature of the observed variability? An initial, superficial consideration of the question usually invites the conclusion that the heterogeneity of the population is the reason for differences in life span. As an illustration, we cite one of the most characteristic pronouncements on the issue: "Obviously a group of individuals of the same age, called a cohort, is qualitatively heterogeneous. The correctness of this statement follows simply from the fact that a cohort dies out gradually, not "all at once", as would be the case if it were homogeneous" (Blokhinov, 1982, p.87). Arguments of this kind, despite their "obviousness", can in fact be remote from reality. Indeed, it is well-known that a homogeneous collection of identical atoms of one and the same radioactive isotope will not all decay at once, but gradually, in conformity with the law of radioactive decay. Consequently, it might be the case that lifetime variability is observed even in a totally homogeneous population, conditioned by the probabilistic nature of the process of dying. The whole experience of radiobiology shows that this kind of explanation is not just a formal construct, remote from real biological processes. It has been demonstrated that the "dose-effect" curve cannot be explained by the heterogeneity of the population, but must be interpreted within the framework of the stochastic theory of targets (Timofeeff-Ressovsky and Zimmer, 1947; Luchnik and Livchak, 1963; Hug and Kellerer, 1966; Timofeev-Ressovsky et al., 1968; Kudryashov and Berenfel'd, 1982).

There are therefore two extreme positions in the attempt to explain lifetime variability. These are the heterogeneity hypothesis and the stochastic hypothesis. Finally, there is also a third source of variability, conditioned by variations in the external environment. The task of the experimenter is to determine the contribution of each of these phenomena to the total life span variability in each concrete case.

In a discussion of the heterogeneity hypothesis, the

first question to arise is the contribution of genetic heterogeneity to the observed variability in lifetimes. Indeed, a large number of different genetic diseases are known which lead to a reduction in life span, including such premature-aging syndromes as Down's syndrome, Werner's progeria, Cockayne syndrome, and progeria (Hutchinson-Gilford syndrome). The desire therefore naturally arises to generalise these facts and explain the observed variability in lifetimes by the genetic differences between individuals. In order to test the genetic heterogeneity hypothesis, a large number of experiments have been carried out to evaluate what is called the heritability of life span. It must be noted however that this term "heritability" encompasses three different concepts, often resulting in confusion (Jacquard, 1983). According to the first and simplest concept of heritability, heritability is a measure of the resemblance between related individuals. Indeed, if the life span of children could be simply calculated knowing the life spans of their parents, this would mean that all differences in lifetime are connected with the genetic heterogeneity of the population.

One of the simplest methods of evaluating the heritability of life span is to calculate the correlation coefficient between the lifetimes of parents and their offspring. The first such calculation was performed in 1901 by the founders of the English school of biometry, M.Beeton and K.Pearson (Beeton and Pearson, 1901). The results of their calculations turned out to be discouraging: the correlation coefficient between the lifetimes of children and parents was only 0.05-0.13, i.e. practically no heritability of life span was observed. This very low value for the correlation coefficient may however be connected with the fact that a direct life span comparison between parents and children is not absolutely correct, involving as it does different generations of people living under different conditions. Indeed, if a comparison is made of the life spans of brothers and sisters living at the same time, and consequently under more similar conditions, the correlation coefficient turns out to be somewhat higher: 0.15-0.30 (Beeton and Pearson, 1901). However, even in this case the correlation coefficient is closer to zero than to one, thus demonstrating the low heritability of life span. These results, which were obtained at the very beginning of the century, have been many times subsequently confirmed. For example, 1n 1931 R.Pearl calculated that the correlation coefficient between the life spans of children and parents is only $0.02\pm0.01$ (Pearl, 1931). In one of the more recent works devoted to this topic, it was also shown that the correlation coefficient between the life spans of parents and children over 20 is very small and amounts to only 0.101 (Philippe, 1978).

The low heritability of life span is not only a human characteristic, but also found in other species, for example drosophila (Tantawy and El-Helw, 1970). What is more, in the case of drosophila, the absence of any resemblance in lifetimes between parents and offspring has been verified by direct experiments involving selection for a high life span. In these experiments, the offspring of long-living individuals who had outlived 80% of the population were systematically selected for further mating. Even though this procedure was repeated 8 times in succession, no selection for an increase in the drosophila life span was observed. By way of comparison, let us note that selection for another quantitative polygenic characteristic (the number of sternopleural bristles) appeared after the first cull (Lints et al., 1979).

Subsequently, an attempt was made to explain the absence of selection for an increased life span in Lints's experiments by the "artificial introduction of strong gene-environment interaction through the use of a competition-free environment" (Luckinbill and Clare, 1985). Luckinbill and Clare showed that selection for an increase in the drosophila life span is nevertheless possible, but is only observed when the drosophila larvae develop in overcrowded conditions, with a density of around 70 larvae per vial (Luckinbill and Clare, 1986). The fact that such conditions are extremely unfavourable for drosophila can be judged by analysing the authors' data (Luckinbill and Clare, 1985): the mean life span of drosophila raised under conditions of overcrowding turns out to be 30% less than that of drosophila raised under normal conditions in conformity with Lints's techniques. It therefore seems to us at least premature to consider the results of Lints's experiments as an artefact. Rather, on the contrary, Luckinbill and Clare's experiments can be treated not as selection for life span, but for resistance to stressful overcrowding during development.

Do these facts mean that individual differences in lifetime are basically non-genetic in origin? The well-known French geneticist and demographer Albert Jacquard replies to this question in the affirmative, citing the following calculations as an illustration (Jacquard, 1982). According to Jacquard's data, the heritability coefficient for the human life span is 0.16. This means that even if the life spans of both parents exceed the mean life span by 20 years, their offspring will gain on average only 0.16 x 20 = 3.2

extra years of life from their "inheritance"! What is more, if we know the life spans of the parents, this reduces the indeterminateness (dispersion) of their offspring's life span by only $0.16^2$ or 2.6%. On the basis of these illustrative calculations, Jacquard (1982) comes to the conclusion that "environmental factors are so important as regards duration of life that genetic factors seem insignificant in comparison".

A similar conclusion is reached by Edmond Murphy (1978) on the basis of the results of the huge Baltimore study into the connection between the life spans of children and parents. It turns out that children "inherit" only one year of extra life for every 10 extra years of their parents' lives, and even this "inheritance" may be conditioned not by genetic, but "purely cultural or environmental factors" determined by family traditions (Murphy, 1978).

According to another point of view, the facts presented above indicate only that the additive genetic component of the total dispersion of life span is indeed small. However lifetime variability may be connected not only with the additive action of the genes, but also with the interaction of the genes amongst themselves (Kogan, 1984; 1986). It is therefore proposed to evaluate heritability not in a narrow, but in a wide sense of the word, i.e. to evaluate a coefficient of genetic determination. It must however be noted that the techniques for evaluating this coefficient are based on a large number of unrealistic assumptions which are painstakingly dismantled in Jacquard (1983).

It seems to us that the data on the low heritability of life span represent a powerful argument against the genetic heterogeneity hypothesis, but nevertheless do not refute it. It is indeed possible to imagine a scenario in which all lifetime differences are genetic in nature, but nevertheless selection for an increase in life span is ineffective and the resemblance between related individuals in respect of this characteristic is insignificant. In order to illustrate this statement, let us consider the following hypothetical example.

Suppose that in a population of diploid organisms the only long-lived individuals are heterozygous with the genotype Aa, while the homozygous individuals with the genotypes AA and aa hardly survive to an adult state and leave almost no offspring. Then, despite the rigorous selection in favour of only one genotype (Aa), the embryos will on each occasion belong to all three genotypes in the previous proportions. The frequency of alleles will invariably be 50%, and only a half of the embryos will live to an old age, even though all of them are the offspring of long-lived individuals. Thus despite the totally genetic nature of the lifetime differences, even the most rigorous selection for longevity will be ineffective. What is more, the frequency with which long-lived offspring are spawned will be the same for both long-lived and short-lived parents.

Of course, this simple example is merely an illustration, although it is known that heterozygotes frequently live longer than homozygotes (the so-called heterosis effect) (Bileva et al., 1978; Bileva and Malinovsky, 1981; Nekrasova and Shakhbazov, 1981; Shakhbazov and Nekrasova, 1980). There is however also some direct experimental evidence for the existence of a genetic variability in lifetimes which is resistant to selection. For example, in a study of three weakly linked diallel loci in the Zvenigorod population of the pine *Pinus sylvestris* L., it turned out that in adult trees (age 50-170 years), only a few of the 27 possible genotypes at these loci are actually found (Zhivotovsky, 1984). At the same time, all 27 genotypes occur in the embryos. The question arises: what is low diversity of genotypes in the adult section of the population related to? Does the appearance of new genotypes in the embryos indicate a change in the genotypical composition of the population from generation to generation? The answer turns out to be in the negative. This cyclic change starts and finishes within the bounds of a single generation: the genotypical variability is sharply increased at the early stages of ontogenesis, while at the reproductive age it again decreases through the preferential elimination of individuals with particular genotypes. Thus genetic diversity, relatively low in the reproducing section of the population, sharply increases in the zygotes of the subsequent generation, then again decreases with age through the elimination of individuals with non-adaptive genotypes. It is important to emphasise that the elimination of non-adaptive individuals does not lead to a change in the frequency of alleles in the population: these remain practically unaltered at around 50%. This is connected with the fact that, because of the phenomenon called gametic integration, there remain in the reproducing section of the population individuals with genotypes which are capable of producing any kind of zygote (Zhivotovsky, 1984). These facts in some respects resemble the simplified hypothetical scenario presented above. In both cases, there indeed exists a genetically determined variability in lifetimes which is resistant to selection.

Another observation which arises in the analysis of data on the heritability of life span is the following. An evaluation of the heritability of life span over the entire age interval may be too crude, since the contribution of genetic variability is not identical at different stages of ontogenesis. From the example quoted above, as well as from a whole range of other data (see Jacquard, 1982), it might be expected that genetic diversity plays a significant role in the early stages of life, while at the end of life its contribution may be insignificant.

The first indication of a change in the heritability of life span with age was obtained by R.Pearl (Pearl and Pearl, 1934). Information was collected for people of different ages on the life span of their parents, and also their four grandparents (Pearl and Pearl, 1934). Pearl summed the six life span figures obtained, and studied how this sum depends on the age of the individuals in the survey. Naturally, if the life span of the parents and grandparents had no influence on the survival of their offspring, no dependency would be observed. The results obtained by Pearl are presented below, the only difference being that the sums he calculated have been divided by six in order to determine the mean life span of the progenitors:

| Age of people in the survey | Mean life span of progenitors |
|---|---|
| years | years |
| 40 | 66.0 |
| 50 | 66.8 |
| 60 | 70.5 |
| 70 | 74.8 |
| 90 | 74.3 |
| 95 | 74.3 |
| 100 | 74.8 |
| 105 | 73.8 |

It is not difficult to observe that up to age 70 there really is some correlation between the life span of progenitors and the survival of their offspring, however this correlation then disappears.

The results presented above have been independently confirmed by later investigations into the life span of monozygotic and dizygotic twins aged 60 and above (see Jacquard, 1982). Analysing the figures, Jacquard notes that although the difference in the life spans of monozygotic twins (36 months) is significantly less than the difference in the life spans of dizygotic twins (74.6 months), these differences decrease with age and finally disappear towards age 80. Thus the contribution of genetic heterogeneity to the observed variability in lifetimes seems to operate only during the early stages of life and strongly decreases with age.

It must also be noted that the traditional techniques used in the genetics of quantitative characteristics, in particular the techniques used for evaluating heritability, may turn out to be unsuitable for studying life span variability. These techniques presuppose that the total dispersion of the life span characteristic can be separated into genetic and environmental components, and the component connected with the interaction of the environment and the genotype is usually ignored, since it is difficult to evaluate. However this kind of simplification can hardly be considered justified, since there is no evidence for a strictly additive cumulation of environmental and genotypical effects (Jacquard, 1983). Besides, returning to the example of radioactive decay, we clearly see that variability in life span might exist in spite of the total absence of any environmental component in the dispersion (environmental conditions have no effect whatsoever on the parameters of radioactive decay), and the obvious "genetic" uniformity of the population. Why does life span variability exist in such a case, and why do the atoms not decay simultaneously? In order to answer this question, it is essential to view life span variability as the result of a survival *process*, i.e. to use a kinetic approach, and also to employ some elements of the theory of stochastic processes (Luchnik and Livchak, 1963; Sacher, 1977).

Thus in the analysis of life span, in addition to the two traditional sources of variation (environment and genotype), it is also necessary to take into account an additional third source of variation: the stochastic (kinetic) nature of the process. However the methodology of this new approach has not yet been worked out, and it is possible that the solution to the problem lies in taking not life span itself but the parameters of its distribution as the characteristic to be studied, as Sacher (1977) has attempted to do. In order to illustrate the possible magnitude of this additional variability, connected neither with the environment nor the genotype, let us note that the coefficient of variation for the life span of nematodes, calculated on the basis of Johnson and Wood's (1982) data, is 52-73%, despite the rigorously controlled environmental conditions and the genetic uniformity of the hermaphroditic nematode strains.

In discussing the hypothesis that life span variability is stochastic (kinetic) in nature, we have come across the objection that such a model cannot explain the existence of high risk groups, i.e. the heterogeneity of the population with respect to the death risk. In fact, this lack of correspondence exists only when the death of organisms is the result of a one-stage process of destruction. If the process which leads to death has many stages, then even in an initially homogeneous population organisms will gradually appear which are at different stages of destruction and consequently have different death risks (Kozlovsky and Gavrilov, 1983). Consequently, the idea that variability in lifetimes is largely determined by a multi-stage process of destruction not only agrees with the well-known facts about the heterogeneity of the population with respect to the death risk, but also permits an explanation of the possible reasons why this heterogeneity might arise.

Summarising this discussion of the possible nature of variability in lifetimes, we can present the following conclusions:

(1) Possible causes of the high variability in life span are: (a) the initial heterogeneity of the population, including its genetic heterogeneity, (b) variations in environmental conditions, and (c) the stochastic (kinetic) nature of the survival process. This last source of variability has until now escaped the attention of many investigators.

(2) Contrary to wide-spread opinion, there is no convincing evidence that the observed individual lifetime differences are primarily genetic in nature. What is more, a multitude of data indicate rather that the contribution of genetic heterogeneity to the observed differences is not great, and in any event decreases with age.

(3) A large variability in lifetimes is maintained even in populations of genetically identical organisms living under strictly controlled laboratory conditions. In order to understand the nature of this variability, it is essential to study the kinetics of the survival of organisms more deeply, and to construct and test the corresponding mathematical models. It is therefore of interest to attempt a formal description of lifetime variability, and this is what will be done in the following section.

## 2.4. In search of the life span distribution law

One approach to investigating the nature of individual lifetime differences is to analyse the characteristics of observed life span distributions. This task can be solved in two ways: either by testing ready-made theories and models for their correspondence with factual data, or by processing observational results with subsequent generalisation of the perceived regularities. The first path is the standard one for the majority of the exact sciences, and its techniques are well worked out. The second path presupposes that the investigator has a developed intuition and a deep knowledge of what is special to the problem. Above all, the question arises as to what index characterising life span distribution should form the basis for such an investigation.

*The problem of choosing the "correct" index.* As is well known, any life table contains a number of indices. The most important of these are: $l_x$ - the probability of surviving to age $x$ (usually multiplied by 100000)*, $d_x$ - the number of those dying in the age interval from $x$ to $x + \Delta x$, $q_x$ - the probability of dying in the same interval, and $e_x$ - the expectation of life at age $x$. So for each age there are four indices. The question arises: which should be chosen for subsequent investigation? At first sight, this question might seem an idle one, since each of the indices contains the same information. Converting the elements of one column into the elements of another is no more than an arithmetical exercise. Consequently, the four indices reflect four different ways of recording the same information. However for the purposes of our investigation these indices will turn out not to be of equal value. Of all the age-related indices, preference should be given to the one which reflects events taking place solely in the age group under investigation, and which would not change inevitably if the mortality at other ages were occasionaly altered. For example, the slaughter of the innocents by King Herod would inevitably have altered all the values of the first two life table indices ($l_x$ and $d_x$) at that age, even if the mortality of all the remaining age-groups in the population had remained unaltered. On the other hand, doing away with old

---

* In the case of survival tables for laboratory animals, sometimes simply the number of individuals surviving to a given age is indicated.

people, a practice of certain wild tribes and ancient societies (see Rosset, 1979), must lead to a reduction in the values for the expectation of life ($e_x$) for all age-groups in the population. Thus of the four life table indices, only one - the probability of death ($q_x$) - is elementary in the sense that its magnitude can reflect a situation which is solely specific to the age-group under investigation. It is therefore sensible to give preference to this index, since its magnitude is determined by the smallest number of factors, a matter of prime importance in the search for the laws of mortality.

At the same time, the probability of dying is not the most convenient index for analysis. Values for the probability of death depend on the length of the age interval ($\Delta x$) for which they are calculated. In the case of human beings, this calculation is usually based on age intervals of 1 year or 5 years. Any conversion of the values for the probability of death from one age interval to another, with the aim of comparing results, must be carried out in conformity with the algebra of probability theory, and not by simple multiplication or division of the figures. Thus calculations which make use of the probability of death must be constantly controlled for their conformity with the algebra of probability theory. In this respect, the problem of the choice of age interval is the one which constantly arises. In a discussion of this problem, Herve Le Bras cites the following example (Le Bras, 1976). Suppose that the probability of death, calculated for an age interval of one year, grows geometrically with age (the Gompertz law). It turns out that the probability of death calculated for any other age interval cannot follow this law. It is easy to convince oneself that this is the case by considering the formula based on probability theory used in calculating the probability of death for a five-year age interval from the values for the probability of death for one-year age intervals:

$$ {}_5q_x = 1 - \prod_{i=0}^{4} (1 - {}_1q_{x+i}), \qquad (7) $$

where ${}_1q_{x+i}$ is the probability of death in the course of the year at age $x + i$ , and ${}_5q_x$ is the corresponding probability for a five-year age interval. Using this formula for his calculations, Le Bras showed that even when the probability of death grows with age over a year in strict conformity with Gompertz's law, the probability calculated for a 5-year interval grows with age significantly more slowly than the law predicts. It therefore turns out that the form of the age-dependence of the probability of death is determined by the choice of the age interval. At the same time, we have no principled basis for preferring one age interval to another as being the more correct.

Finally, since the probability of death cannot be greater than unity, the use of a probability scale when the mortality figures are large can lead to erroneous conclusions. Studying the growth with age of the probability of death, we will almost inevitably discover a reduction in the rate of this index's growth as it approaches its upper limit. Since this upper limit is always unity, we will also "discover" that the mortality differences in the populations under comparison are blotted out. It is clear, however, that such "discoveries" reflect the nature of the index rather than the nature of the phenomenon.

Therefore instead of the probability of death, which cannot exceed unity, it is better to use the force of mortality (synonyms: hazard function, hazard rate, failure rate, intensity of risk of death, instantaneous mortality rate), which is not bounded from above. This parameter is determined in the following way:

$$\mu(x) = \lim_{\Delta x \to 0} \frac{l(x) - l(x + \Delta x)}{\Delta x \cdot l(x)} = - \frac{dl(x)}{l(x)\, dx} \qquad (8)$$

In order to estimate the force of mortality at age $x$, it is possible to use the following formula proposed by Sacher (1956; 1966):

$$\mu(x) = \frac{1}{2\Delta x} \ln \left(\frac{l(x - \Delta x)}{l(x + \Delta x)}\right) \qquad (9)$$

This formula can be used for intervals $\Delta x$ which are sufficiently small for the change in the force of mortality to be ignored or considered close to being linear.

There are also other methods for estimating the force of mortality. For example, in the statistical analysis of survival, the formula proposed by Cutler and Ederer (1958) is often used:

$$\mu(x) = \frac{2q_x}{\Delta x \cdot (2-q_x)} = \frac{2}{\Delta x}\left[\frac{l(x) - l\ (x - \Delta x)}{l(x) + l\ (x - \Delta x)}\right] \qquad (10)$$

Gehan and Siddiqui (1973), using the Monte-Carlo method, came to the conclusion that Cutler and Ederer's formula is preferable to Sacher's, since Cutler and Ederer's formula gives less bias. Subsequently, this conclusion has been widely quoted and has served as the basis for the preferential use of Cutler and Ederer's formula in most publications and even in application program packages developed for the analysis of survival (for example in the BMDP package). However, if Gehan and Siddiqui's (1973) work is carefully analysed, it emerges that the formula for the force of mortality which they call Sacher's formula does not in fact coincide with the formula given above and proposed by Sacher (1956), but rather has the following form:

$$\mu(x) = \frac{1}{\Delta x} \ln\left[\frac{l(x + \Delta x)}{l(x)}\right] \qquad (11)$$

For aging systems with a monotonically increasing force of mortality, it can easily be seen that this formula, attributed to Sacher, will always lead to biased (reduced) values for the force of mortality, since these values relate to the beginning of the age interval, and not to the middle of the age interval as in the case of the real Sacher formula. Thus it is clear, even without using the Monte-Carlo method, that the formula tested by Gehan and Siddiqui will give biased valued for the force of mortality. However, the question as to which formula is better - the real Sacher formula or Cutler and Ederer's formula - remains open. If we compare the two formulae as they apply to large values for the force of mortality, it becomes obvious that the formula proposed by Sacher is much better than the one proposed by Cutler and Ederer. Cutler and Ederer's formula has the disadvantage that the resulting values cannot exceed the upper limit of $2/\Delta x$, whereas the actual force of mortality may be growing unrestrictedly. There is therefore every reason to use Sacher's (1956) formula to estimate the force of mortality, and this is what we have done in the present book.

The force of mortality, like the probability of death,

reflects mortality only in the age group under investigation, and does not change inevitably when the mortality at other ages is occasionally altered. However, in distinction to the probability of death, the force of mortality does not depend on the length of the age interval, and in addition, calculations using the force of mortality are exceptionally simple and do not necessitate the use of the algebra of probability theory. Since the force of mortality, unlike the probability of death, can in principle assume arbitrarily high values, it well reflects the dynamics of high mortality during old age and infancy. Finally, it must be noted that the force of mortality is determined in exactly the same way as the failure rate in the mathematical theory of reliability. The use of this index is therefore a significant aid in applying the ideas and methods of the theory of reliability to the construction and testing of mathematical models of mortality. All these considerations make the force of mortality the most successful and "correct" index in the search for lifespan distribution laws.

*A short survey of functions describing life span distribution.* An analysis of the works dedicated to the search for mortality and life span "laws" confirms that the force of mortality is indeed the index chosen.

One of the first and most successful attempts to express the dependency between mortality and age mathematically was that of the English actuary Benjamin Gompertz, dating to 1825 (Gompertz, 1825):

$$\mu(x) = -\frac{dl(x)}{l(x)dx} = R\cdot\exp(\alpha x), \qquad (12)$$

where $\mu(x)$ is the force of mortality at age $x$; $l(x)$ is the number of those surviving to age $x$, and $\alpha$ and $R$ are the parameters of the equation. This formula, which describes the mortality of people older than 20, was called the Gompertz law, and its parameters were named the Gompertz parameters. Subsequently, the Gompertz law began to be used widely for describing the mortality of laboratory animals.

Gompertz suggested the following theoretical justification for his empirical law. Let us suppose that the rate at which the "resistivity to death" decreases is proportional to the resistivity itself. Since the force of mortality $\mu(x)$ acts as a measure of the human susceptibility to death, Gompertz took its reciprocal $1/\mu(x)$ as a measure of resistivity, thereby deriving the equation:

$$\frac{d}{dx}\left(\frac{1}{\mu(x)}\right) = -\alpha\left(\frac{1}{\mu(x)}\right) \qquad (13)$$

where $\alpha$ is a non-negative constant. After integration and simplification, this equation turns into formula (12).

In his work, Gompertz noted that, in addition to the mortality which grows exponentially with age, there can also exist a component of mortality which is independent of age: "It is possible that death may be the consequence of two generally coexisting causes: the one chance, without previous disposition to death or deterioration, or increased inability to withstand destruction" (Gompertz, 1825). However, for the analysis of the life tables which were then available, Gompertz considered it possible to restrict himself solely to the exponential component of mortality. Not until 35 years later, in 1860, did another actuary William Makeham add the age-independent component to the Gompertz formula (Makeham, 1860). This component, usually denoted by the letter $A$, received the name of the Makeham parameter. Thus the formula appeared which we now know as the Gompertz-Makeham law:

$$\mu(x) = A + R\cdot\exp(\alpha x) \qquad (14)$$

Subsequently, there were many attempts to modify the Gompertz law. For example, several works use the quadratic form of the Gompertz equation (El Shaarawi et al., 1974):

$$\mu(x) = A + R\cdot\exp(\alpha x + \beta x^2) \qquad (15)$$

Risser proposed instead of the quadratic dependency a polynomial one (see Le Bras, 1976):

$$\mu(x) = A + R\cdot\exp(a_0 + a_1 x + a_2 x^2 + \cdots + a_n x^n) \qquad (16)$$

Makeham himself later supplemented the Gompertz-Makeham

formula with a component linearly dependent on age (see Henderson, 1915):

$$\mu(x) = A + Bx + R \cdot \exp(\alpha x) \qquad (17)$$

Another modification to the Gompertz-Makeham formula looks as follows (see Henderson, 1915):

$$\mu(x) = ma^{x} + nb^{x} \qquad (18)$$

Another way of modifying the Gompertz function is to use what are called logistic equations. The most widely known of these is the Perks equation (Perks, 1932):

$$\mu(x) = \frac{A + B \cdot \exp(bx)}{1 + D \cdot \exp(bx)} \qquad (19)$$

It is interesting to note that this formula can be theoretically derived as a special case of the model for the avalanche-like chain-destruction of the organism in aging (Gavrilov, 1987). Beard (1959) proposed a simplified version of the Perks formula:

$$\mu(x) = \frac{B \cdot \exp(bx)}{1 + D \cdot \exp(bx)} . \qquad (20)$$

A distribution which is different in principle was proposed by Weibull for describing the variability of technical systems with respect to their "lifetimes" (Weibull, 1951). This distribution, now known as the Weibull law, is widely used in reliability theory. The rate of failure (the analogue of the force of mortality) in this case is a power function of age:

$$\mu(x) = B \cdot x^{c} . \qquad (21)$$

Recently, the Weibull distribution has also begun to be applied in descriptions of the lifetime variability of organisms (Rosenberg et al., 1973; Slob and Janse, 1988).

In some publications, a generalised form of the Weibull law is used (see Gavrilov, 1980):

$$\mu(x) = A + Bx^c . \tag{22}$$

It seems to us useful to add to the formulae listed above another one which we have called the generalised binomial mortality law:

$$\mu(x) = A + (b + cx)^n . \tag{23}$$

For some relationships between the parameters, this formula is close to the Gompertz-Makeham formula, while for others it resembles the generalised Weibull law, thus unifying two different classes of distribution. Indeed, if the constant $b$ turns out to be much less than the constant $c$, the generalised binomial law of mortality coincides with the generalised Weibull law. If, on the other hand, the constant $b$ turns out to be much greater than the constant $c$, the generalised binomial law of mortality coincides with the Gompertz-Makeham law, with $R = b^n$, and $\alpha = \frac{nc}{b}$. We have discovered that the binomial mortality law can be theoretically derived from models customarily leading to the Weibull law, as long as additional account is taken of the heterogeneity of the population with respect to the number of defects initially present in each organism (see ch.6).

A significantly more complicated formula generalising the Gompertz and Weibull laws was proposed by Brillinger (1961):

$$\mu(x) = \sum_i H_i(x-B_i)^{c_i} + \sum_j \frac{A_j}{(B_j-x)^{c_j}} + \sum_k E_k d_k^{\,x} \tag{24}$$

Sometimes the basis chosen is not the force of

mortality, but other indices. For example, in the investigations of certain actuaries, formulae like (25) have been used:

$$\frac{q_x}{1-q_x} = \exp\ (F(x)) \tag{25}$$

where $q_x$ is the probability of death, and $F(x)$ is a polynomial of the required degree (Keyfitz, 1982).

H.A.Johnson and M.Pavelec (see Economos, 1980a) proposed the following formula for the number of survivors:

$$l(x) = 1 - (1 - e^{-kx})^p \tag{26}$$

Economos (1980a) suggests approximating the dependence of mortality on age with two curves. The first of these describes the increased proportion of deaths at early ages:

$$m(x) = m_0 e^{\alpha x} , \tag{27}$$

where $m(x)$ is the proportion of deaths. The second graph describes the decreased proportion of survivors at later ages:

$$l(x) = l_p \cdot e^{-\alpha'(x-x_p)} , \tag{28}$$

where $l(x)$ is the proportion of survivors, and correspondingly $l_p$ and $x_p$ are the number of survivors and the age at which the graph starts. In semi-logarithmic coordinates, therefore, these graphs take the form of two straight lines: at the beginning a rising straight line for the proportion of deaths, and then a falling straight line for the proportion of survivors.

Some investigators prefer to use formulae which describe the change in the expectation of life with age. For example, G.H.Hardy proposed the following formula (see Keyfitz, 1982):

$$e(x) = \exp(a + bx + cx^2 + dx^3 + fx^4) . \tag{29}$$

J.F.Steffenson used another dependency (see Le Bras, 1976):

$$e(x) = \frac{1}{A + Bc^x} \tag{30}$$

Most of the formulae quoted above are suitable only for desribing the mortality of adult, sexually mature individuals. There have however been attempts to describe mortality over the entire age interval. The first attempt of this kind seems to be the Wittstein formula (see Henderson, 1915):

$$q_x = a^{-(M-x)^n} + \frac{1}{m} a^{-(mx)^n} \tag{31}$$

An analysis of this formula quoted in Henderson's (1915) book shows that the first component describes the mortality of adults, and the second describes the "additive mortality of infancy".

At the present time, the best known of all the formulae which describe mortality over the entire age interval is the formula proposed by Heligman and Pollard (1979):

$$\frac{q_x}{1 - q_x} = A^{(x+B)^C} + De^{-E(\ln x - \ln F)^2} + GH^x , \tag{32}$$

where $q_x$ is the probability of death during the year. The first component describes child and infant mortality, and the last describes the mortality of old people. The second component approximates the mortality peak which is observed in the region of 20 years and which is basically linked to accidents.

The list of formulae proposed as approximations to the life span distribution function could be extended (Henderson, 1915; Le Bras, 1976; Keyfitz, 1982; Hsieh, 1985). However, even from the examples quoted it is obvious that there is at present no lack of formulae for describing this function. The problem is to choose from all the possible formulae the one which actually reflects the essence of the phenomenon under investigation and aids our understanding of the mechanisms behind variable lifetimes. At the same time, searching for the life span distribution formula by simply scrutinising all the possible variants is a thankless task performed in the hope of hitting on the right one by good fortune. It stands the same chance of success as solving a problem by substituting all the possible answers. It is therefore above all necessary to formulate some methodological principles which would enable us to reach the necessary formula by the shortest possible route.

*Methodological principles for selecting the life span distribution law.* Let us formulate the general methodological principles which usually guide researchers solving problems of this kind.*

(1) The principle of theoretical justification. According to this principle, only those equations should be used which have theoretical justifications. Then the recording of data using such an equation is simultaneously the first step to its interpretation. Starting from this principle, special attention should be devoted to formulae derived from various theoretical hypotheses rather than to the empirical formulae used for life insurance.

(2) The principle of universality. The aim of revealing general regularities which are valid for the widest possible range of natural phenomena is the very essence of the scientific world-view. In conformity with this principle, special value should be attached to general life span distribution laws which are valid for the greatest variety of organisms, including man.

(3) The principle of the best approximation with the fewest parameters. A formula satisfying this principle allows data to be recorded in the most compact form, thereby permitting the distribution to be recovered with the minimal number of observations (Keyfitz, 1982). This principle is a particular case of the idea that "entitities are not to be multiplied beyond necessity", known as Occam's rasor. As applied to the problem of life span, this principle points us not towards an absolutely exact description of the observed lifetime distributions using formulae with many parameters, but towards the use of models which reflect the

---

* These principles, unfortunately, are rarely employed simultaneously in one and the same investigation.

most prominent characteristics of those distributions. In this connection, especially promising is a factor analysis of mortality, permitting a determination of the minimum number of parameters necessary to describe it.

(4) The principle of local description. Since many systems have critical periods in their development when they qualitatively change their properties and behaviour (Zhirmunsky and Kuz'min, 1980), we should not try to describe the whole extent of the process in one go. The history of science demonstrates that the local description of a process is the most efficient path to take, with the subsequent "dovetailing" of the various scientific approaches in the framework of a new, more general conception. Therefore if a proposed life span distribution law is valid only for a restricted age interval, this is not in itself a basis for being critical towards it. The restricted applicability of the law does not demonstrate that it is incorrect, but merely that it is only a special case of another, more general and as yet unknown law.

If we are guided by the principles cited above and analyse a sufficiently large body of factual data, it turns out that the Gompertz-Makeham law is still in many respects better than most other well-known formula. We must therefore tarry in greater detail on this law and the arguments in its favour.

## 2.5. The Gompertz-Makeham law

We shall begin our discussion with an analysis of the data on the life span of the traditional subject of genetics - the fruit fly *Drosophila melanogaster*. When an analysis is made of the survival data from a large group of simultaneously born, genetically identical individuals which are kept under standard laboratory conditions, an interesting regularity emerges. It turns out that over a significant age interval the force of mortality grows with age according to the law of geometric progressions (exponentially, in conformity with the Gompertz formula). In Fig.2 we present a graph of the logarithm of the force of mortality against the age of the drosophila, which is as straight as a ruler. In fact, the correlation coefficient between the logarithm of the force of mortality and the age reaches 0.999 at 11 points on the graph. The same regularity is valid for male rats of the Wistar strain (see Fig.3).

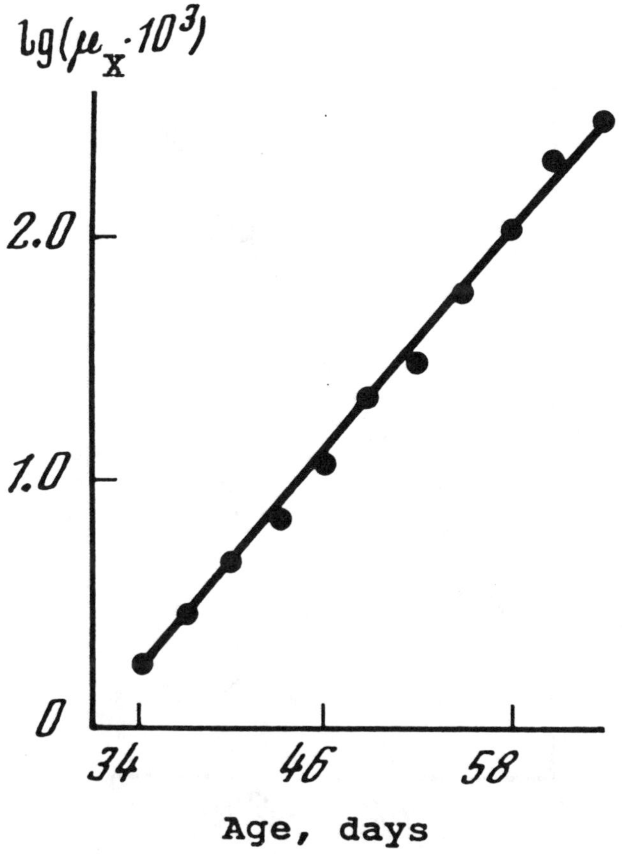


**Figure 2** The dependence of the logarithm of the force of mortality on the age of drosophila. Calculated and constructed on the basis of the life table for 2400 *Drosophila melanogaster* females of the Canton-S strain published by Hall (1969). In calculating the force of mortality, a three-day age interval was selected.

abscissa: age, days

ordinate: $\lg(\mu_x \cdot 10^3)$

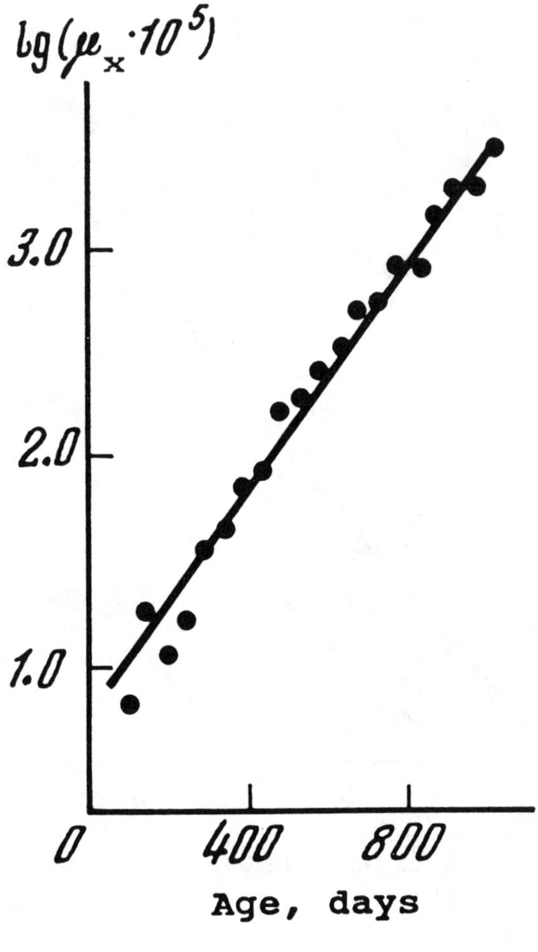


**Figure 3** The dependence of the logarithm of the force of mortality on the age of laboratory rats. Calculated and constructed on the basis of the life table for 2113 male rats of the Wistar strain published by Schlettwein-Gsell (1970). In calculating the force of mortality, a 50-day age interval was selected.

abscissa: age, days

ordinate: $\lg(\mu_x \cdot 10^5)$

Sometimes we hear it objected that this linearity is not surprising, since many dependencies look straight on a logarithmic scale. Fig.4 contains an answer to this comment. It shows the dependence of the logarithm of the force of mortality on the age of male rats of the Wistar strain. The same graph also shows as a dashed line the theoretical curve calculated on the basis that the distribution of the lifetimes of laboratory rats follows the normal law, with the same mean and dispersion as is observed experimentally. It is clear that the experimental points can be fitted to a straight line graph much better than to the dashed theoretical curve, which once again confirms the lack of any justification for using the normal law in a description of lifetime distributions.

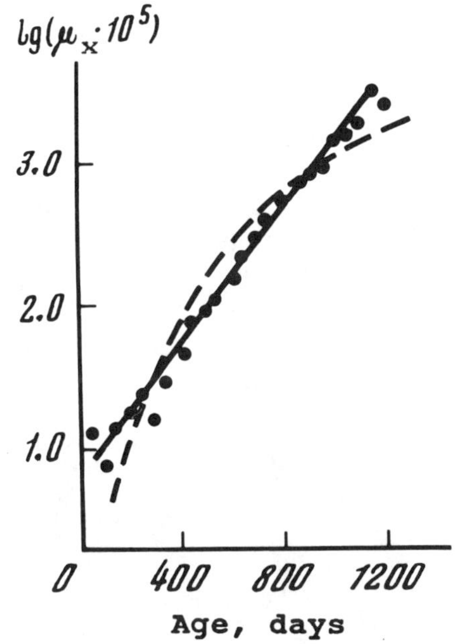


**Figure 4** The dependence of the logarithm of the force of mortality on the age of laboratory rats; the theoretical dependence in the case of a normal life span distribution (dashed curve). Calculated and constructed on the basis of the life table for 2050 female rats of the Wistar strain published by Schlettwein-Gsell (1970). In calculating the force of mortality, a 50-day age interval was selected.

abscissa: age, days

ordinate: $\lg(\mu_x \cdot 10^5)$

The examples cited above of the exponential growth of the force of mortality with age are far from being isolated. For example, for the very same laboratory drosophila we have found 8 life tables constructed for populations with initial numbers greater than 1000 individuals (Hall, 1969; Pearl and Parker, 1921). An analysis of these tables shows the linear growth of the logarithm of the force of mortality with age: in particular, this can be seen from the high values for the correlation coefficient between the variables ($r$ = 0.97-0.99, see Table 3).

**Table 3** Characteristics of the life span distribution of *Drosophila melanogaster*.*

| Initial population | $\alpha \cdot 100$, days$^{-1}$ | $\ln(R \cdot 10^4)$ | Correlation coefficient $r$ | Number of points in the dependence |
|---|---|---|---|---|
| | | Males | | |
| 1000 | 4.47 | 0.71 | 0.98 | 27 |
| 1407 | 6.35 | 0.40 | 0.98 | 6 |
| 1200 | 7.80 | -0.04 | 0.99 | 10 |
| 2400 | 9.31 | -0.29 | 0.99 | 10 |
| | | Females | | |
| 1000 | 3.90 | 0.86 | 0.99 | 27 |
| 1415 | 5.93 | 0.37 | 0.97 | 6 |
| 1200 | 9.32 | -0.44 | 0.99 | 10 |
| 2400 | 13.07 | -1.43 | 0.98 | 10 |

* Calculated on the basis of life tables published by Pearl and Parker, 1921; 1924a; Hall, 1969. The values of $R$ and $\alpha$ were determined by the least squares method in the coordinates $\ln(\mu_x) \div x$. In calculating the force of mortality, a six-day interval was selected. The age range for the linearisation was 28-55 days; $r$ is the correlation coefficient between the logarithm of the force of mortality and age.

Of course, the correlation coefficient is not the best measure of the linearity of the relationship under investigation, since its departure from unity may be connected as well with the random scattering of data as with systematic divergences from linearity. Typical of the majority of laboratory animal life tables is a low initial population (less than 1000 individuals) and, as a consequence, a large statistical scatter of data. In these circumstances, the correlation coefficient between the logarithm of the force of mortality and age will be small even when the deviations from the Gompertz law are purely random in character. Consequently, in order to test mortality laws on data with a large statistical scatter, it is essential to use other methods and indices.

In 1979-1980, a method was proposed for testing the adequacy of mortality laws when the data are imprecise (Gavrilov, 1980; Gavrilova et al., 1979). When applied to the Gompertz formula, this method consists of the following. If the life span distribution is really described by the Gompertz formula, then the graph of the logarithm of the force of mortality plotted against age ought to be linear. In this case, the ratio between the tangent of the angle of the slope over the initial portion of the graph and the tangent of the angle of the slope over the final portion is equal to one. This tangent ratio has been called the K-criterion (Gavrilov, 1980). For each graph taken individually, the value of the K-criterion may differ significantly from unity, but if the divergence from the Gompertz formula is not systematic and has a random nature, the centre of distribution of the K-criterion will tend to unity as the number of observations is increased.

Thus testing the Gompertz formula by this method amounts to determining the centre of distribution of the values of the K-criterion (the median, mode or arithmetic mean), calculated for as large a number of life tables as possible.

The authors of the present volume, together with V.G.Semyonova, analysed 129 previously published drosophila life tables. The centre of distribution of the K-criterion, calculated as the arithmetic mean of a distribution truncated at the outliers, took the value $1.03 \pm 0.19$, i.e. exactly coincided with the theoretical value (1.0) which would be expected if the Gompertz law were correct (Gavrilov, 1980).

Of course this approach, like any other statistical technique, cannot strictly speaking prove the validity of this or any other law, let alone its uniqueness. In the best case, it is possible to say that the proposed formula does not contradict the factual data. However the use of the K-criterion provides a quick and easy test of the adequacy of the competing formulae, and the justified rejection of many of them. As an example, we can cite the results of testing the adequacy of the already mentioned Weibull law.

We have given a proof that for *any* positive values of the parameters in the Weibull formula, the theoretically expected value of the K-criterion is equal to the inverse ratio of the ages for which the tangents are calculated (Gavrilov, 1980; Gavrilova et al., 1979). In the case described above, this ratio has the value 1.89-2.25 (the spread is connected with the fact that in the 129 drosophila life tables the age intervals did not always coincide). It is not difficult to see that the observed value of the centre of distribution of the K-criterion (1.03±0.19) significantly and strongly differs from the theoretical values (1.89-2.25) which would be expected if the Weibull law were correct. Thus the Weibull law, unlike the Gompertz law, does not agree with the observed facts on the life span of drosophila. In a similar manner, it is possible to test other formulae simply and quickly for their correspondence with the factual data.

The example just cited shows that the volume of data accumulated in the scientific literature is already sufficient for a strong test of the competing formulae, and of the corresponding conceptions of the mechanisms which determine life span. In the light of the new data, the "old" Gompertz law has not only maintained its importance, but has also turned out to be more able to beat off the competition than a whole range of "younger" and more fashionable formulae. The validity of the Gompertz law has been noted not only for drosophila and rats, but also for nematodes (Johnson, 1987), human lice (Gavrilov, 1984b) (Fig.5), mosquitoes (Gavrilov, 1980), mice (Kunstýř and Leuenberger, 1975), horses (Strehler, 1962) and mountain sheep (Gavrilov, 1980). Naturally, the question arises: why does the Gompertz law have such wide applicability to such varied species as drosophila and horses? Perhaps analogies of the Gompertz law also exist in the non-living world? It turns out that such analogies do indeed exist. In particular, the Japanese researcher Kase (1953) studied the "survival" of two hundred specimens of rubber as loads were increased. It turned out that the intensity of snapping in the rubber grows exponentially with the increase in the load, expressed in kg/cm$^2$. This same regularity is observed in investigations into the electrical breakdown of oil as the intensity of an electrical field is increased (see Gumbel, 1958). Thus the regularity which has been brought to light is so wide-spread that we must seek to provide it with a maximally general

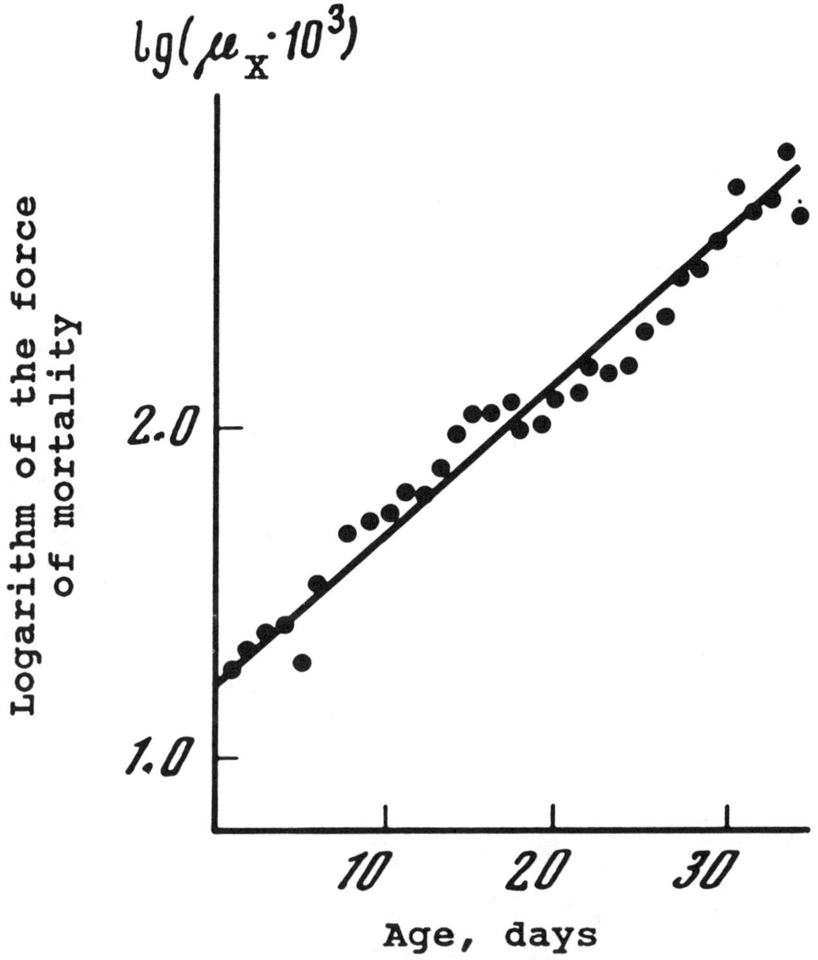


**Figure 5** The dependence of the logarithm of the force of mortality on the age of human lice. Calculated and constructed on the basis of the life table for 400 male human lice (*Pediculus humanus* L.) published by Evans and Smith (1952). In calculating the force of mortality, a four-day age interval was selected.

abscissa: age, days

ordinate: logarithm of the force of mortality, $\lg(\mu_x \cdot 10^3)$.

theoretical basis.

It turns out that this basis already exists and is provided by the statistics of extreme values. In this subdivision of probability theory, a distribution with an exponentially growing intensity of failures (in the particular case - force of mortality) is treated as an extreme distribution and is called the first asymptotic function of the distribution of least values (Gumbel, 1958). Thus the regularity in question has a theoretical basis which is just as rigorous as, for example, the well-known normal distribution law. With its wide applicability and rigorous theoretical basis, it can consequently be recognised as fundamental.

In those cases when a deviation from the Gompertz law is observed, the following must be borne in mind. In addition to mortality factors whose effect depends on age, there are situations in which a lethal outcome is inescapable for all, even the most healthy organisms (for example, catastrophes, accidents, acute infections and poisonings). In other words, in addition to the exponentially growing component of mortality, which is conditioned by aging, there must also exist an age-independent component connected with extreme situations.

This notion can be formulated in formal mathematical terms as the principle that the two types of mortality are to be summed. According to this principle, the total force of mortality from all causes is the sum of two non-negative components, one of which does not depend on age:

$$\mu(x) = A + f(x); \qquad \frac{\partial A}{\partial x} = 0, \quad A \geq 0, \; f(x) \geq 0, \qquad (33)$$

where $\mu(x)$ is the force of mortality at age $x$ (meaning the relative rate of dying); $A$ is the age-independent component of mortality, which we call the background component of mortality (Gavrilov and Gavrilova, 1979a; Gavrilova and Gavrilov, 1983); $f(x)$ is the age-dependent component of mortality (Gavrilov and Gavrilova, 1979a). As can be seen from what has been said, the age-dependent component of mortality is an exponential. In the particular case in which the background mortality can be ignored (for example under good laboratory conditions), the total force of mortality grows exponentially with age, i.e. in accordance with the Gompertz law.

For the cases in which the background component of

mortality cannot be ignored, we have proposed a method for linearising the data which is based on preliminary numerical differentiation (Gavrilova and Gavrilov, 1983; Gavrilov et al., 1978). Indeed, on differentiating, the constant component (the background component of mortality) disappears, and then the logarithm of the increment in the force of mortality must be a linear function of age:

$$\ln(\Delta\mu_x) = \alpha x + \ln[R\cdot(e^{\alpha\Delta x} - 1)]; \quad \Delta x = \text{const.} \qquad (34)$$

Fig.6 contains an example of the use of the proposed method. It can be seen that the logarithm of the force of mortality in males of the confused flour beetle *Tribolium confusum* is not linear, but a concave function of age (dependency 1). It can be shown however that this divergence from the Gompertz law is connected with a failure to take the background component of mortality into account. Indeed, it can also be seen from Fig.6 that the logarithm of the increment in the force of mortality increases with age in a strictly linear fashion (dependency 2). This means that the mortality law is the sum of an exponential and a constant component (i.e. follows the Gompertz-Makeham law). In addition, the constant component is greater than zero, as is demonstrated by the concave nature of dependency 1 in Fig.6. Thus taking into account the background component of mortality can explain the observed deviations from the Gompertz law and add to our knowledge of the regularities governing the distribution of organisms' life spans.

The variability of organisms' lifetimes can therefore in many cases be described sufficiently precisely with the aid of the Gompertz-Makeham formula:

$$\mu(x) = A + R\cdot\exp(\alpha x) \ . \qquad (35)$$

It can be shown that this formula also satisfies all the principles formulated above for seeking out and choosing amongst the competing life span distribution laws.

For example, the Gompertz-Makeham formula satisfies the principle of theoretical justification, since it can be derived from a whole range of mathematical models (Gavrilov, 1978; Gavrilova et al., 1978; Skurnick and Kemeny, 1978) and is an asymptotic distribution in the statistics of extreme values (Gumbel, 1958).

The Gompertz-Makeham law also conforms to the principle

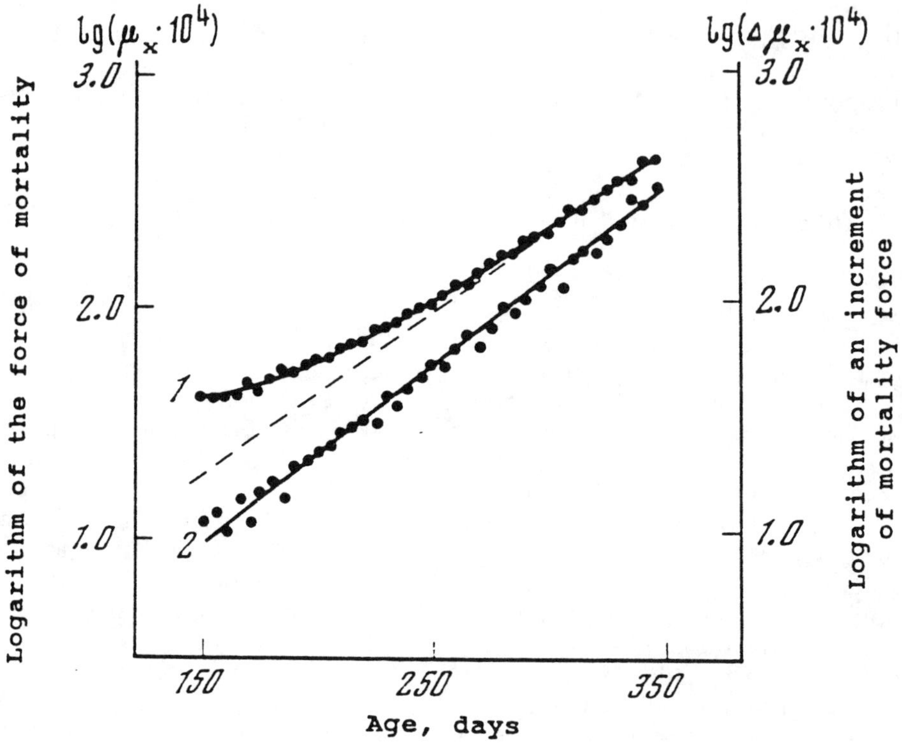


**Figure 6** The dependence of the logarithm of the force of mortality (1) and the logarithm of the increment in the force of mortality (2) on the age of flour beetles. Calculated and constructed on the basis of the life table for 400 female confused flour beetles *Tribolium confusum* Duval published by Pearl and Miner (1941). In calculating the force of mortality, a 5-day age interval was selected. For the subsequent calculation of the increments in the force of mortality, an age interval of 30 days was selected.

abscissa: age, days

ordinate (left): logarithm of the force of mortality, $\lg(\mu_x \cdot 10^4)$

ordinate (right): logarithm of the increment in the force of mortality, $\lg(\Delta\mu_x \cdot 10^4)$.

that the two types of mortality are to be summed, since it consists of two factors, one of which does not depend on age (parameter *A*). What is more, direct calculation has shown that both factors in the Gompertz-Makeham formula are indeed as a rule non-negative (Gavrilova, 1982; Gavrilov et al., 1983). Consequently, the two factors can indeed be taken as the constituent components of mortality.

The Gompertz-Makeham formula satisfies the principle of universality, since it describes the life span distributions of a wide variety of biological species (drosophila, mosquitoes, flour beetles, mice, rats, horses and mountain sheep), including man (see ch.3).

Finally, it must be noted that the Gompertz-Makeham law conforms to the principle of best approximation with the fewest parameters. If we consider the family of three-parameter formulae, which includes the generalised Weibull law and also the normal and log-normal life span distribution laws, it turns out that the Gompertz-Makeham formula gives the best approximation. What is more, it has been established that the three-parameter Gompertz-Makeham equation is a far better approximation to the survival curve of laboratory drosophila than a fourth-degree polynomial containing five parameters.

Of course, the facts and arguments we have adduced are in no way a proof of the general applicability of the Gompertz-Makeham law, and even less are they a demonstration of its uniqueness as the fundamental law of mortality. However there is every reason to state that the use of this law is valid as a research tool whenever it turns out to be in good agreement with the observed life span distribution.

## 2.6. Unresolved issues and problems

A discussion of the problem of the variability of organisms' lifetimes would be incomplete if we were to pass over in silence the obstacles in the path of future investigations.

Above all, it must be recognised that the law governing the life span distributions of organisms has unfortunately not yet been established. Despite all the virtues of the Gompertz-Makeham formula, this formula (which is valid in a limited age interval) can only be considered as a special case of a more general and as yet unknown distribution law. This gives rise to a whole range of problems and restrictions on future research.

Firstly, information on organisms' life spans which is recorded in a compact form, using indices such as the mean, median and maximal life span, is defective since it does not permit the original distribution to be retrieved. Therefore the results of survival experiments must be published in a maximally full form, as detailed life tables. In this regard, the works of R.Pearl can serve as a model (Pearl and Miner, 1935; 1936; 1941; Pearl et al., 1927; Pearl and Parker, 1921; 1922a; 1922b; 1922c; 1922d; 1924a; 1924b; Pearl et al., 1923). The data in these works is still used in contemporary investigations (Gavrilov, 1984a). Any attempt to cut the volume of published information on survival risks causing irreversible damage to its value for future research. As an extreme measure, it is possible to restrict publication to the parameters of a multi-parameter empirical formula, if it has been shown that the discrepancy between the calculated and observed dependencies can be considered random. This device was used by R.Pearl (Pearl and Parker, 1924a), and is sometimes encountered in modern investigations (Kunstýř and Leuenberger, 1975).

The second consequence of our lack of knowledge of the life span distribution law is the fact that we are not justified in using the techniques of parametric statistics (usually based on the hypothesis, inapplicable in this case, of a normal distribution), but are instead forced to use the less powerful techniques of non-parametric statistics (for example, the Kolmogorov-Smirnov lambda test or the chi-square test) in order to establish the significance of the effect of particular factors on lifetimes. This means that many investigators must turn away from the classic biometric methods and assimilate the techniques of non-parametric statistics. At the present time, a number of non-parametric statistical techniques have been developed specially for survival analysis (Kalbfleisch and Prentice, 1980). Although these techniques were initially created in order to analyse the survival of oncological patients, they have turned out also to be useful for research in the biology of life span (Mode et al., 1984); Muenchow, 1986; Ermakov and Gavrilova, 1987). Since the techniques of non-parametric statistics are relatively unpowerful, it is essential to recognise also that economising on the number of animals used can make an experiment pointless.

Finally, the third consequence of our lack of knowledge of the life span distribution law is the lack of a theoretical basis for the correct interpretation of the obtained data. Therefore any conclusions which are drawn on the basis of survival experiments must be considered to be only one of the possible interpretations of the results under discussion.

In conclusion, it is essential to emphasise once more that the problem of the distribution of organisms with respect to their lifetimes is the key question in the biology of life span. Until the life span distribution law is established, researchers will have to operate with bulky amounts of data on survival, set up experiments with a deliberately augmented number of organisms and constantly experience a feeling of doubt when interpreting their experimental results.

# CHAPTER 3. THE HUMAN LIFE SPAN

## 3.1. Specific questions connected with research into the human life span

Of all the problems in the biology of life span, the problem of research into the biological aspects of the human life span is simultaneously the most interesting, the most important, and unfortunately the most complex. A priori at least four reasons can be named why the human life span distribution has to assume a complex form which is in principle different from the one we observe in other biological species.

First of all, the conditions in which people live have little similarity with unchanging laboratory conditions. When we therefore consider the mortality of a simultaneously developing group (a cohort), the age-related increase in the intensity of mortality conditioned by the aging process can be overlain by a reduction in mortality connected with progress in medicine and health care. Consequently, we can sometimes even observe an age-related reduction in the intensity of mortality amongst adults. This problem can be solved in part if we do not study cohort life tables, constructed for a generation of simultaneously born individuals, but rather the mortality of different age groups in the population living at one and the same time. On the basis of this information, demographic life tables are constructed for a hypothetical cohort. These then serve as the basis for calculating the value of the mean life span.

Life tables for a hypothetical cohort are also called current life tables (Chiang, 1978). A distinction is made between complete life tables, in which the values of the indices are listed for every year of age, and abridged life tables, in which the values of the indices are usually given at five-year age intervals.

As a source of data for the construction of current life tables, we can use census results and information about the number and age of those dying in the census year. In order to increase the accuracy of the calculations, information about the number of deaths in the years adjacent to the census year is also frequently used. On the basis of this statistical information, when a complete current life table is constructed, a series of age-specific death rates are calculated in the following manner (Chiang, 1978):

$$M_x = D_x/P_x \, , \tag{36}$$

where $D_x$ is the number of those dying in the age group $(x, x + 1)$ in the course of the calendar year under investigation (the census year), and $P_x$ is the number of the population in the same age group $(x, x + 1)$, taken in the middle of the calendar year. The figure $P_x$ characterises the sum of the person-years lived in the course of the calendar year in the given age interval $(x, x + 1)$.

It is then possible to use several techniques to convert the age-specific death rates $(M_x)$ into life tables (Chiang, 1978; Keyfitz and Flieger, 1971). The simplest of these, and one which is also reasonably accurate, is the technique proposed by Chiang (1978). In this technique, the calculation of the probabilities of dying is carried out using the formula:

$$q_x = \frac{M_x}{1 + (1 - a_x)M_x} \, , \tag{37}$$

where $a_x$ is the fraction of the last year of life for age $x$ which the people dying in the calendar year under investigation managed to live. The value of $a_x$ for ages greater than four years is as a rule close to 0.5, corresponding to a uniform distribution of the number of deaths in the age interval $(x, x + 1)$ (Chiang, 1978).

Current life tables are calculated for a hypothetical cohort whose initial numbers are usually taken to be equal to 100,000. This then constitutes the initial value for the number of survivors - $l_0$. Then all the remaining values for the numbers of survivors are calculated using the formula:

$$l_{x+1} = l_x \, (1 - q_x) \, . \tag{38}$$

The tabular number of deaths is then given by:

$$d_x = l_x q_x \tag{39}$$

The number $L_x$ of years lived in the age interval is obtained using the formula:

$$L_x = (l_x - d_x) + a_x d_x \ , \quad x = 0,\ 1,\ \ldots \qquad (40)$$

Then the total number of years lived beyond age $x$ is calculated:

$$T_x = L_x + L_{x+1} + \ldots + L_w, \quad x = 0,1,\ \ldots\ ,\ w. \qquad (41)$$

And finally, the expectation of life at age $x$ is calculated using the formula:

$$e_x = T_x / l_x \ . \qquad (42)$$

Thus a current life table reflects the way in which a cohort of 100,000 new-born individuals would die out if the levels of age-specific mortality observed at a particular moment in time were maintained throughout these individuals' lives.

The use of life tables for a hypothetical cohort enables us, at a first approximation, to separate the influence on human mortality of age and time (more precisely, the change in living conditions with time).

It must however be noted that this approach is nevertheless not a complete solution to the problem. People of different ages do indeed belong to different generations with different histories. If the risk of dying were determined solely by age and the current situation, then the data obtained using this approach would not differ in principle from the cohort data for constant conditions. However, there is no a priori basis for assuming that people's pasts have no influence on the risk that they will die in the future.

The second difference between human living conditions and the constant conditions in which laboratory animals are kept is connected with the phenomenon of age discrimination in the population (Gavrilov, 1984a; 1984b). For example, the living conditions of people of call-up age, especially at times of war, differ significantly from the living conditions of the remaining population groups, including pensioners. Therefore the dependence of mortality on age will inevitably be distorted by differences in the living conditions of different age groups in the population.

The third feature of human populations which contrasts them with genetically uniform groups of laboratory animals kept under identical conditions is the significant heterogeneity of the population, both in its biological (for example, genetic) and its social characteristics. Since many of these characteristics influence the risk of dying, the total dependence of mortality on age may take on an exceptionally complicated form.

Finally, the fourth and last feature, which tends to be continually emphasised, is connected with the conviction that the human life span is determined by specifically human social laws. This notion that the status of humans is exceptional enjoys such popularity that it has even been reflected in fiction. As a particularly clear example we cite the musings of the hero in A.Kron's novel, Insomnia, doctor of biology Yudin: "Man ages and dies in a fundamentally different way than animals. This truth is rather banal..." (see Kron, 1980, p.176).

Thus there are a whole range of reasons and considerations why the human life span distribution might have an exceptionally complex form which makes it different from the analogous distributions for other organisms. Let us see whether this is in fact the case.

## 3.2. Regularities of human mortality

> "I said in mine heart concerning the estate of the sons of men, that God might see that they themselves are beasts.
>
> For that which befalleth the sons of men befalleth beasts; even one thing befalleth them: as the one dieth, so dieth the other; yea, they have all one breath; so that a man hath no preeminence above a beast..."
>
> Ecclesiastes 3, verses 18-19

Turning from general considerations to the analysis of real data, we are surprised to discover that there are no principled differences between human life span distributions and the analogous distributions for other biological species.

What is more, it turns out that the dependence of human mortality on age, as in the case of laboratory animals, consists of the following three periods: a period of high child mortality, when the force of mortality decreases with age; a period of sexual maturity, when the force of mortality grows with age, usually following the Gompertz-Makeham law (Gavrilov, 1984a); and finally, a senile period, when the force of mortality is very high and grows comparatively slowly with age (Gavrilov, 1980; Gavrilova and Gavrilov, 1982). Thus although there is a strong difference in magnitude between the life spans of humans and, for example, laboratory drosophila, the general shape of the survival curves turns out to be totally identical.

To exemplify this, Fig.7 contains the results of an analysis of the mortality of Italian women. Our attention is drawn to the striking similarity which exists between this figure and the figure presented above on the mortality of female confused flour beetles (Fig.6).

In both cases the force of mortality grows according to the Gompertz-Makeham law. This means that no biologist or demographer is capable of distinguishing between human life tables and the analogous tables for laboratory animals if the units of age are not specified. In other words, such tables are fundamentally indistinguishable, not what one would have expected if man really aged and died "in a fundamentally different way than animals". The congruity between the life span distributions of humans and laboratory animals means that none of the reservations expressed above are in fact of decisive importance. Otherwise, this congruity would not be observed. Of course, past events can affect the risk of dying, but their contribution can be ignored in comparison with the effect of age and the current situation. The same can be said for age discrimination and the heterogeneity of human populations. All these factors doubtlessly affect the dependence of mortality on age, but their effect is an order of magnitude weaker than the effect of age and the current situation.

Of course, the example just cited is only an illustration of the applicability of the Gompertz-Makeham law to the study of the human life span. In order to submit the adequacy of the law to a thorough test, in 1979 we analysed 285 abridged human life tables from all the world's geographical regions: Africa, America, Asia, Europe, the USSR, Australia and Oceania (Gavrilov and Gavrilova, 1979b). Our test of the Gompertz-Makeham law consisted in checking

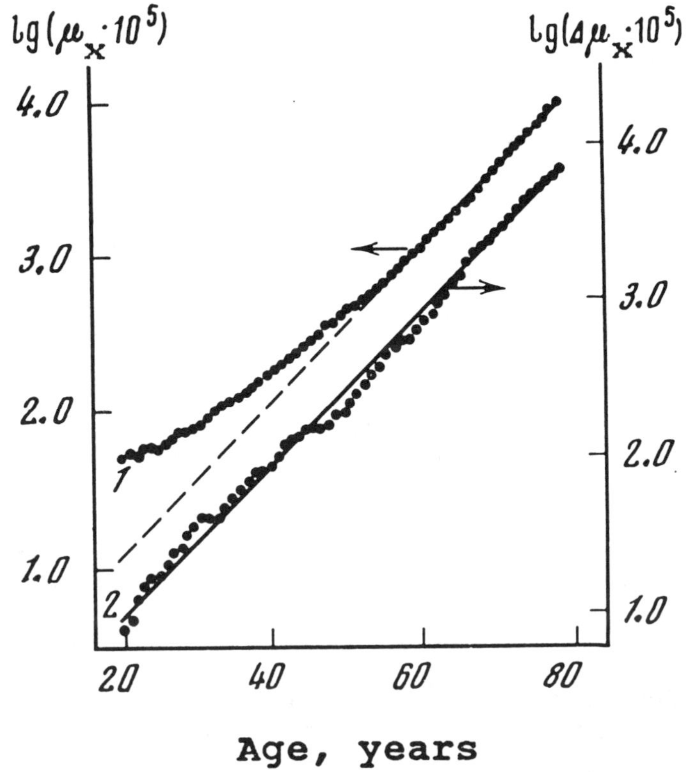


**Figure 7** The dependence of the logarithm of the force of mortality (1) and the logarithm of the increment in the force of mortality (2) on the age of human beings. Calculated and constructed on the basis of the life table for Italian women during 1964-1967 (for the source of the data, see Gavrilova et al., 1983). In calculating the force of mortality, an age interval equal to 1 year was selected. In the subsequent calculation of the increments in the force of mortality, a five-year age interval was selected.

abscissa: age, years

ordinate (left): $\lg(\mu_x \cdot 10^5)$

ordinate (right): $\lg(\Delta\mu_x \cdot 10^5)$.

whether there was a linear dependency between the logarithm of the age-specific increment in the force of mortality and the age of people in the interval 35-75 years. Each dependency contained 9 points at 5-year intervals. It turned out that in 242 cases out of 285 (85%) the dependencies were straight lines with a correlation coefficient $r \geq 0.98$. Straight lines with $r \geq 0.99$ were found in 74% of all the cases under analysis. Since the square of the correlation coefficient reflects the proportion of explained dispersion, the figures just presented demonstrate that the residue of dispersion "not explained" by the Gompertz-Makeham law in most cases does not exceed 2-4%. These results show that the dependence of mortality on age for adult humans can in the overwhelming majority of cases be described with sufficient accuracy by the Gompertz-Makeham law. By way of comparison, if the force of mortality were for example to increase with age linearly rather than exponentially, then the correlation coefficient between the logarithm of the increment in the force of mortality and age would be zero (since in this case the magnitude of the increment in the force of mortality would be constant).

As has already been noted, the correlation coefficient is not the best measure of the linearity of the dependency under investigation, since its divergence from unity may just as well be connected with the random scatter of data as with systematic deviations from linearity. For this reason, we also used another method for testing the adequacy of the Gompertz-Makeham formula. The principle behind this method was described in the previous chapter: it involves testing for linearity by calculating the ratio between the tangents of the angles of the slope at the beginning and end of each dependency. In each concrete case, this ratio may deviate significantly from unity, but if the deviation is random rather than systematic in character, the centre of the distribution of the ratios for a large series of analysed tables must tend towards unity as the number of observations increases.

To test the adequacy of the Gompertz-Makeham law using this technique, we checked, by calculating the corresponding tangent ratios, whether there was a linear dependency between the logarithm of the age-related increment in the force of mortality and human age. For each life table, we analysed the dependency and calculated the ratio between the tangent of the angle of the slope in the age interval 40-50 years and the corresponding tangent in the age interval 60-70 years. This ratio, named the C-criterion (Gavrilova et al., 1979), was calculated for 290 life tables published by the UN. It turned out that the centre (median) of the C-criterion's distribution was equal to 0.98 ± 0.03 (Gavrilova et al., 1979).

The results just presented were obtained for a combined analysis of male and female life tables (Gavrilova et al., 1979; Gavrilova, 1982). Subsequently, Pakin and Hrisanov (1984) repeated these calculations with some modifications to the previous technique, and most importantly, with a separate analysis of male and female mortality. As a result, they discovered a significant tendency towards deviation from the Gompertz-Makeham formula: in men they observed a tendency for the force of mortality to grow less steeply with age, and in women they observed a tendency for the force of mortality to grow more steeply with age.

Since Pakin and Hrisanov did not carry out a quantitative evaluation of the observed deviations, merely pointing out their existence, we have tried to do this using the C-criterion.

The C-criterion calculations carried out separately for men and women do indeed confirm the validity of Pakin and Hrisanov's conclusions, and permit a quantitative characterisation of the observed deviations. It turns out that the centre (median) of the C-criterion's distribution is 1.18 for men (1.04-1.34, $P > 0.99$), and 0.76 for women (0.68-0.84, $P \geq 0.99$). This means that in the age interval 45-65 years the angular coefficient of the dependency between the logarithm of the increment in the force of mortality and age decreases on average by 18% in men and increases on average by 24% in women (Semyonova et al., 1985).

In order to evaluate how significant these deviations from the Gompertz-Makeham law are, let us compare the Gompertz-Makeham formula with some other competing life span distribution laws. For example, it was shown above that if the generalised Weibull law is correct, the centre of the C-criterion's distribution ought to be equal to the inverse ratio of the ages for which the tangents are calculated (Gavrilov, 1980). In our case, this theoretically expected value is 65/45 years, i.e. 1.44. Now let us compare the theoretically expected and the observed values of the C-criterion:

| | |
|---|---|
| Theoretically expected value of the C-criterion if the Gompertz-Makeham law is valid: | 1.00 |
| Theoretically expected value of the C-criterion if the generalised Weibull law is valid: | 1.44 |

| | |
|---|---|
| Observed value of the C-criterion's centre of distribution for men and women: | 0.98 |
| Observed value of the C-criterion's centre of distribution for men: | 1.18 |
| Observed value of the C-criterion's centre of distribution for women: | 0.76 |

Comparing these results, it is easy to see that the Gompertz-Makeham law is significantly more preferable than the generalised Weibull law. Thus despite the existence of a significant tendency towards systematic deviations from the Gompertz-Makeham law, the results obtained nevertheless demonstrate that it should be preferred to the other mortality laws. If we consider the whole range of individual characteristics in human populations (different experiences for different generations, the phenomenon of age discrimination, genetic and social heterogeneity), what is surprising is not the existence of deviations from the Gompertz-Makeham law, but the negligibility of these deviations from such a simple formula in the description of such a complex phenomenon as human mortality. What is more, it has become clear that even these small deviations are not historically stable, but fluctuate around the trajectory corresponding to the Gompertz-Makeham formula (Pakin, 1988). Consequently, the Gompertz-Makeham law can be used in human life span investigations if its adequacy is checked in each concrete case, and the validity of any conclusions is verified by other methods.

### 3.3. The biological characteristics of the human life span

As was demonstrated above, the life span distribution of human beings and laboratory animals is described by one and the same formula. A question which naturally arises is how social factors affect the human life span if they do not essentially alter the shape of the life span distribution. The only reasonable answer to this question is the proposition that social factors have no direct effects, but act indirectly by altering human ecology. This leads to a change in the numerical values of the distribution's parameters without altering its shape. Therefore the first stage in an analysis of the biosocial structure of life span is to study the extent to which each of the parameters in the Gompertz-Makeham distribution depends on social and biological factors.

*Methods of estimating the parameters of the Gompertz-Makeham formula.* In order to study the influence of social and biological factors on the parameters of the life span distribution, we must first be able to calculate them.

Two methods can be proposed for estimating the parameters of the Gompertz-Makeham formula. The first, simplified method is based on linearisation of the data using the following coordinates: (i) the natural logarithm of the age-specific increment in the force of mortality, (ii) age. As a result, the problem is reduced to the standard procedure for evaluating the parameters of a linear regression by the least squares method:

$$y = ax + b, \tag{43}$$

where $y = \ln(\Delta\mu_x)$ is the logarithm of the age-specific increment in the force of mortality, and $x$ is age.

As was shown earlier, the parameters of this linear equation are connected with the Gompertz parameters in the following manner:

$$a = \alpha \qquad b = \ln R + \ln(\exp(\alpha\Delta t) - 1), \tag{44}$$

where $\Delta x$ is the fixed step of numerical differentiation chosen in calculating the value of the age-specific increment in the force of mortality. Thus it turns out that the slope coefficient of the linear regression coincides with the unknown parameter $\alpha$, and the parameter $R$ can be calculated in the following manner:

$$R = \frac{\exp(b)}{\exp(\alpha\Delta t) - 1} \tag{45}$$

Then, knowing the values of the parameters $R$ and $\alpha$, it is possible to estimate the value of the parameter $A$, which is the difference between the observed force of mortality and its age-dependent component ($R\exp(\alpha x)$). For a more accurate estimate of the value of parameter $A$, it is possible to calculate the arithmetic mean of these differences at various ages:

$$\bar{A} = \frac{1}{n} \sum_{i=1}^{n} (\mu(x_i) - R \cdot \exp(\alpha x_i)). \qquad (46)$$

This method of estimating the parameters is accessible to any investigator, since it does not require the use of complex computational techniques. What is more, even at the first stage in the calculation (the construction of the graph of the linear dependency), it is possible to estimate how adequate the Gompertz-Makeham formula is in the given case, and to determine the range of ages to which it is applicable (by observing where the graph is linear). It is worth mentioning also that this method led in its time to a series of fundamental results (Gavrilov and Gavrilova, 1979a; 1979b), which were subsequently verified by other more sophisticated methods (Gavrilov, 1984a; 1984b; Gavrilov et al., 1983). At the same time it must be recognised that this particular method has little statistical justification and gives biassed estimates of the parameters, especially when the data are significantly scattered (overestimation of parameter $\alpha$ and underestimation of parameter $R$). Finally, the method turns out to be inapplicable when the scatter of data results in negative values for the age-specific increment in the force of mortality (the logarithm cannot be calculated).

Another method of determining the parameters of the Gompertz-Makeham formula consists in the use of standard programs for evaluating the parameters of non-linear regressions. One of these programs can be found in the well-known BMDP package, and another was written in Fortran by V.N.Nosov (Biological Department of Moscow State University).

As many years of experience working with computers have shown, calculations of the parameters of the Gompertz-Makeham law are significantly accelerated, simplified and made more reliable if it is possible to make reasonably accurate initial estimates. With this aim in mind, a method was devised which permits the parameters to be estimated on the basis of the number of survivors at four equally-spaced ages (for example at 20, 40, 60 and 80 years).

Let $l_{x_0}$, $l_{x_0+\Delta x}$, $l_{x_0+2\Delta x}$ and $l_{x_0+3\Delta x}$ be the numbers of those surviving to the given age in the life table under analysis. Then in order to determine the parameters, the following auxiliary values are first calculated:

$$y_1 = \ln\left(\frac{l_{x_0}}{l_{x_0+\Delta x}}\right) \; ; \qquad y_2 = \ln\left(\frac{l_{x_0+\Delta x}}{l_{x_0+2\Delta x}}\right) \; ;$$

$$y_3 = \ln\left(\frac{l_{x_0+2\Delta x}}{l_{x_0+3\Delta x}}\right) \; ; \quad z = y_1 + y_3 - 2y_2$$

$$w = \frac{(y_3 - y_2)}{(y_2 - y_1)} \, . \tag{47}$$

Subsequently it is possible to calculate the parameters of the Gompertz-Makeham equation according to the formulae:

$$A = \frac{y_1 y_3 - y_2^2}{z\Delta x} \; ; \quad R = \frac{(y_2 - y_1)^2 \ln w}{z\Delta x(w - 1)w^{(x_0/\Delta x)}} \; ;$$

$$\alpha = \frac{1}{\Delta x} \ln w \, . \tag{48}$$

Using this method of estimating the parameters, the graph of the dependence of the number of survivors on age, theoretically calculated according to the Gompertz-Makeham law, precisely passes through all four points which correspond to the numbers of survivors chosen for the evaluation. It is also important to note that subsequent refinement of the values of these parameters by the method of least squares on a computer using the the non-linear regression program only adds a minor correction to the initial estimates. For this reason the method is widely applicable.

Let us illustrate the proposed method with a concrete example. The numbers of survivors to age 20, 40, 60 and 80 as given in the life table for Swedish men from 1926-1930 are 88,575, 80,997, 66,825 and 24,197 respectively. The auxiliary values are:

$$y_1 = 0.0894;\ y_2 = 0.1923;\ y_3 = 1.0159;\ z = 0.7206;$$

$$w = 8.0033,$$

which gives the following values for the parameters:

$$A = 3.74 \cdot 10^{-3}\ \text{yr}^{-1},\quad R = 2.73 \cdot 10^{-5}\ \text{yr}^{-1},$$

$$\alpha = 0.104\ \text{yr}^{-1}.$$

It is interesting to note that further refinement of these parameters with 61 values for the number of survivors (age interval 20 to 80 years) and using the most sophisticated mathematical and computational techniques gives essentially the same results:

$$A = (3.76 \pm 0.08) \cdot 10^{-3}\ \text{yr}^{-1},$$

$$R = (2.74 \pm 0.27) \cdot 10^{-5}\ \text{yr}^{-1},$$

$$\alpha = 0.104 \pm 0.001\ \text{yr}^{-1}.$$

It is not difficult to see that between the initial and the final estimates of the parameters there are no significant differences. In addition, the calculation of the initial estimates takes only a few minutes and does not require the use of computational techniques.

Thus there are at least three methods for evaluating the parameters of the Gompertz-Makeham law: the traditional method based on the linearisation of the data, which is very transparent; the modern method based on the use of standard programs for evaluating the parameters of non-linear regressions; and the express method which is useful for preparatory calculations. The investigator is justified in selecting any of these methods, depending on his aims and possibilities, or in searching for other methods of solving the problem, some of which are described in the special literature (Grenander, 1956; Garg et al., 1970; Slob and Janse, 1988).

*The criterion of historical stability.* In order to determine which parameters of the life span distribution basically depend on social factors and which depend on biological factors, the simplest method would be to carry out a comparison of populations which differ solely according to a complex of social factors or solely according to a complex of biological factors. Since different countries and even different regions can simultaneously show significant differences both in the socio-economic conditions and in the ecological characteristics of the populations under comparison, simply contrasting them can do little to help solve the problem at hand. It seems to us that the problem can only be solved by analysing the historical changes in the parameters of the life span distribution over the period in the 20th century when there was a sharp fall in mortality. Indeed, it is well known that falls in human mortality over very short historical periods are caused exclusively by socio-economic transformations. Therefore the biological characteristics of life span must satisfy the criterion of historical stability. In other words, those parameters which changed during the period when there was a sharp fall in mortality are socially regulated parameters, and those which remained unaltered despite the sharp fall in mortality are socially autonomous and reflect deeper (biological) characteristics of human populations.

*The phenomenon of the historical stability of the age-dependent component of mortality.* In 1979, during an analysis of the historical changes in the mortality of the male population of Sweden, the hitherto unknown phenomenon was discovered of the historical stability of the age-dependent component of mortality and the parameters which determine it (Gavrilov and Gavrilova, 1979a). Further more careful investigation confirmed the reliability of the phenomenon (Gavrilov et al., 1983) and permitted the conclusion that it was quite general in character (Gavrilov, 1984a; 1984b; Gavrilov and Gavrilova, 1984b; Gavrilov et al., 1986; Gavrilova, 1982a; Gavrilov et al., 1985).

In table 4 we give the values of the parameters of the Gompertz-Makeham formula for the male population of Sweden during the period 1901 to 1983. It can be seen that parameter $A$ (the background component of mortality) is the only parameter which has significantly changed over the period under investigation. The two other parameters ($R$ and $\alpha$), which determine the value of the age-dependent component of mortality, turn out to be practically unchanged, despite the sharp fall in total mortality in the 20th century.

**Table 4** Historical dynamics of the parameters of the Gompertz-Makeham equation for Swedish men

| Years | Values of the parameters and their confidence intervals*, $year^{-1}$ | | |
|---|---|---|---|
| | $A \cdot 10^3$ | $R \cdot 10^6$ | $\alpha \cdot 10^3$ |
| 1901-1910 | 5.52±0.16 | 33.6±4.7 | 101.3±2.0 |
| 1911-1915 | 5.34±0.15 | 29.0±3.9 | 103.6±1.9 |
| 1916-1920 | 6.58±0.28 | 14.8±4.5 | 112.3±4.3 |
| 1921-1925 | 3.89±0.17 | 25.0±4.2 | 104.8±2.4 |
| 1926-1930 | 3.74±0.08 | 27.4±2.7 | 104.0±1.0 |
| 1931-1935 | 2.93±0.08 | 29.2±2.3 | 103.1±1.1 |
| 1936-1940 | 2.33±0.13 | 35.0±3.9 | 101.3±1.6 |
| 1941-1945 | 2.00±0.14 | 26.4±4.0 | 103.7±2.2 |
| 1946-1950 | 1.11±0.13 | 29.6±3.9 | 102.6±1.9 |
| 1951-1955 | 0.70±0.10 | 28.5±3.0 | 102.7±1.5 |
| 1956 | 0.67±0.09 | 27.4±2.6 | 103.2±1.4 |
| 1957 | 0.57±0.14 | 29.1±4.4 | 102.7±2.2 |
| 1958 | 0.54±0.12 | 23.6±3.3 | 105.1±2.0 |
| 1959 | 0.66±0.09 | 20.6±2.3 | 106.7±1.6 |
| 1960 | 0.53±0.13 | 23.6±3.5 | 105.8±2.2 |
| 1961 | 0.50±0.14 | 24.4±3.8 | 104.7±2.3 |
| 1962 | 0.52±0.10 | 23.1±2.6 | 106.1±1.6 |

*The confidence intervals correspond to the 5% confidence level

Table 4 (continued)

| Years | $A \cdot 10^3$ | $R \cdot 10^6$ | $\alpha \cdot 10^3$ |
|---|---|---|---|
| 1963 | 0.50±0.14 | 23.7±3.9 | 105.3±2.4 |
| 1964 | 0.56±0.11 | 24.5±3.1 | 104.8±1.8 |
| 1965 | 0.65±0.09 | 21.6±2.3 | 106.6±1.5 |
| 1966 | 0.57±0.09 | 24.0±2.6 | 105.0±1.6 |
| 1967 | 0.72±0.07 | 24.1±2.0 | 105.0±1.2 |
| 1968 | 0.83±0.05 | 19.9±1.2 | 107.6±0.9 |
| 1969 | 0.68±0.08 | 22.3±2.2 | 106.4±1.4 |
| 1970 | 0.61±0.08 | 26.4±2.4 | 103.2±1.3 |
| 1971 | 0.66±0.05 | 26.4±1.6 | 103.5±0.9 |
| 1972 | 0.69±0.06 | 25.9±1.9 | 103.8±1.0 |
| 1973 | 0.66±0.05 | 25.2±1.6 | 104.3±0.9 |
| 1974 | 0.62±0.04 | 28.1±1.4 | 102.7±0.7 |
| 1975 | 0.69±0.06 | 27.1±1.8 | 103.2±1.0 |
| 1976 | 0.63±0.05 | 27.7±1.6 | 103.2±0.9 |
| 1977 | 0.65±0.05 | 30.8±1.9 | 101.1±0.9 |
| 1978 | 0.65±0.06 | 28.4±2.0 | 102.4±1.0 |
| 1979 | 0.65±0.07 | 28.6±2.4 | 102.2±1.2 |
| 1980 | 0.60±0.06 | 29.7±2.2 | 101.4±1.1 |
| 1981 | 0.48±0.06 | 28.7±2.1 | 101.8±1.1 |
| 1982 | 0.49±0.05 | 27.6±1.9 | 102.0±1.0 |
| 1983 | 0.48±0.05 | 24.8±1.8 | 103.2±1.0 |

Fig.8 illustrates the meaning of this surprising phenomenon. It can be seen that the decline in human mortality in the 20th century is almost exclusively accounted for by the background component of mortality. The age-dependent component of mortality has remained practically unchanged, despite radical social transformations and advances in medicine and health-care. When the level of background mortality finally approached its ultimate value of zero, the traditional reserve for a decline in mortality was exhausted. At that moment, the mortality of adults practically ceased to fall and has not changed essentially to the present day (see Fig.8).

The constancy of the age-dependent component of mortality is striking. In fact, in the 20th century the human life span has increased almost by a factor of two, and there have been radical changes in life-styles and the structure of causes of death. Nevertheless, despite all the social transformations and the advances in medicine and health-care, the age-dependent component of mortality has remained practically unchanged. Consequently, the age-dependent component of mortality is determined not by social conditions, but by significantly more stable biological characteristics of human populations. Thus, in conformity with the criterion of historical stability, it is necessary to consider the age-dependent component of mortality ($R$ $\exp(\alpha x)$ as a biological characteristic which is of special interest for the biology of life span. The same can be said of the parameters which determine it ($R$ and $\alpha$).

Since this conclusion is fundamentally important for further research into the biological basis of the human life span, it is absolutely essential to verify its reliability. In the first instance, it is important to ascertain the extent to which the historical stability of the age-dependent component of mortality is a universal regularity.

With this aim in mind, 150 complete life tables were analysed for the period from the beginning of the century to the present day and for the populations of the following countries: Austria, England and Wales, Belgium, Bulgaria, Hungary, the German Democratic Republic and the Federal Republic of Germany, Denmark, Spain, Italy, Norway, the USA (white and non-white population separately), Finland, France, Switzerland, Sweden and Japan. The analysis of each table involved the calculation of the parameters of the Gompertz-Makeham equation in the age interval 20-80 years. In order to give a quantitative characterisation of the tendency of each parameter to alter in time, we determined the Spearman rank correlation coefficients between the parameters ($A$, $R$ and $\alpha$) and time (over the 20th century).

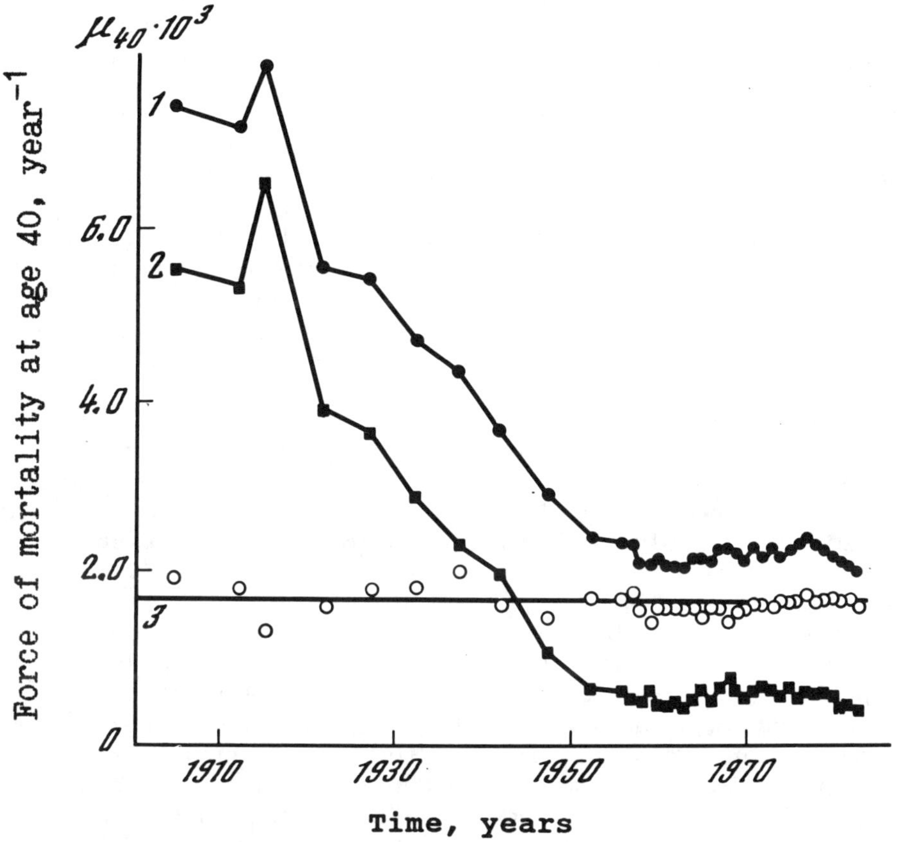


**Figure 8** Historical changes in the total force of mortality (1), its background component (2) and age-dependent component (3) for Swedish men.

The mortality figures correspond to the number dying in the course of the year out of 1000 40-year-old men.

abscissa: time, years

ordinate: the force of mortality of 40-year-olds, $year^{-1}$; $\mu_{40} \cdot 10^3$

In total 32 time dependencies were worked out for each parameter (the data for men and women were processed separately). It turned out that at the 1% confidence level there was a significant trend for parameter *A* to alter with time in 25 of the 32 cases, for parameter *R* only in 6, and for parameter $\alpha$ only in 5. It is important to emphasise that even in the 7 cases when there was no significant trend for parameter *A* to decrease with time, this was caused solely by the fact that the value of the parameter fell extremely sharply, rapidly approached its zero limit and subsequently showed no essential changes. In this case, therefore, the results of investigating the tendencies of the observed trends are more revealing. It turns out that in all 32 cases parameter *A* had a tendency to fall in time (the tendency was significant). However, the historical change in parameters *R* and $\alpha$ showed no significant tendency (this conclusion was reached with the aid of a sign test at the 1% significance level).

The analysis we have carried out has shown that the phenomenon of the historical stability of the age-dependent component of mortality is a reasonably universal regularity, and not a unique peculiarity of the male population of Sweden. To illustrate this conclusion, we can offer a few more examples (see Figs. 9-11). It must however be emphasised that the concept of historical stability in no way entails the absolute constancy of the age-dependent component of mortality in time. It is totally possible for a temporal sequence of the values for this component not to be random, and for a more careful analysis to reveal hidden periodicities or even trends. Historical stability should not be taken to mean absolute constancy, but the insignificance of any changes in comparison with the sharp change in total mortality. The mere fact of a small change in the age-dependent component of mortality cannot be treated as an argument against its biological nature, inasmuch as the biological characteristics of human populations do not remain absolutely unaltered (it suffices to recall the phenomenon of acceleration). In a number of countries one can indeed observe a tendency for the age-dependent component of mortality to decline (Switzerland, England and Wales) or increase (Hungary). The causes of these trends are unknown and merit special study. However in the overwhelming majority of cases the age-dependent component of mortality turns out to be so resistant to social transformations that this fact cannot be considered accidental.

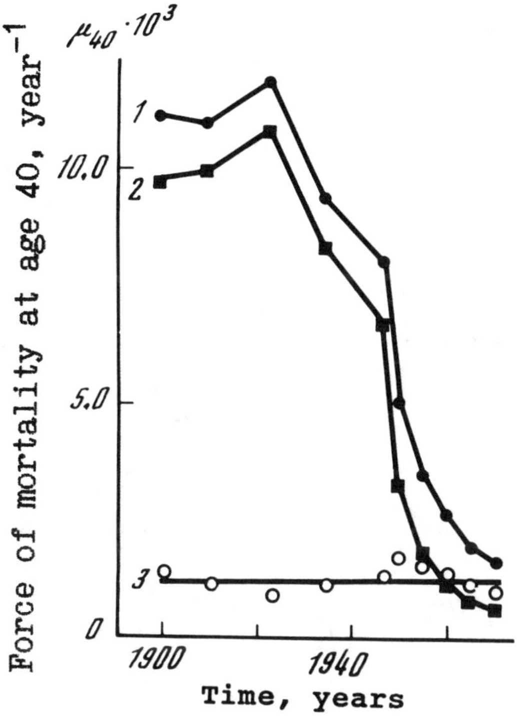


**Figure 9** Historical changes in the total force of mortality (1), its background component (2) and age-dependent component (3) for Japanese women.

The mortality figures correspond to the number dying in the course of the year out of 1000 40-year-old women. Taken from Gavrilov (1984b).

abscissa: time, years

ordinate: the force of mortality of 40-year-olds, year$^{-1}$; $\mu_{40} \cdot 10^3$

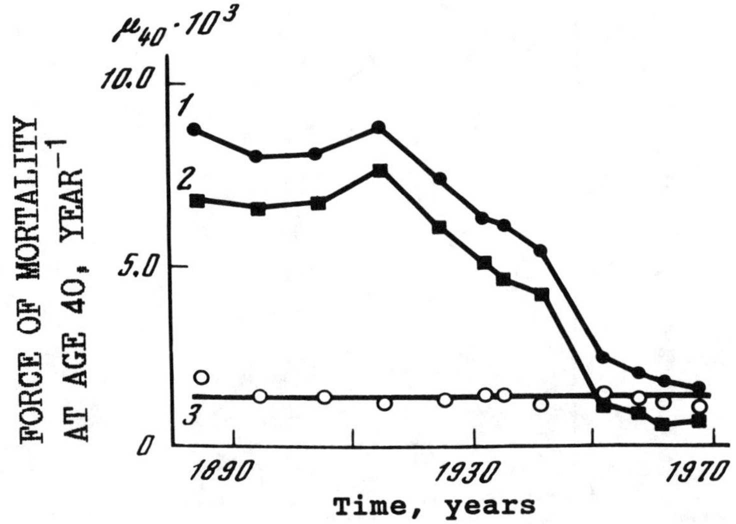


**Figure 10** Historical changes in the total force of mortality (1), its background component (2) and age-dependent component (3) for Finnish women.

The mortality figures correspond to the number dying in the course of the year out of 1000 40-year-old women. Taken from Gavrilov (1984b).

abscissa: time, years

ordinate: the force of mortality of 40-year-olds, year$^{-1}$; $\mu_{40} \cdot 10^3$

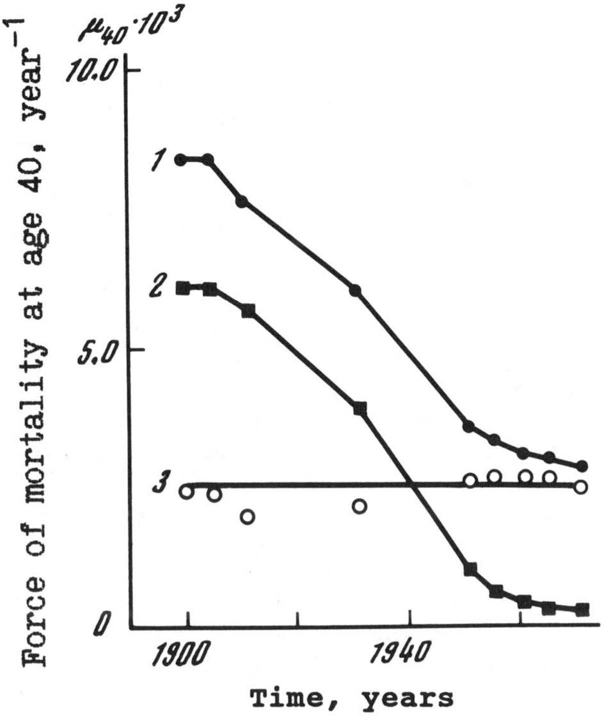


**Figure 11** Historical changes in the total force of mortality (1), its background component (2) and age-dependent component (3) for Italian men.

The mortality figures correspond to the number dying in the course of the year out of 1000 40-year-old women. Taken from Gavrilova et al. (1982).

abscissa: time, years

ordinate: the force of mortality of 40-year-olds, year$^{-1}$; $\mu_{40} \cdot 10^3$

### 3.4. The epidemiological approach to studying the biology of the human life span

Amongst the methods for studying the biological bases of the human life span, a special role belongs to epidemiological investigations, since scientists are restricted in the experiments they can perform on human beings, but nature knows no limits. Epidemiology, which was born as a subdiscipline of medical geography studying the foci of infectious diseases and the laws governing the development of epidemics, later dramatically expanded its sphere of research. There were rapid and fruitful advances in the epidemiology of noninfectious diseases, including the epidemiology of cancer (Doll and Peto, 1981). Traditional methods of epidemiological investigation, which were based on a comparison of different geographical regions, were supplemented by an analysis of the characteristics of different population groups living in the same area. As a result, modern epidemiology has transcended the bounds of medical geography and purely medical research, turning into a powerful tool for the investigation of fundamental medico-biological problems. One of these problems is the determination of the biological foundations of the human life span.

Up to recent times, epidemiological investigations into the problem of life span have essentially been restricted to an analysis of the phenomenon of long-living (Zubov and Kozlov, 1982). However, such an approach has two essential drawbacks. Firstly, the age of long-lived people is usually exaggerated (Abrams, 1985), and it is therefore unclear what is being investigated: the epidemiology of extreme longevity or the epidemiology of tales and legends. Secondly, the geography of concentrations of long-lived people may be determined not by modern conditions, but by circumstances unknown today which obtained many years ago. Therefore it is significantly more fruitful to analyse reliable demographic data concerning adult mortality (20-80 years), with the aim of finding out the biological characteristics of life span, and subsequently investigating the regional differences.

At the same time, the search for fundamental biological differences in the mortality of human populations is complicated by the influence of social factors. Therefore, a comparison of the mortality levels of people living in different socio-economic conditions can in itself contribute little to an investigation of the biological foundations of life span. Theoretically, it might be possible to solve this problem by taking into account each of the confounding factors, deducting its effect, and thereby obtaining the remaining biologically determined differences. In practice, however, this exclusion technique represents an almost hopeless task, since it is difficult even to enumerate all the possible confounding factors, let alone to quantify their effects. For this reason, epidemiological investigations into the biology of human life span have been developing exceptionally slowly.

This situation can be radically altered, now that we know the biological characteristics of the human life span as described in the previous section. As has already been noted, when we analyse the dependence of human mortality on age, we can isolate a component of mortality which is historically stable despite the sharp reduction in total mortality and the radical socio-economic transformations which have occurred in the 20th century. The historical stability of this component of mortality shows that it is determined by the fundamental (genetic and ecological) characteristics of human populations, which evolve significantly more slowly than socio-economic living conditions. Thus we have the unique opportunity to study the biologically determined differences in the human life span against the complex background of the socio-economic differences of the populations under comparison.

Before turning to an analysis of regional differences in the biological characteristics of life span, it is essential to decide which population groups should be compared. Above all, the problem arises: how should we analyse data concerning the mortality of males and females, together or separately? Indeed, there are usually significant differences in mortality levels between these two large population groups (male mortality is greater than female mortality). On the other hand, if these differences are basically social in nature, as some demographers suppose, the biological characteristics of the life spans of each sex ought to be close. At the same time, the experience of previous research (c.f. the history of the calculation of the C-criterion on pages 70-72) shows that a mixed analysis of male and female data is often an extreme simplification. Therefore, we must first of all determine which parameters of the life span distribution are linked to the differences in male and female mortality.

*A comparative analysis of the male and female life span distribution.* In order to study the gender differences in human mortality, we used the technique of paired comparisons for each of the three parameters of the Gompertz-Makeham formula. In total, 255 paired comparisons were carried out for each parameter in geographical and historical cross-sections. The magnitude of the background component of the force of mortality (parameter *A*) was as a rule similar for men and women, within the limit of measurement errors. If these data are compared as a whole, we do not find even a tendency for differences in this parameter: in half the cases the value of parameter *A* for women was somewhat greater than the corresponding value for men (53±6% for $P \geq 0.95$), whereas in the remaining half (47±6% for $P \geq 0.95$) the opposite was observed. Consequently, gender differences in life span are connected not with the background component, but with the age-dependent component of the force of mortality. Indeed, the age-dependent component of the force of mortality for men turned out, as a rule, to be significantly greater than the age-dependent force of mortality for women. What is more, these differences represent a reliable tendency when the data are compared in total: in 97.6% of cases (95.2-99.0% for $P \geq 0.95$), the value of the age-dependent component of the force of mortality for 40-year-old men was greater than the corresponding value for women. This regularity is determined both by differences in parameter *R* and in parameter $\alpha$. The value of *R* for men was in 94.7% of cases (91.5-97.1% for $P \geq 0.95$) greater than the corresponding value for women. The opposite was observed for parameter $\alpha$ : in 92.3% of cases (88.7-95.3% for $P \geq 0.95$) the value of $\alpha$ for women was higher. Thus, the age-dependent component of the force of mortality and its doubling period are smaller for women than for men. Table 5 illustrates the observed regularities with concrete examples. It shows the median values of the parameters of the distribution and age-dependent force of mortality for age 40, as well as confidence intervals for these median values (in brackets, corresponding to 97-99% confidence limits). The figures in table 5 were obtained by non-parametric statistical techniques (Campbell, 1967), since the distribution of the values of the parameters might not be a normal one. It can be noted that gender differences in life span are most prominent in Finland and France, and minimal in Sweden.

Thus, gender differences in the mortality of adult males and females are basically connected with historically stable (biological) characteristics of the human life span. Consequently, an analysis of regional differences in these characteristics must be carried out separately for men and women.

**Table 5** Comparison of the life span distributions of men and women (values for the parameters of the distributions, $year^{-1}$)

| Country | Years | Number of tables processed | $\alpha \cdot 10^3$ | $R \cdot 10^6$ | $R \cdot e^{40\alpha} \cdot 10^3$ |
|---|---|---|---|---|---|
| | | | Men | | |
| Bulgaria | 1920-1977 | 8 | 102.8 (96-105) | 32.5 (28-53) | 1.98 (1.8-2.5) |
| Italy | 1899-1979 | 12 | 95.1 (93-107) | 57.7 (32-68) | 2.59 (2.2-2.7) |
| Finland | 1881-1980 | 15 | 87.1 (85-95) | 137.1 (79-155) | 4.33 (3.5-4.7) |
| France | 1856-1981 | 45 | 87.2 (86-88) | 110.5 (104-118) | 3.61 (3.5-3.9) |
| Sweden | 1901-1983 | 38 | 103.7 (103-105) | 26.4 (24-28) | 1.64 (1.6-1.7) |
| Japan | 1891-1983 | 31 | 103.7 (102-105) | 31.1 (26-39) | 2.03 (1.7-2.4) |
| | | | Women | | |
| Bulgaria | 1920-1977 | 8 | 118.4 (117-122) | 7.8 (7-10) | 0.91 (0.9-1.0) |
| Italy | 1899-1979 | 12 | 115.2 (110-129) | 10.8 (6-11) | 1.02 (1.0-1.1) |
| Finland | 1881-1980 | 15 | 115.1 (115-117) | 12.2 (9-14) | 1.23 (1.1-1.4) |
| France | 1856-1981 | 45 | 104.0 (103-106) | 18.8 (17-21) | 1.29 (1.2-1.4) |
| Sweden | 1901-1983 | 38 | 112.5 (110-115) | 9.6 (8-11) | 0.82 (0.8-1.1) |
| Japan | 1891-1983 | 31 | 112.9 (112-115) | 10.7 (8-12) | 1.00 (0.8-1.1) |

*The epidemiology of the biological characteristics of the human life span.* As can be seen from table 5 above, the magnitude of the age-dependent component of mortality, which is historically stable, is strongly dependent on sex and shows significant regional variation.

Figs. 12 and 13 contain cartograms of the levels of the biological (age-dependent, historically stable) component of mortality for the male and female population of Western Europe at age 40. We shall subsequently show that the selection of a given age affects only the absolute level of the biological component of mortality, and does not alter the ranking of regions according to their values for this index (see chapter 4, p.150). It must be noted that the level of the biological component of mortality for women turned out to be so much lower than the corresponding level for men that in constructing the cartogram we had to use a different absolute scale to grade the levels of the index.

Comparing the cartograms, we note that they have a certain similarity. For instance, in both cases low levels of the biological component of mortality are observed in Sweden, Norway, Greece, Bulgaria, Spain, and the south of Italy. On the other hand, north-western Europe (Great Britain, northern France and Belgium) is characterised by a high level of the biological component of mortality for both men and women.

At the same time, it is possible to observe, as well as a significant similarity, certain differences between the cartograms. For instance, the male population of Finland is characterised by the highest level of the biological component of mortality in Europe. Simultaneously, the female population of Finland occupies an intermediate position for the indicator under investigation.

Comparing this fact with other epidemiological data confirms the special status of the male population of Finland. Indeed, it is well-known that one of the highest mortality levels in the world from cardiovascular disease is to be found in Finland, predominantly amongst men (Salonen et al., 1982). What is more, an especially high mortality from this kind of disease is recorded in the eastern regions of Finland (North Karelia) (Salonen et al., 1982), where, according to our calculations, the highest level of the biological component of mortality is also to be observed. Thus, the observed regularity (the high level of the biological component of mortality in Finnish men) is directly supported by other epidemiological investigations.

In discussing the possible mechanisms underlying this phenomenon, it must be noted that the blood cholesterol content in Finnish men is one of the highest in the world

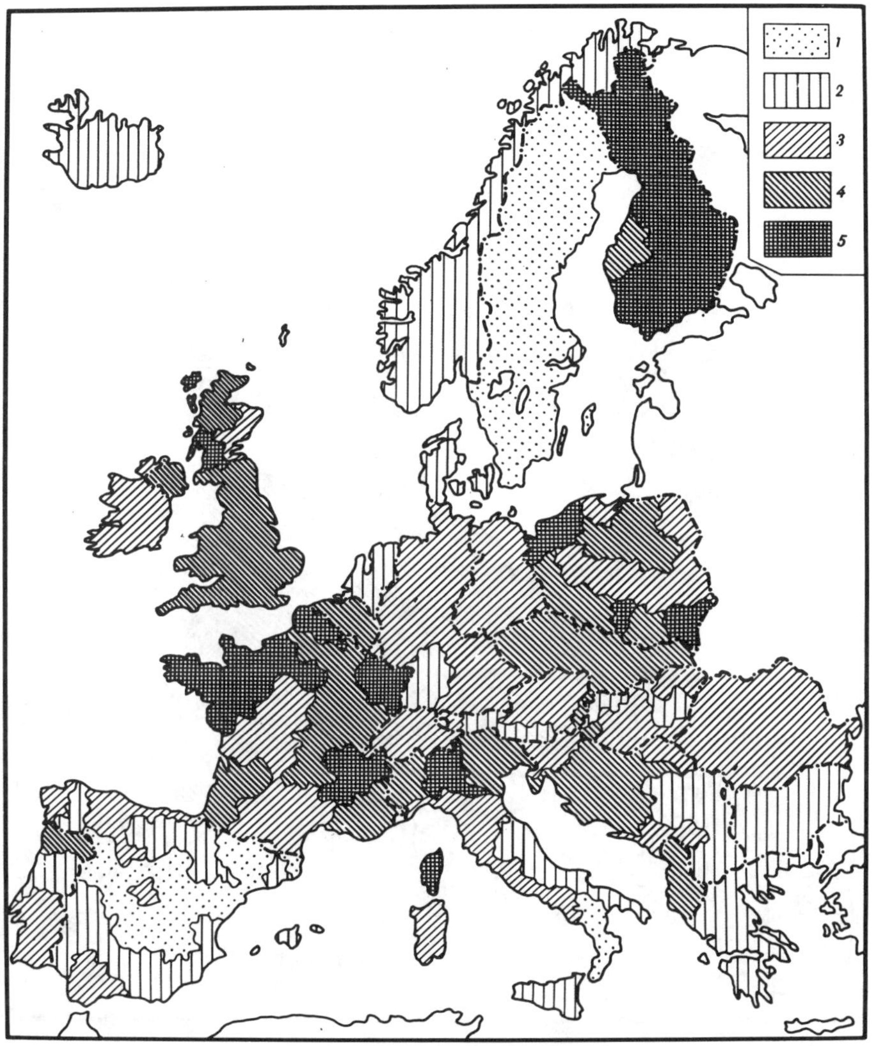


**Figure 12** Cartogram showing the levels of the age-dependent (biological) component of male mortality for age 40 years in Western Europe.

1 - low level: less than 1.6 · $10^{-3}$ $year^{-1}$
2 - reduced level: (1.61-2.20) · $10^{-3}$ $year^{-1}$
3 - medium level: (2.21-2.80) · $10^{-3}$ $year^{-1}$
4 - heightened level: (2.81-3.40) · $10^{-3}$ $year^{-1}$
5 - high level: higher than 3.41 · $10^{-3}$ $year^{-1}$

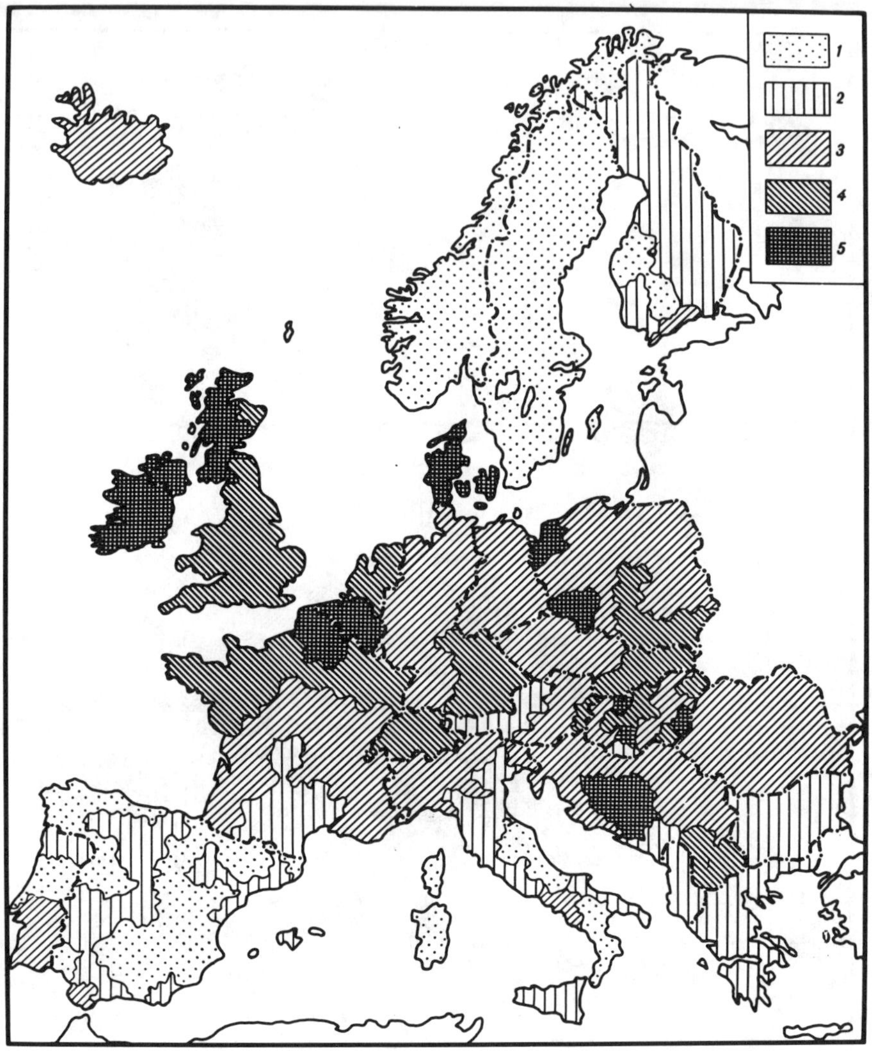


**Figure 13** Cartogram showing the levels of the age-dependent (biological) component of female mortality for age 40 years in Western Europe.

1 - low level: less than 0.83 · $10^{-3}$ $year^{-1}$
2 - reduced level: (0.84-1.03) · $10^{-3}$ $year^{-1}$
3 - medium level: (1.04-1.23) · $10^{-3}$ $year^{-1}$
4 - heightened level: (1.24-1.43) · $10^{-3}$ $year^{-1}$
5 - high level: higher than 1.44 · $10^{-3}$ $year^{-1}$

(Knuiman et al., 1982). An especially high blood cholesterol content is recorded in men living in the eastern part of Finland (6.3-6.4 mmol/l), where the highest levels of the biological component of mortality (Gavrilov and Gavrilova, 1986) and mortality from heart disease (Salonen et al., 1982) are observed. Men in the western part of Finland have blood cholesterol levels which are somewhat lower (5.9-6.2 mmol/l), but still significantly greater than those which are characteristic of most other countries (for example, the blood cholesterol concentration of Italian men is 5.1-5.3 mmol/l) (Knuiman et al., 1982).

The significance of these facts is obvious when one takes into account that a high level of cholesterol in the blood is considered to be an important risk factor in the development of cardiovascular disease (Knuiman et al., 1982).

Another interesting characteristic of Finland is the fact that the soils and waters of the country contain unusually small quantities of the trace element selenium (Mutanen and Koivistoinen, 1983). As a result, the blood concentration of selenium for Finns is significantly lower than the levels characteristic of other countries. For example, the blood concentration of selenium for the inhabitants of Helsinki is only 0.081 μg/ml, almost three times lower than for the populations of most other countries in the world. By way of comparison, we note that the average level of the blood selenium level for the US population is 0.206 μg/ml, and for the inhabitants of Venezuela it is 0.355 μg/ml (Levander, 1982). It is interesting to observe that an especially low blood selenium level is again found in the population of eastern Finland. For example, for the inhabitants of the town of Lappeenranta (North Karelia), the blood selenium content is only 0.056 μg/ml (Levander, 1982).

In discussing the biological consequences of a deficiency of selenium in the organism, the following facts must be borne in mind. It is well-known that a selenium deficiency in the diet of experimental animals leads to the development of heart disease and a number of other disorders (Levander, 1982). Epidemiological investigations have confirmed that regions with a low selenium content show increased mortality from a whole series of diseases, including cardiovascular ones (Shamberger et al., 1978). What is more, in the course of these epidemiological investigations it became clear that the male population is especially sensitive to a selenium deficiency. For example, the correlation coefficient between mortality from circulatory diseases and the blood selenium content for the population of 19 US states is -0.79 for men and only -0.53

for women. The analogous coefficients for mortality from hypertension were -0.84 for men and only -0.54 for women (Shamberger et al., 1979). What is more, it turns out that a selenium deficiency increases the risk of developing malignant tumours by a factor of almost 3 in men, but has no significant effect on the risk of developing cancer in women (Kok et al., 1987). The role of a selenium deficiency in the occurrence of circulatory diseases was also confirmed in the course of a long-term investigation carried out in the regions of Finland with a low selenium content (Salonen, 1987). It became clear that in men with a serum selenium content of less than 0.045 μg/ml the risk of death from cardiovascular diseases was 2.7 times higher, and the risk of death from ischaemic heart disease was 3.6 times higher.

Thus, the "selenium" hypothesis under discussion can explain a whole series of facts: the high level of biological mortality in Finnish men (especially in regions with a low selenium content), the preferential affectedness of the male part of the population, and a pathology which is characteristic of selenium deficiency (circulatory diseases).

Further more detailed discussion of the role of selenium in the metabolism of living organisms lies beyond the bounds of this book. We merely note that selenium is a component of the enzyme glutathione peroxidase, entering the active centre of this enzyme in the form of selenocysteine (Levander, 1982). This enzyme restores hydrogen peroxide to water, and organic hydroperoxides to the corresponding alcohols, thereby preventing the peroxidation of biological membranes and damage to other cell structures (Harman, 1981). A selenium deficiency leads to a reduction in the activity of this enzyme, assisting the development of a number of pathological processes including heart diseases and accelerated aging (Harman, 1981; 1985). The greater resistance of women to selenium deficiency seems to be connected with the fact that the female organism has more efficient defence mechanisms against damage by free radicals (Harman, 1981; 1985).

To conclude this review of the "selenium" hypothesis, it is worth noting that 1969 saw the enactment of a resolution by the Finnish Ministry of Agriculture and Forestry that selenium should be added to cattle fodder with the intention of increasing the content of this trace element in meat and dairy products (Mutanen and Koivistoinen, 1983). Therefore, if the "selenium" hypothesis is correct, we should expect a reduction in the biological component of mortality for the male population of Finland. In the 1970s, there was indeed some increase in the blood selenium content of Finns

(Salonen, 1987), accompanied by a reduction in total mortality and CHD (coronary heart disease) mortality (Werkö, 1987). This forces us to pay more attention to the "selenium" hypothesis. It is also striking that in the 1970s, as our calculations have shown (Gavrilova, 1982), the biological component of mortality for Finnish men began to decrease. In 1976-1980, the level of this component for 40-year-old men stood at 79% of the level observed in 1966-1970.

This example shows that the separation of mortality into social and biological components is to a significant extent conventional and historically transient. In fact, the biological component is that part of mortality which man has not yet learnt to fight. It cannot be doubted that as science advances an ever-growing proportion of deaths will be placed under social control and eliminated. Nevertheless, the separation of mortality into social and biological components is useful for the problem under review.

Another example of the lack of correspondence between cartograms showing the male and female levels of the biological component of mortality concerns Denmark. It can be observed that the male population of Denmark is characterised by a comparatively low level of the biological component of mortality. At the same time, the female population of Denmark has one of the highest levels of this index in Europe. It remains the case, however, that the absolute values of the levels of biological mortality for Danish men and women are close. In other words, the highest level of female biological mortality is still somewhat less than the lowest level of male biological mortality. Nevertheless, the question arises as to what the very high biological mortality of Danish women, compared with the female population of other European countries, is related to.

In this regard, it is interesting to note that the observed level of biological mortality is in good agreement with the results of epidemiological investigations cited above. It turns out that the female population of Denmark is characterised by one of the highest levels of oncological morbidity and mortality in the world, including mortality from breast cancer (Doll and Peto, 1981). The causes of this phenomenon have not yet been established, but it is well-known that Denmark traditionally has one of the highest levels of fat consumption in Europe. The level of fat consumption in the country amounts to 170g per person per day, almost 40% greater than in Finland and almost 6 times greater than in Japan (Doll and Peto, 1981). Many epidemiological investigations (Knox, 1977; Doll and Peto,

1981; Hopkins and Carroll, 1985), in addition to animal experiments (Hopkins and Carroll, 1985), have shown that excess consumption of fats may be an important factor in the development of a number of diseases, and may facilitate the development of senility (Harman, 1981). Of course, this explanation is merely a hypothesis which illustrates the possible use of such cartograms in determining the biological foundations of the human life span.

In concluding this discussion of cartograms, we must turn our attention to the magnitude of the observed differences. As a whole, the female population of Europe is characterised by less variation in the level of the biological component of mortality. Thus, at age 40 the differences between the best and the worst regions amount to 200% in women and 300% in men. These are very large differences, especially when it is borne in mind that no evident tendency can be observed for them to decrease. Determining the causes of the observed regional differences would therefore permit a two-fold or three-fold reduction in the mortality of the working population, a matter of great practical as well as scientific interest. In this respect, the study of regions where there are "jumps" from very low to very high levels of mortality is particularly interesting, for example the border of Finland with Norway and Sweden (for men), or the border of Denmark with Norway and Sweden (for women). Of course, most interesting are "jumps" in mortality within neighbouring regions of the same country which are characterised by similar living conditions. An investigation of the mortality of migrants seems promising, since it would discriminate between the effects of genetic and environmental factors on the biological component of mortality.

Despite the great variety of all the possible explanations for the observed regional differences, they can be reduced to four general hypotheses: the environmental determination hypothesis, the genetic determination hypothesis, the socio-cultural determination hypothesis, and the adaptational determination hypothesis.

According to the environmental determination hypothesis, the biological component of mortality is determined by historically stable environmental factors. An example of such a hypothesis is the suggestion discussed above that selenium deficiency plays a role in increasing the biological component of mortality in Finnish men. In conformity with hypotheses of this type, it would be expected that the biological component of mortality in migrants should be determined by their new place of residence.

According to the genetic determination hypothesis, the biological component of mortality is inherited in the same way as many other racial characteristics. If this hypothesis is correct, the biological component of mortality in migrants should be basically determined by their ethnic and racial affiliation.

According to the socio-cultural determination hypothesis, the biological component of mortality is determined by fundamental properties of life-style, such as diet and a variety of nationally specific modes of behaviour. At this point it must be emphasised that what we have in mind are socio-cultural characteristics which are so stable that they can be treated as components of the ecology of human populations. An example of this hypothesis is the suggestion made above that the high consumption of fats in Denmark acts as a factor in the high biological component of mortality for women. In conformity with the socio-cultural determination hypothesis, we would expect that in assimilated migrants the level of biological mortality should be determined by their new environment, but that it should remain unaltered in migrants who preserve their national charactersitics and customs.

Finally, the adaptational determination hypothesis proposes that biological mortality is determined by the degree to which a human population is adapted to its environment. This hypothesis predicts that the biological mortality of migrants should generally be higher than in the regions from which they come, and higher in comparison with the native population of the area. According to this hypothesis, we should also expect exceptionally low levels of the biological component of mortality in populations which have lived for ages in the same place. In this regard, it is interesting to note that a low level of the biological component of mortality is observed in such countries as Sweden and Norway, despite the fact that the climate in these countries is not especially favourable. This hypothesis can be considered as a special case of the idea that there are adaptive types of human beings which have arisen as a consequence of a long history of adaptation by human populations to different living conditions (Alekseeva, 1986).

*The reliability of the revealed regional differences.* Although the above discussion of cartograms was fragmentary and at times speculative in nature, it is clear that the epidemiological approach can in principle be used to investigate the biological foundations of the human life span. Since the scientific and practical significance of these kinds of investigation cannot be overestimated, we must again return to the question of the reliability of the observed regional differences.

What we understand in this case by reliability is not the statistical significance of the regional differences, of which one can easily convince oneself by turning for example to table 5, but the correctness of final results which are based on tested, although not absolutely rigorous, assumptions (the correctness of the Gompertz-Makeham formula, the historical stability of the age-dependent component of the force of mortality). In other words, before going more deeply into an investigation of the nature of the observed differences, it would be useful once again to check independently whether they really exist, or are an artefact of the method used for processing the data.

It must also be noted that the very fact that there are biological differences in the mortality of human populations is not self-evident. For example, some demographers suppose that all people are originally equal in respect of their possible life span, and that the existing regional differences in mortality are connected exclusively with the effect of social factors. Since the results we have obtained are in contradiction with this wide-spread opinion, their reliability must be especially rigorously tested.

In order to test the reality of the observed differences, let us see whether we can use them as a basis for correct and non-trivial predictions. Below we cite several examples of the most paradoxical mortality predictions and how they have been tested.

The first example involves a comparison of the mortality of women in Denmark and Norway. At the beginning of the 20th century, the total mortality of women in Norway was significantly higher than in Denmark (see Fig.14), without any indications that this situation might alter. However, from the same life tables belonging to the beginning of the century, it was calculated that the biological (age-dependent) component of mortality in Norway was not greater, but rather significantly smaller than in Denmark (Fig.14). The expectation was therefore that as living conditions improved and the social (background) component of mortality decreased, a historical mortality *inversion* should take place: Norway and Denmark should change places according to the level of female mortality. As can be seen from Fig.14, this prediction was totally fulfilled.

Let us consider another example. At the beginning of the century, the total mortality of women in Bulgaria was much higher than in Italy (Fig.15), and likewise there was no basis for assuming that this situation might change.

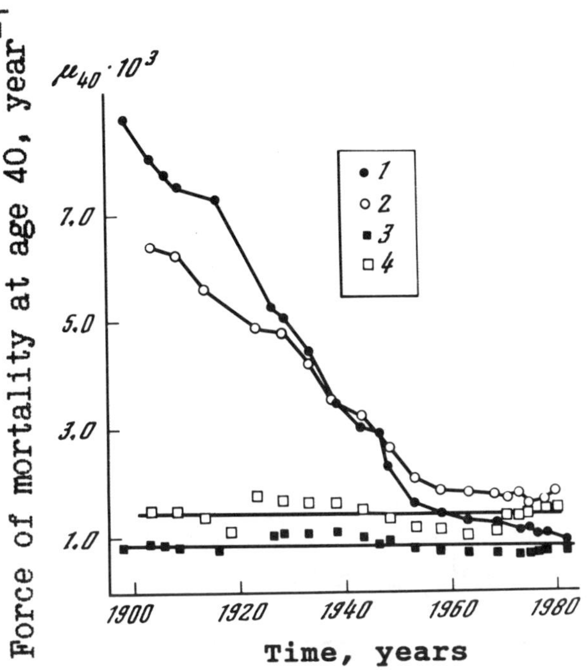


**Figure 14** Comparison of the historical changes in the mortality of Norwegian and Danish women.

1 - total force of mortality for Norwegian women at age 40
2 - total force of mortality for Danish women at age 40
3 - age-dependent component of the force of mortality for Norwegian women at age 40
4 - age-dependent component of the force of mortality for Danish women at age 40.

Calculated and constructed on the basis of statistical data for which the references are given in Gavrilova et al., (1983).

abscissa: time, years

ordinate: force of mortality at age 40, year$^{-1}$; $\mu_{40} \cdot 10^3$

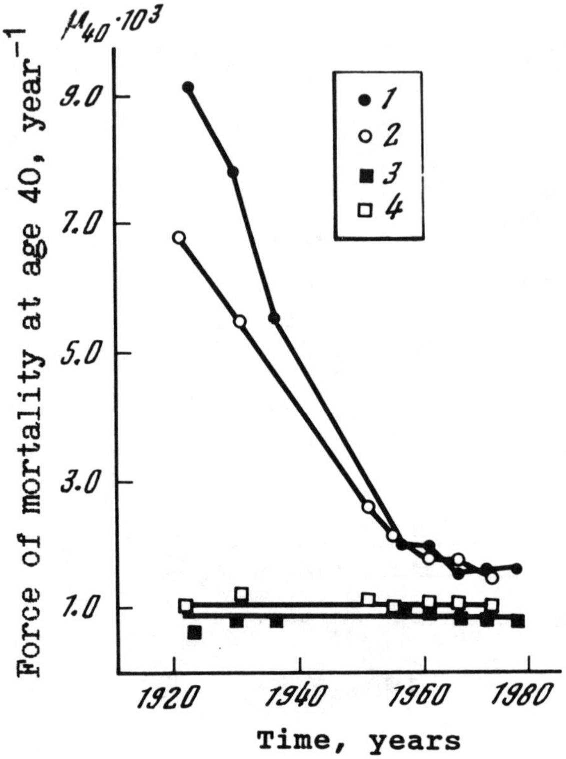


**Figure 15** Comparison of the historical changes in the mortality of Bulgarian and Italian women.

1 - total force of mortality for Bulgarian women at age 40
2 - total force of mortality for Italian women at age 40
3 - age-dependent component of the force of mortality for Bulgarian women at age 40
4 - age-dependent component of the force of mortality for Italian women at age 40.

Calculated and constructed on the basis of statistical data for which the references are given in Gavrilova et al., (1983).

abscissa: time, years

ordinate: force of mortality at age 40, year$^{-1}$; $\mu_{40}\cdot 10^3$

However, a calculation of the biological (age-dependent) component of mortality showed that there are no significant differences between Bulgaria and Italy in the level of this component for women (Fig.15). Consequently, the prediction could be made that as the socially determined (background) component of mortality decreased, a historical mortality *convergence* should take place: the differences in the mortality of Bulgarian and Italian women should become significantly smaller. This prediction was totally confirmed (see Fig.15).

The third example involves a comparison of the mortality of people of different sexes in different countries. At the beginning of the century, the mortality of Finnish women greatly exceeded the mortality of Danish men (Fig.16). At first sight, it might seem a hopeless task to make any predictions about the differences between such diverse population groups. Nevertheless, a calculation of the biological (age-dependent) component of mortality leads to the confident prediction of a historical mortality *inversion*, since the age-dependent component in Finnish women is much smaller than in Danish men (Fig.16). Modern data fully confirm the correctness of this prediction (Fig.16).

Thus, the model that we have been testing makes correct predictions about the levels at which mortality will be historically stabilised, and the magnitude of the differences between these levels for different populations. Especially to be noted is the fact that these predictions require no information about causes of death or temporal mortality trends. All that is needed is data about total mortality in the form of traditional, almost 100-year-old life tables! As far as we know, there are no other methods in modern demography and biology which can be used to perform a similar task. We are therefore justified in maintaining that the cited calculations are not a meaningless computational exercise, but an objective evaluation of real phenomena. In addition, it is an evaluation with a practical purpose (new possibilities for predicting mortality).

The approach we have been developing also permits us to explain and make quantitative predictions about the increasing differences in the life spans of men and women. As is well-known, the gap between the life spans of men and women at the beginning of the century was only one or two years. Now these same differences have reached 8-10 years (Lopez, 1983). There is a wide-spread opinion that since these differences have emerged comparatively recently, their origin must be sought in the present, i.e. in the properties

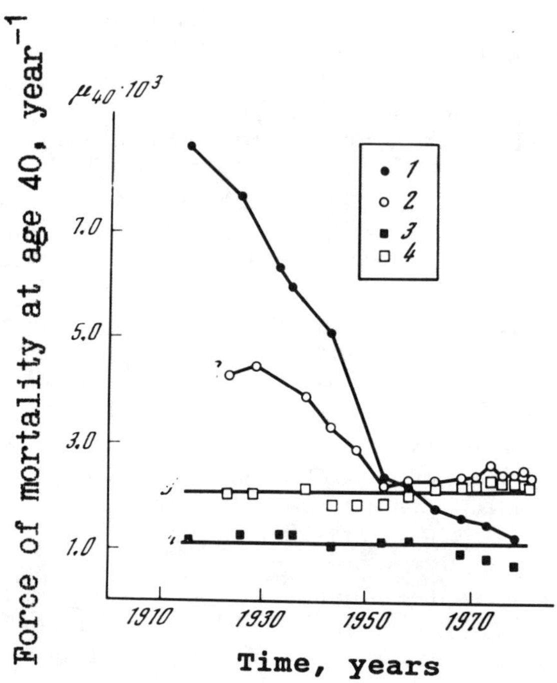


**Figure 16** Comparison of the historical changes in the mortality of Finnish women and Danish men.

1 - total force of mortality for Finnish women at age 40
2 - total force of mortality for Danish men at age 40
3 - age-dependent component of the force of mortality for Finnish women at age 40
4 - age-dependent component of the force of mortality for Danish men at age 40.

Calculated and constructed on the basis of statistical data for which the references are given in Gavrilova et al., (1983).

abscissa: time, years

ordinate: force of mortality at age 40, year$^{-1}$; $\mu_{40} \cdot 10^3$

of modern life-styles (Lopez, 1983). However, the results we have obtained give grounds for the diametrically opposite conclusion. Fig.17 shows the historical changes in the mortality of Danish men and women. It can be seen that at the beginning of the century female mortality was close to male mortality, and even somewhat exceeded it. However, from the very same tables it can be calculated that the biological component of mortality in women is rather smaller. These differences at the beginning of the century were simply masked by the high level of background mortality. Then, as social progress took place, the background component of mortality tended towards zero, and the previously hidden biological differences could be fully revealed (see Fig.17).

This regularity is even more obvious in the case of Italy (Fig.18). At the beginning of the century, male and female mortality in this country was practically equal. However, a calculation of the biological component of mortality reveals large differences (Fig.18). It is therefore possible to predict that in the course of time a mortality *divergence* should take place between Italian men and women, and that the gender differences in mortality should be greater than the analogous differences observed in Denmark. In order to convince oneself of the correctness of this prediction, it is sufficient to compare Figs. 17 and 18.

Thus, within the approach we have been developing, it is possible not only to explain the growing gap between male and female life spans, but also to make quantitative predictions about the magnitude of these differences in particular countries. It must again be emphasised that all we need to make the calculations are traditional demographic life tables from the beginning of the century. The predictions need no additional information about the structure of causes of death or temporal mortality trends.

The examples we have cited show that the observed sexual and regional differences in the levels of biological mortality are an objective reality and can be used in practice to predict the mortality of the population. In particular, an evaluation of the biological component of mortality permits us to make reliable predictions about the historical *convergence*, *divergence*, and *inversion* of mortality in the human populations under comparison. And although the conclusion that even in similar social conditions different human populations can have different life expectancies might seem debatable to some demographers, it is a conclusion which is completely natural from the point of view of population genetics and modern

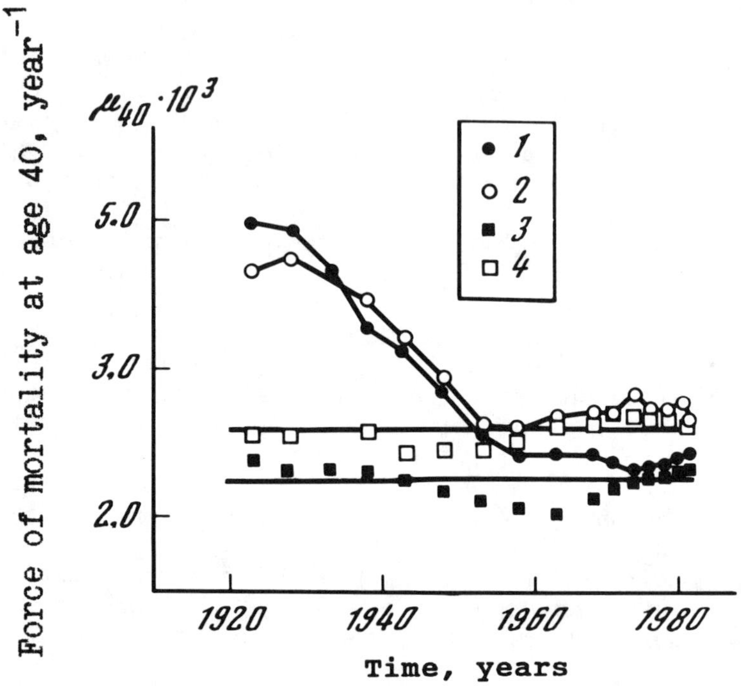


**Figure 17** Comparison of the historical changes in the mortality of Danish women and men.

1 - total force of mortality for Danish women at age 40
2 - total force of mortality for Danish men at age 40
3 - age-dependent component of the force of mortality for Danish women at age 40
4 - age-dependent component of the force of mortality for Danish men at age 40.

Calculated and constructed on the basis of statistical data for which the references are given in Gavrilova et al., (1983).

abscissa: time, years

ordinate: force of mortality at age 40, year$^{-1}$ ; $\mu_{40} \cdot 10^3$

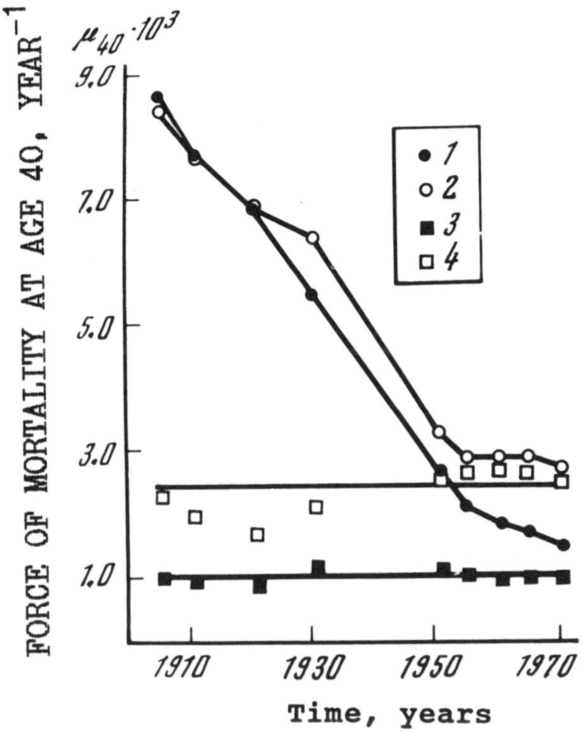


**Figure 18** Comparison of the historical changes in the mortality of Italian women and men.

1 - total force of mortality for Italian women at age 40
2 - total force of mortality for Italian men at age 40
3 - age-dependent component of the force of mortality for Italian women at age 40
4 - age-dependent component of the force of mortality for Italian men at age 40.

Calculated and constructed on the basis of statistical data for which the references are given in Gavrilova et al., (1983).

abscissa: time, years

ordinate: force of mortality at age 40, year$^{-1}$; $\mu_{40} \cdot 10^3$

epidemiology.

### 3.5. Prospects for human life span extension

To conclude this chapter, it is natural to turn to a discussion of the methods and prospects for increasing human life expectancy. This topic has attracted the unfailing attention of a wide range of specialists: biologists, medics, demographers, philosophers, futurologists, etc. In the USSR, it is being pursued within the framework of the special scientific programme "The Extension of Life", supported by the Ministry of Health (Chebotarev, 1979; 1986). It is therefore sensible not to rehearse problems which have already been discussed on many occasions, but to limit ourselves to a deeper analysis of a comparatively small number of questions and hypotheses. In particular, it seems that it is possible to judge the correctness of hypotheses concerning the prospects and methods of life span extension according to how they explain current mortality trends. In other words, in order to make a deduction about the future, it is essential to have a good understanding of the past and present. Current ideas about the observed mortality trends can be summarised in the form of three hypotheses. Let us examine these hypotheses in turn, with the conventional names: (a) *the ecological crisis hypothesis*, (b) *the endogenous causes of death hypothesis*, and (c) *the limited reliability of the organism hypothesis*. Each of these hypotheses, which has a different explanation for the observed mortality trends, leads to different ideas about the prospects and methods for extending human life.

*The ecological crisis hypothesis*. In the opinion of some biologists, current mortality trends are not related to the biological aspects of human life span extension. Their conception involves the existence of an "apparent biological limit" to the human life span, equalling 100-110 years, "that has remained unchanged across time, races and civilisations" (Economos. 1985). Consequently, the basic task of the biology of life span is to explain the mechanisms which determine the magnitude of this limit, and to find the means whereby it can be overcome. It is supposed that since a certain proportion of people survive to this limit, all other people might attain it if they did not fall victim to random circumstances (wars, diseases, hunger, accidents, etc.). In this approach, human mortality is seen as a phenomenon of interest solely to demography, and not to biology. In answering why there has nevertheless been a reduction in the rate at which adult mortality has fallen and average life expectancy has increased, the supporters of this hypothesis use the idea of ecological crisis (Abrams, 1985; Economos, 1985). In their opinion, the end of the fall in mortality is connected with historically transient factors, and not with the biological characteristics of human populations. Correspondingly, the levels at which the mortality figures have stabilised reflect only the level of prosperity which has been established in the given country, and do not contain any information for biologists. To the question why the undoubted advances in medicine have had no significant effect on mortality, the supporters of the ecological crisis hypothesis reply that the beneficial influence of medicine is counterbalanced by the negative consequences of modern life styles (stress, hypodynamia, obesity, smoking, etc.), the destruction of the ecological balance, and environmental pollution.

When discussing this hypothesis, it has to be admitted that it has a number of positive attributes: it is simple, clearly formulated, and optimistic. Indeed, it has the consequence that people's life expectancy might be increased to 100-110 years without any changes in human biology. The only drawback to the hypothesis is that it does not correspond to reality. In fact, our results cited above demonstrate that the levels at which mortality stabilises are not unforeseen, but can be predicted on the basis of life tables relating to the beginning of the century. In addition, as has already been observed, mortality stabilisation levels are different for different sexes and for different regions of the world, and this is in good agreement with presently observed mortality data. The hypothesis under examination cannot explain this fundamental fact. Thus, the ecological crisis hypothesis is an extreme simplification, and is not suitable for scientific research.

*The endogenous causes hypothesis*. In 1952, the well-known French demographer Jean Bourgeois-Pichat attempted to predict population mortality using the idea of endogenous and exogenous causes of death. In the exogenous causes of death he included infectious and parasitic diseases, respiratory diseases, accidents, poisonings, and violence. The endogenous causes of death included malignant neoplasms, circulatory diseases, and the remaining causes of death (Bourgeois-Pichat, 1952).

In Bourgeois-Pichat's opinion, the evolution of human mortality can be likened "to the erosion of soil composed of two kinds of rock: soft rock and hard rock." At first, the "soft rock" is quickly eroded (exogenous causes of death), then the "hard rock" slowly erodes (endogenous causes of death). On the basis of these ideas, it was predicted that medical advances in eliminating the exogenous causes of death would lead to the endogenous causes of death coming to the fore. The search for effective methods of fighting the exogenous causes would take a long time (Bourgeois-Pichat, 1952; 1979). Thus, Bourgeois-Pichat formulated "the concept of a temporary limit on mortality decline", and even calculate the level of this "temporary limit" for each age, calling it "the biological limit of mortality decline" (Bourgeois-Pichat, 1952; 1979).

It is interesting to note that the biological component of mortality calculated in this way has grown almost exponentially with age, has shown significant historical stability (against the background of a sharp drop in total mortality), and has been materially dependent on sex (the endogenous mortality of the male population has been much higher than the female population) (Bourgeois-Pichat, 1952; 1979). If in addition we bear in mind that the levels for the biological component of mortality calculated using the Gompertz-Makeham formula are reasonably close to the levels for endogenous mortality calculated using Bourgeois-Pichat's method, the coincidence of the results is striking.

Without doubt, the coincidence of these conclusions, reached by two independent methods on the basis of different original data, is a strong argument in their favour. Nevertheless, the observed coincidence does not imply the equivalence of the two approaches, nor the validity of the endogenous causes hypothesis.

Indeed, the technique of separating mortality into its background and age-dependent components allows one to determine the scope for a reduction of mortality on the sole basis of demographic life tables, without any knowledge of the causes of death. On the other hand, in order to calculate exogenous and endogenous mortality, it is essential to have reliable statistics for causes of death, but the quality of such statistics even in the developed countries of today leaves much to be desired (Geller, 1982). For example, in patients dying from cancer, the cause of death on the death certificate may be a totally different disease, such as pneumonia (in primary or metastatic cancer of the lung), or other infectious diseases (in various leukaemias and lymphomas) (Doll and Peto, 1981). In some cases, it is simply impossible to establish whether death is exogenous or endogenous. For example, a patient may be suffering from several diseases, none of which alone would lead to death, but which are lethal in combination.

Finally, the endogenous causes of death hypothesis encounters a number of fundamental objections. For example, taking into account modern advances in epidemiology, it is impossible to agree that malignant neoplasms and circulatory diseases should simply be treated as purely endogenous causes of death. On the contrary, recent research has shown that a significant contribution to death from these causes is made by external factors which can be eliminated (it suffices to mention environmental carcinogens) (Doll and Peto, 1981). Further, the very principle of dividing deaths into exogenous and endogenous types seems to be an unjustifiable simplification of reality. Of course, it will always be possible to find a clear example of a purely exogenous death and, with some difficulty, to select a corresponding example of a purely endogenous death. However, in the overwhelming majority of cases, an intermediate situation is what is observed, with a complex interaction of exogenous and endogenous factors. Bourgeois-Pichat's method does not take this interaction into account, and represents total mortality as a simple sum of exogenous and endogenous components.*

When discussing the endogenous causes of death hypothesis, it has to be admitted that it is significantly more profound and substantial than the ecological crisis hypothesis reviewed above. Indeed, the endogenous causes of death hypothesis provides a rational explanation for the observed mortality trends, and even allows for quantitative predictions. According to this hypothesis, by no means all cases of death in youth and middle-age need be considered premature and relatively easy to eliminate. On the contrary, it is supposed that each age is characterised by its own biological limit to a reduction in mortality. This limit increases with age, reflecting the aging of the organism. Without doubt, this claim is in closer correspondence with the real data than do the conceptions of the ecological crisis hypothesis. Nevertheless, this promising approach needs serious revision, since total mortality is in reality determined not by the sum, but by the interaction of exogenous and endogenous factors.

The inadequacy of the endogenous causes of death

---

*In this connection, attempts have been made by A.G.Vishnevsky and E.M.Andreev to supplement Bourgeois-Pichat's approach with the concept of so-called quasi-endogenous causes of death (Vishnevsky, 1982; Andreev, 1987).

hypothesis also becomes clear when it is used to determine the prospects and methods of human life span extension. For example, the hypothesis leads us to expect that a further increase in life expectancy will be achieved by a reduction in mortality from individual "endogenous" causes. However, this claim seems far from obvious when modern data are subjected to careful analysis. What is more, a body of evidence has accumulated in recent years which shows that although the fight against individual "endogeneous" causes of death may lead to some reduction in mortality from these causes, it has practically no effect on total mortality. In other words, a reduction in mortality from one set of causes is frequently compensated by a growth in mortality from another set of causes, with the result that the expected increase in life expectancy does not occur. We will cite just a few examples to illustrate this statement.

At the end of the sixties, the World Health Organisation, using three European centres as a base (Edinburgh, Budapest and Prague) carried out an investigation into the effectiveness of primary prevention of ischaemic heart disease with the aid of clofibrate, a drug which reduces the blood cholesterol content (Committee of Principal Investigators, 1978). The investigation, which lasted five years, involved 10,627 essentially healthy, middle-aged men with an increased blood cholesterol content. The individuals who took part in the investigation were randomly divided into two groups, one of which received clofibrate. The result was that in the group receiving clofibrate there did indeed occur a reduction in the blood cholesterol content, and a significant decrease in CHD incidence (by 20%). At the same time, the expected reduction in total mortality compared with the control group did not occur, because of a growth in mortality from causes not connected with heart disease, including oncological diseases.

In the Lipid Research Clinics Coronary Primary Prevention Trial, another drug was used which lowers the blood cholesterol level: cholestyramine (Lipid Research Clinics Program, 1984). This multicentre, randomised, double-blind trial, which lasted seven years, involved 3,806 asymptomatic men with hypercholesterolaemia. In the trial group which received cholestyramine, a reduction in the blood cholesterol content did indeed occur, and CHD incidence and mortality decreased. However, because of an increase in mortality from other causes, including accidents and violence, there was no statistically significant decrease in total mortality.

In one of the recent investigations into the primary

prevention of ischaemic heart disease, the Helsinki heart study, another drug was used which lowers the blood cholesterol level: gemfibrozil (Frick et al., 1987). This experiment, which lasted five years, involved 4,081 asymptomatic middle-aged men with primary dyslipidemia. In the trial group which received gemfibrozil, there was a significant reduction in the content of cholesterol and triglycerides in the blood. However, despite a significant reduction in the incidence of heart diseases, a reduction in total mortality did not occur because of a rise in mortality from other causes (intracranial haemorrhage, accidents and violence).

A rather different method of preventing cardiovascular diseases was used in American Physician's Health Study. In this case, the prophylactic agent was aspirin, well-known for its antithrombic action. The study involved 22,000 healthy volunteers randomly divided into two groups, one of which received aspirin for a period of five years, and the other a placebo (Young et al., 1988). In the group which received aspirin, the total number of myocardial infarctions, as well as of fatal myocardial infarctions, was significantly lower than in the control group. However, because of a growth in mortality from other causes, including strokes, there was no statistically significant decrease in total mortality (Relman, 1988; Young et al., 1988).

The list of examples could be augmented, but even these facts are sufficient to cause one to reflect whether the connection between total mortality and mortality from particular causes is really as simple as it seems at first sight. In 1986, when we first drew attention to the fact that total mortality is surprisingly constant in intervention trials for the prevention of particular diseases (Gavrilov and Gavrilova, 1986), our generalisation seemed premature to many in the field. Subsequently, however, other investigators (Klevay, 1987; Oliver, 1988) have also independently arrived at the same conclusion after analysing a wide number of different intervention trials.

The strange interaction between total mortality and its component parts is also manifested in the very fact that the age-dependent component of human mortality is historically stable. Indeed, as we have already noted, this phenomenon permits us to predict the mortality of a population on the sole basis of demographic life tables, through an analysis of the dependence of total mortality on age. In other words, the limit to a reduction in adult mortality can be calculated on the basis of life tables relating to the beginning of the century, which contain no information about the structure of the causes of death. It turns out that causes of death are superfluous: they can be totally ignored in the prediction of human mortality. What is more, it turns out that the phenomenon of the historical stability of the age-dependent component of mortality cannot be reconciled with the notion of causes of death. Indeed, it is well-known what causes of human mortality have declined in the 20th century. In the first instance, these are pneumonia, influenza, tuberculosis, enteritis and other infectious diseases (Bourgeois-Pichat, 1979). It is also well-known that the mortality from each of these causes alters with age. Therefore the elimination of these diseases ought inescapably to have altered the dependence of total mortality on age, and the magnitude of the age-dependent component of mortality. However, as we have demonstrated above, the age-dependent component of mortality either alters insignificantly, or does not alter at all. The only way of resolving this contradiction is to admit that causes of death are not independent of each other, but act in concert in such a way that the age-dependent component of mortality remains historically stable, despite sharp changes in the structure of the causes of death. However, the following question then arises: in what manner do the causes of death "agree" amongst themselves so as to leave the age-dependent component of mortality unaltered?

Since for a long time no sensible answer to this question could be offered, even in the form of a hypothesis, a mystic quality was attributed to many related facts and observations. These were therefore usually ignored or given different explanations in each individual case. For example, the well-known geneticist A.Jacquard, noticing that monozygotic twins show great similarity in causes of death, but little similarity in lifetimes, restricted himself to the conclusion that the given phenomenon "leaves little justification for any biological interpretation" (see Jacquard, 1982, p.310).

However, all these facts receive a relatively easy explanation in the framework of the limited reliability of the organism hypothesis, to which we now turn.

*The limited reliability of the organism hypothesis.* According to this hypothesis, proposed in 1978, an organism is a multiply redundant system with a high, but not infinitely high, reliability (Gavrilov, 1978a; Gavrilov et al., 1978). There is therefore always a certain probability that failures in the working of individual elements of the organism will coincide in time, and the organism will pass into a state of nonspecific vulnerability. Such a breakdown

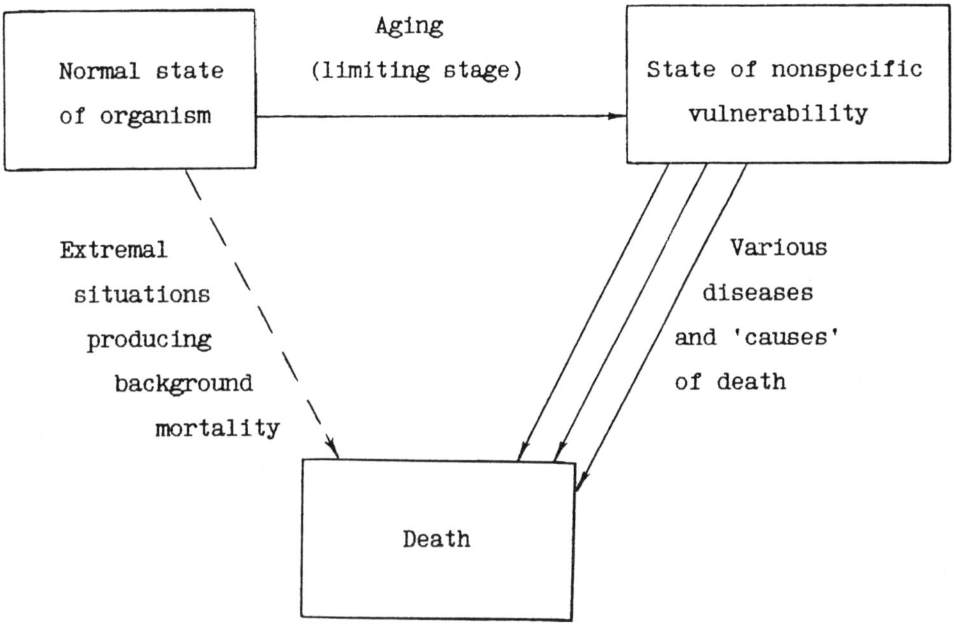


**Figure 19** Simplified representation of the survival model for organisms, illustrating the hypothesis of an intermediate state of nonspecific vulnerability.

"causes a whole cascade of dependent breakdowns in the organism's remaining systems, and therefore there are a large number of direct causes of death" (see Gavrilov, 1978b, p.490). Fig.19 is a simplified schematic representation of the hypothesis. According to this diagram, an organism in the normal state can perish only in extreme situations, those which would be fatal for any, even the heathiest organisms (this corresponds to the background component of mortality, already close to zero in the developed countries). Equally, if there is a breakdown in one of the organism's systems, it can enter into a state of nonspecific vulnerability, then being described as "having one foot in the grave" (Gavrilov, 1984b; Gavrilov et al., 1982). It should be noted that this state has a completely clear biological interpretation. For example, breakdowns in the immune system, whose frequency sharply increases with age, create a nonspecific vulnerability to a very broad spectrum of both "endogenous" and "exogenous" diseases (Walford et al., 1978). Once it has fallen into a state of nonspecific vulnerability, an organism will quickly perish from the first cause to arise, no matter what this may be.

As can be seen from the diagram, the age-dependent component of mortality is determined by the speed of the first, limiting stage in the organism's transition from the normal state to the "one foot in the grave" state. This means that the age-dependent component of mortality is not *composed* of individual causes of death, but, on the contrary, is *distributed* between them. In other words, the speed of the first, limiting stage determines the size of the *death quota*, which is then distributed between the various concrete manifestations which are tendentiously called "causes" of death. The proposed mechanism explains why the elimination of individual "endogenous" causes of death cannot significantly alter the magnitude of the biological component of mortality. In fact, any reduction in the rate at which organisms perish when they are in a state of nonspecific vulnerability inevitably leads to an increase in the proportion of individuals to be found in this state, and hence to the restoration of the previous level of mortality through an increase in mortality from other causes (Gavrilov, 1984a; Gavrilov et al., 1982).

Thus, the limited reliability of the organism hypothesis can explain the phenomenon of the historical stability of the age-dependent component of mortality, as well as the "autonomous" behaviour of total mortality in relation to its component causes. What is more, the hypothesis provides a justification of the Gompertz-Makeham formula, based on simple notions of the nature of aging as an age-related reduction in the redundancy of the organism's systems (Gavrilov, 1978; Gavrilov et al., 1978). Therefore, the concept of the limited reliability of the organism is sufficiently well-founded and natural to be used as a working hypothesis in determining the methods and prospects for extending human life.

According to this concept, at any age there is always a certain risk of death, determined by the limited reliability of the organism. This view is diametrically opposed to the above-cited opinion of certain gerontologists that cases of phenomenal longevity demonstrate how long everyone could live given favourable circumstances. From the perspective of the limited reliability of the organism hypothesis, this seemingly plausible argument is just as improbable as, for example, the statement that it is possible for major prizes to be handed out to everyone in a lottery because some people have been lucky. Thus, this hypothesis links the prospects for prolonging human life not to the total elimination of so-called premature death, but to the significantly more practicable task of gradually reducing the risk of death at each age.

In distinction to the endogenous causes of death hypothesis, the hypothesis under discussion asserts that the problem of human life prolongation cannot be reduced to a fight against individual causes of death. What is more, the limited reliability hypothesis predicts that a reduction in mortality from particular causes will only lead to a significant reduction in total mortality when the initial stage in the destruction of the organism (the transition to a state of nonspecific vulnerability) has ceased to be a limiting stage in the whole process. Therefore, the fight with individual "endogenous" causes of death must mainly lead to a reduction in mortality in old-aged and senile people, amongst whom many will already be in a state of nonspecific vulnerability. This conclusion is diametrically opposed to the opinion of some investigators, who believe that as individual causes of death are eliminated, the survival curve will become more and more "rectangular" (Fries, 1980). However, as the analysis of real data on human mortality has shown, this process, sometimes called the rectangularisation of the survival curve, is absent at ages above 60 years (Myers and Manton, 1983). What is more, factor analysis of the time series of mortality confirms the preferential reduction in the mortality of old-aged and senile people which is predicted by the limited reliability hypothesis (Gavrilov and Nosov, 1985). This process, called the *derectangularisation* of the survival curve, has started to emerge especially intensively in the last decade, giving rise to an increase in the proportion of old-aged and senile people in the total population (Gavrilov and Nosov, 1985). The observed reduction in the mortality of old-aged people is not accompanied, however, by an improvement in their health; this is demonstrated by the growth in the number of diseases accompanying the basic cause of death (Myers and Manton, 1983).

Thus, the limited reliability hypothesis predicts that the traditional approach, based on the fight against individual causes of death, has no prospects (Gavrilov, 1984a; 1985). According to the hypothesis, the strategy for the future struggle to prolong human life must be fundamentally altered. Apparently, the future belongs to another strategy founded on explaining the mechanisms which determine the reliability of the organism and its nonspecific resistance to a broad spectrum of harmful factors. If there is any success in this direction, a simultaneous reduction in mortality from a wide variety of diseases is bound to occur.

This formulation of the prospects and methods for prolonging human life may seem at first sight too abstract and fundamental. However, concrete research in this direction is already under way. For example, the World Health Organisation has approved and is coordinating the so-called Integrated Programme for the Prevention of Noncommunicable Diseases (Glasunov et al., 1983). According to this Programme, the most significant research is not that which is directed at the struggle with a single cause of death, but with a whole range of causes of death. Indicators have already been found (for example, an increased blood sugar level), which are markers of the nonspecific vulnerability of the organism to a very broad spectrum of causes of death, up to and including accidents (Grabauskas et al., 1985). Therefore, the development of this new integrated health strategy is being carried out at an international level, and has reached the stage of concrete research programmes. Also of especial interest in this regard is the work of Professor V.M.Dilman (1981; 1986; 1987), who demonstrates the possible existence of unitary mechanisms in the origin of many diseases, and develops approaches to the elimination of whole groups of "endogenous" diseases which form the basis for a new, so-called integrated medicine. Thus, the methods of prolonging life we have been discussing have a completely realistic scientific foundation.

# CHAPTER 4. SPECIES-SPECIFIC LIFE SPAN

## 4.1. Review of ideas about species-specific life span

Amongst the fundamental problems of the biology of life span, the problem of inter-species differences in lifetimes occupies a special position. Indeed, until we clarify the reasons for the enormous diversity in the life spans of different biological species, and until we understand the mechanisms which cause life spans to change in the course of biological evolution, our comprehension of the biological bases of life span must be considered at least incomplete. In order to convince oneself how substantial the inter-species differences in lifetimes are, it is not even necessary to make a comparison of exotic species: it suffices, for example, to compare the life spans of human beings and rats, species which sometimes live next to each other and consume the same food. Thus, a rat which lives for 5 years is considered by research scientists to be a case of such improbable longevity that it is even excluded from consideration in the statistical processing of results (see Sacher, 1977). It is clear that these significant inter-species differences are at the same time a problem and a key to the solution of a whole range of issues in the biology of life span. Indeed, comparing the species-specific life span with other species-specific characteristics is now a very popular approach to the investigation of the biological bases of life span (Cutler, 1984; Economos, 1980b; Hart aand Setlow, 1974; Kato et al., 1980; Schwartz and Moore, 1979).

The large scale of inter-species differences in lifetimes has led to the point where for a long time the question of the exact measurement of these differences has had no decisive significance. Indeed, if we compare drosophila, rats and human beings, it does not matter what life span index we select, the result will not be fundamentally different: drosophila lives least of all (months), then comes the rat (years), and man lives longest of all (decades). Moreover, the results of the comparison are not fundamentally altered if we abandon figures altogether and replace them with a rank-ordering (who occupies which position on the scale of life spans). In this case, using the simplest techniques of non-parametric statistics (the calculation of the Kendall or Spearman coefficient of rank correlation), it is possible to confirm the existence of all the presently established inter-species correlations: for example, the positive correlation between life span and body size or relative brain weight, and the negative correlation with metabolic rate. The use of rank indicators in this case is especially appropriate, since it does not create an illusion of great accuracy and rigour in the results.

Thus, the analysis of inter-species differences has for a long time not required a knowledge of the absolute magnitudes of these differences. As a result, a paradoxical situation has arisen: the comparison of the life spans of species is now carried out with great accuracy, but a *measure* of the life span of a species is lacking. Values of *species-specific life spans are not measured, but prescribed* by investigators without an indication of substantiated confidence intervals, since the reliability of such expert estimates is unknown (see Cutler, 1984a). Therefore, despite the complexity of the computational procedures which are followed, the extent to which the conclusions are justified is very dubious.

It has to be admitted that the problem of determining species-specific life span is not a simple one, despite the intuitive obviousness of the concept. In fact, most investigators agree that what is meant is the characteristic life span for individuals of the species. But how is this to be determined if there is always a particular distribution of lifetimes depending on external circumstances and the genetic structure of the populations of individuals which belong to the given species? Which figure should be chosen, and how can this choice be justified? Of course, the task would be somewhat simplified if we were to succeed in finding a particular point on these distributions which had specific properties and corresponded to a particular age (within the bounds of the given species), independently of the particular circumstances and characteristics of the populations under investigation. Thus, we again return to the problem of individual differences in lifetimes.

As was demonstrated above, an analysis of life span distributions is a fairly laborious and complicated task, without any guarantee of success in the sense that some special point will be distinguished corresponding to a species-specific life span. What is more, the force of mortality usually grows smoothly with age, without any sharp breaks which would point to the existence of any special periods in aging. The task therefore seems at first sight a rather hopeless one.

One of the ways of "solving" this problem is to postulate the existence of a special age and to attribute it with the desired properties. The most suitable age for such a role is the maximal life span, about which any assertion can be made without a great risk of contradiction, given the inadequacy of reliable statistical data concerning extremely old ages. It seems that in this way there has arisen a notion, which has become established through a long period of mutual citation, that there exists an "apparent biological limit" to life span, which in the case of man "has remained unchanged across time, races and civilisations" (Economos, 1985). The uncompromising nature of this assertion is striking in the total absence of any convincing proof.

This kind of notion is quite wide-spread. For example, some authors maintain that "the maximal life span has remained unaltered for many centuries" (see Vanyushin and Berdyshev, 1977, p.80), recognising, however, that "the limit to longevity has not yet been established" (Vanyushin and Berdyshev, 1977, p.79). The question naturally arises: how can one be convinced of the historical immutability of the maximal life span if its magnitude remains unknown to this day?

The same question arises in a study of many other publications, including, for example, the article by H.L.Abrams (1985). On the one hand, he correctly observes that "precisely how long humans can possibly live is still unknown" (p.81). However, in the same article, Abrams asserts: "Although life expectancy in the United States has increased greatly since 1900, the life span - the biological age limit - remains the same as in ancient times" (p.78). It remains a puzzle what facts and what evidence have been used to justify this exceptionally strong conclusion, and how one can in general convince oneself of the immutability of what "is still unknown".

The subjective nature of estimates of the species-specific life span limit is also revealed in the significant discrepancies between specialist opinions (see table 6). In this list, which is reminiscent of an auction report, the high frequency of round figures is notable, in particular the frequent use of the figure 100. Pointing out this comic circumstance, B.Ts.Urlanis (1978, p.252) wrote: "In assertions of a 100-year limit we see the magnetism of round numbers, which itself is a consequence of man's adoption of the decimal system. This in its turn is based on the number

**Table 6** Estimates of the biological limit to the length of human life

| Human maximum life span, years | Reference |
|---|---|
| 60 | Harvey et al. (1987, see p.185) |
| 70 | Harvey and Clutton-Brock (1985, see p.566) |
| 80 | Moses: "The days of our years are threescore and ten; and if by reason of strength they be fourscore years..." (The Book of Psalms, Psalm 90, A Prayer of Moses) |
| 86 | De Moivre, 18th century Quoted in Thatcher (1982), p.348 |
| | Cutler (1984a, see p.94) |
| 90 | Cutler (1984a, see p.71) |
| 91 | Cutler (1984a, see p.95) |
| 93.1 | Sacher (1978): 34,000 days |
| 94 | Cutler (1984a, see p.32) |
| 95 | Hart and Setlow (1974) |
| | Tolmasoff et al. (1980) |
| | von Hahn (1983) |
| | Cutler (1984a, see p.6) |
| | Witten (1985) |
| 95.5 | Hofman (1983) |
| 96 | Cutler (1984a, see p.21) |
| 100 | Flourens (1856) |
| | Tarkhanov (1891) |

Table 6 (continued)

| Human maximum life span, years | Reference |
|---|---|
| 100 | Sacher (1959) |
| | Davidovsky (1962) |
| | Zepelin and Rechtschaffen (1974) |
| | Economos (1980a; 1980b; 1982) |
| | Bremermann (1982) |
| | Cutler (1984a, see pp.4, 127-128) |
| | Rothstein (1986) |
| | Nakano et al. (1990) |
| 100.3 | Hofman (1984) |
| 103 | Strehler and Mildvan (1960) |
| 110 | Hayflick (1975) |
| | Stanley et al. (1975) |
| | Schwartz and Moore (1977) |
| | Kato et al. (1980) |
| | Röhme (1981) |
| | Clarke (1982) |
| | Cutler (1984a, see p.23) |
| | Masoro (1988) |
| 113 | Walford (1983) |
| | Cutler (1984a, see p.123) |
| | Sunlin (1985) |

Table 6 (continued)

| Human maximum life span, years | Reference |
|---|---|
| 113.586 | Fries (1983): 113 years and 214 days |
| 114 | Fries (1980) |
| 115 | Benjamin (1982) |
| | Cutler (1984a, see p.75) |
| | Schneider (1985) |
| | Pioli et al. (1986) |
| | Hayflick (1988) |
| 115.1 | Tréton and Courtois (1989) |
| 120 | the Lord: "My spirit shall not abide in man for ever, but his days shall be a hundred and twenty years" (Genesis 6:3) |
| | Cramer (1981) |
| | Solomon (1983) |
| | Comfort (1984) |
| | von Ardenne (1985) |
| | Lestienne (1988) |
| 120.649 | The Guinness Book of Records 1990: 120 years and 237 days. |
| 121 | Berdyshev and Asadov (1985) |
| 125 | Barnett (1982) |
| 160 | Prokof'ev et al. (1982) |
| 167 | Zhirmunsky and Kuz'min (1980) |
| 200 | Hufeland (1796) |

of fingers on both hands or the number of toes on both feet. What is the basis for asserting that the length of a man's lifetime is determined by the number of his fingers, multiplied by 10?"

It is also interesting to note that the well-known gerontologist R.Cutler in the course of a single article quotes first one then another value for the species-specific life span of human beings, finally using ten figures: 86, 90, 91, 94, 95, 96, 100, 110, 113, 115 (see Cutler, 1984a and table 6).

The accuracy of these estimates is unknown, and this is revealed, in particular, by the absence of substantiated confidence intervals. Nevertheless, some experts are careful to quote not a value for the species-specific life span, but a supposed interval within which they believe this value to be found:

| | |
|---|---|
| 90-95 years | (Zubov and Kozlov, 1982) |
| 90-100 years | (Chebotarev, 1978) |
| 100-110 years | (Kurtzman and Gordon, 1976; Economos, 1985) |
| 110-115 years | (Strehler and Mildvan, 1960; Spector, 1974) |
| 112.5-137.5 years | (Barnett, 1982) |
| 115-150 years | (The New Encyclopaedia Britannica, 1989, vol.7, p.347) |
| 120-130 years | (Vanyushin and Berdyshev, 1977) |
| 125-150 years | (A.A.Bogomolets, quoted in Urlanis, 1978) |
| 150-180 years | (P.P.Lazarev, quoted in Urlanis, 1978) |

It can be seen, however, that even these estimates hardly overlap.

The figures we have quoted have involved estimates of the species-specific life span of human beings. Regarding the life spans of other biological species, there are fewer discrepancies, apparently a consequence of the smaller number of investigators who consider themselves experts in the particular field (Cutler, 1984). However, since the most reliable life span data are data relating to the human life span, the existing discrepancies in expert opinion are especially alarming. It seems that the only conclusion that can be drawn from a survey of the estimates that have been made is that they are unsuitable for scientific research.

It might seem that a way out of this state of affairs would be simply to record any reliably attested cases of longevity. Then, if we are to believe the latest information, the species limit to the human life span is 120 years and 237 days (The Guinness Book of Records, 1990). However, there is no certainty that this record will not be broken in the future. Thus, what is a scientific problem takes on the flavour of a chronical of sporting records. Indeed, in discussions of the species-specific life spans of other organisms, the analysis often sinks to the level of hunting stories. It is clear that on the basis of such data it will be difficult to achieve an understanding of the biological basis of life span.

Another approach to the problem is the use of techniques from the statistics of extremes (Gumbel, 1958). Following this approach, what should be recorded is not the maximal, but ***the largest characteristic life span***, equal to the age of the last organism ***remaining alive*** out of a sample of known size. In this case, it is possible to process the data statistically and compare the values of the largest characteristic life spans for different species. It is essential, however, to take into account the fact that the magnitude of the largest characteristic life span grows with an increase in the number of observations, and that this growth may not be limited from above. Consequently, in a comparison, it is necessary to take into account the size of the sample. At the present time, no-one makes this adjustment to the number of observations. Therefore, it is naturally to be expected that for human beings, given the huge population of the Earth and the long history of recording cases of longevity, the magnitude of the largest characteristic life span will be significantly greater than the corresponding figure for a species with the same longevity, but represented by a dozen individuals in a zoo. After this, it is hardly surprising that in the majority of inter-species correlations the human data are out of line, i.e. man lives longer than would be expected on the basis of his anatomical, physiological and biochemical characteristics. Sometimes, an attempt is made to attribute these deviations to the special status of man as a biosocial creature (Arshavsky, 1982). However, before discussing this interesting hypothesis, we must first exclude the possibility of another, simpler explanation, connected with the differences in the number of observations.

Of course, if an absolute upper life span limit existed for every species, the maximal life span, given a sufficiently large number of observations, would be a good estimate of the magnitude of this limit. However, before starting to measure the species-specific life span limit, we must first check whether it really exists.

## 4.2. Is there a species-specific life span limit?

The existence of a species-specific life span limit has for a long time seemed so obvious that the question whether its existence can be reasonably asserted has not even been raised. Doubts about this issue have seemed as ridiculous as doubts about the very fact of the limited life spans of organisms, and the correctness of the ancient thesis that "all men are mortal". Belief in the existence of a species longevity limit has been absolute, given the absence of a reasonable alternative.

However, a more careful examination of this point of view shows that it is far from being obvious. Indeed, the hypothesis of a species limit contains hidden within it the proposition that there exists a maximal age which has special properties. This rather strong assertion is not at all equivalent to the fact that life spans are limited. For example, the life span of the atoms of a radioactive element is limited: this is shown, in particular, by the existence of a finite mean lifetime. However, it is impossible in this case to point to any maximal time which has special properties. Of course, it is always possible to determine the half-life of a radioactive element, as well as the time by which 95%, 99% or 99.9% of the original atoms will have decayed. In this way it is possible to calculate a multitude of values, none of which has any essential priority over the others. As is well-known, there is no absolute limit to the life span of the atoms of a radioactive element. Naturally, in a sample of finite size some atom will always be the last to decay, but the magnitude of this "maximal life span" grows without limit with an increase in the number of observations. In the statistics of extremes, it has been shown that in this case the largest characteristic life span increases proportionately to the logarithm of the initial number of atoms (Gumbel, 1958).

Thus, the hypothesis of a species longevity limit is not at all equivalent to the fact that the life spans of organisms are limited, and is not self-evident. What is more, there are even doubts about its theoretical coherence. One observation of this kind is made in a well-known probability theory textbook written by the distinguished mathematician W.Feller. This book, which has supported three editions, is a standard textbook for students. However, as emerged in the course of a scientific discussion in which one of the authors of the present book took part (see Gavrilov, 1984), the extent to which the supporters of the species limit hypothesis are acquainted with issues in probability theory is far from unlimited, and therefore it becomes necessary to cite a short extract from this well-known textbook.

At the very beginning of the book, W.Feller poses the following question: "The question then arises as to which numbers can actually represent the life span of a person. Is there a maximal age beyond which life is impossible, or is any age conceivable? We hesitate to admit that man can grow 1000 years old, and yet currently actuarial practice admits no bounds to the possible duration of life. According to formulas on which modern mortality tables are based, the proportion of men surviving 1000 years is of the order of magnitude of one in $10^{10^{36}}$ - a number with $10^{27}$ billions of zeros. This statement does not make sense from a biological or sociological point of view, but considered exclusively from a statistical standpoint it certainly does not contradict any experience. There are fewer than $10^{10}$ people born in a century. To test the contention statistically, more than $10^{10^{35}}$ centuries would be required, which is considerably more than $10^{10^{34}}$ lifetimes of the earth. Obviously, such extremely small probabilities are compatible with our notion of impossibility. Their use may appear utterly absurd but it does no harm and is convenient in simplifying many formulas. Moreover, if we were seriously to discard the possibility of living 1000 years, we should have to accept the existence of maximum age, and the assumption that it should be possible to live $x$ years and impossible to live $x$ years and two seconds is as unappealing as the idea of unlimited life" (see Feller, 1968, pp. 7-8).

A similar opinion is expressed by the French demographer and geneticist A.Jacquard, who considers that "it is true that, however old we are, our probability of death within the next hour is never equal to 1; there is therefore no final barrier that we all encounter (Jacquard, 1982, p.303).* The very notion of a species-specific life span limit is called by A.Jacquard "an extremely vàgue concept" (Jacquard, 1982, p.303).

Thus the species longevity limit hypothesis, which seems obvious at first sight, turns out on closer examination to be highly contentious and even dubious. Therefore, the question naturally arises as to how to test its correctness.

It must be noted that testing the correctness of ideas about a species limit is in principle a practicable exercise. Indeed, if such a limit exists, at the maximal age the number of survivors ($l_x$) should become zero, the probability of death ($q_x$) should equal unity, and the force of mortality at this age ($\mu_x$) should tend towards plus infinity. Thus, in order to test the hypothesis in question, it is sufficient to investigate the characteristics of the life span distribution in the maximal age region and check whether at these ages there really is a catastrophic increase in the force of mortality.

Fig.20 is a plot of the force of mortality against age in Swedish men. The topmost curve, which is theoretically calculated, corresponds to the case when the species-specific life span limit really exists. This curve has been calculated on the assumption that the *probability of death* grows exponentially with age. Then at a certain age the probability of death becomes unity, corresponding to the absolute life span limit. The second curve, also theoretical, has been calculated for the case when the *force of mortality* grows exponentially (the Gompertz law). In the semi-logarithmic coordinates used in the figure, this corresponds to a straight line. It can be seen that initially (in the region of small ages), these theoretical curves practically coincide. For this reason, many investigators often fail to distinguish between the concepts of probability of death and force of mortality, believing that the Gompertz law equally corresponds to an exponential growth in the probability of death. It can be seen from the figure, however, that the divergence between the two theoretical curves grows swiftly. The fundamental difference between them is that in the first curve the absolute life span limit exists, and in the second it is absent. It is not difficult to convince oneself of this, once the Gompertz-Makeham equation is integrated:

* Jacquard has in mind a barrier on the time axis corresponding to a maximal age. (Authors' note)

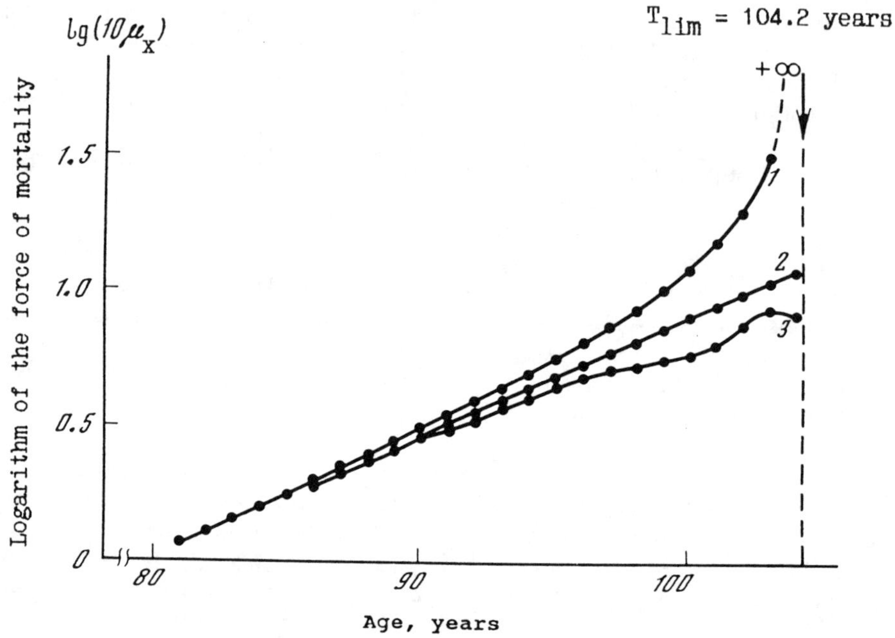


**Figure 20** Testing the hypothesis of a species-specific limit to the human life span

1 - theoretical dependence of the logarithm of the force of mortality on age, corresponding to the case when the species-specific life span limit exists

2 - theoretical dependence corresponding to exponential growth in the force of mortality with age (the Gompertz law with parameters $R = 3.46 \cdot 10^{-5}$ year$^{-1}$ and $\alpha = 0.101$ year$^{-1}$)

3 - real dependence of the logarithm of the force of mortality on age for Swedish men. The statistical data in the interval 81-85 years relate to the mortality of Swedish men in 1956-1960 (see Gavrilova et al., 1983), and in the interval 85-100 years to mortality in 1945-1967 (Depoid, 1973).

$$l_x = l_0 \exp[-Ax - \frac{R}{\alpha}(e^{\alpha x} - 1)], \quad (49)$$

where $l_x$ is the number of those surviving to age $x$, $l_0$ is the size of the original population at the initial moment of time. In the particular case when parameter $A$ is equal to zero, this curve corresponds to the Gompertz law. Analysing the formula, it can easily be seen that whatever the values of the parameters are, the number of survivors becomes zero only at an infinitely large age, i.e. there is no absolute life span limit. It can also be seen from the figure that the real curve (Fig.20, *3*) lies significantly closer to the theoretical curve corresponding to the Gompertz law (Fig.20, *2*). Thus the observed dependence of mortality on age does not support the hypothesis that there is a species-specific life span limit. Moreover, it can be noted that in the region of large ages, the real curve involves even somewhat lower mortality values than the Gompertz law predicts. Therefore the number of people surviving to older ages even somewhat exceeds what is theoretically expected from a calculation based on the Gompertz formula. These divergences only strengthen the conclusion that the hypothesis of a species-specific life span limit does not fit the actual data.

Of course, this example is not unique. In order to test the species longevity limit hypothesis thoroughly, a study was made of 26 tables describing the mortality of long-lived people. These tables were constructed separately for men and women in France (6 tables), the Netherlands (6 tables), Sweden (8 tables) and Switzerland (6 tables) (Depoid, 1973). It turns out that in the age interval 90-103, the force of mortality increases smoothly according to the Gompertz law, and moreover, the rate of this increase turns out to be somewhat lower than at earlier ages (the doubling period of the force of mortality in long-lived people was about 9 years (Gavrilov et al., 1984). The catastrophic growth in the force of mortality predicted by the species longevity limit hypothesis is not observed.

It is interesting to note that many demographers have already reached the conclusion that there is no upper limit to the human life span. For example, in 1961, the leading demographer of our century, Alfred Sauvy, who is the head of the French school of demography and who has for a long period been the chairman of the UN Commission on Population, dedicated a whole book to the problem called "The limits of human life" ("Les limites de la vie humaine"). Having carefully analysed all the currently known facts, A.Sauvy ended his book with the following conclusion: "Thus it seems illusory to put a precise ceiling on human life" (Sauvy, 1961, p.131).*

It must however be noted that the adherents of the species longevity limit hypothesis are surprisingly impenetrable to any arguments. Their faith in the existence of an absolute life span limit is comparable only to the strength of a religious conviction. This is especially surprising, given that the opposing point of view might bring them greater consolation. In response to the arguments we have cited, supporters of this hypothesis suggest that the species longevity limit lies in a region as yet inaccessible to statistical observation. Indeed, it is difficult to test anything in this area, although some material on the mortality of people aged over 100 does exist (Greenwood and Irwin, 1939). An analysis of this material shows that at extreme ages the force of mortality practically ceases to increase with age. The result is that the kinetics of the mortality of long-lived people coincides with the kinetics of radioactive decay, with a "half-life" corresponding to approximately one year (Gavrilov, 1980; Gavrilov and Gavrilova, 1982). As has already been noted, in this event there can be no absolute life span limit. The conclusion that mortality at exceptionally large ages follows a first order kinetics has also been made by other authors (Greenwood and Irwin, 1939), including A.Economos, who demonstrated the correctness of this law for human beings and a number of laboratory animals (Economos, 1980; 1983). It is surprising that A.Economos, who insists on his own precedence as the discoverer of the new kinetics of the mortality of old organisms (the law of radioactive decay), maintains, contrary to his own results, that an absolute life span limit exists and that for human beings it is equal to 100-110 years (Economos, 1985).

Another device used by adherents of the species longevity limit hypothesis is to invent several such limits. Thus, when it became clear that the species limit hypothesis does not fit the actual data, the new proposition was introduced that 95% of people in the population have a maximal life span of 100 years, 4.75% - 110 years, and 0.25% - 120 years (Koltover, 1984). The baseless and artificial nature of this assumption is already obvious from the use of round numbers. However, even with such an arbitrary choice of parameters, no correspondence can be achieved with the

* "Ainsi, il paraît illusoire d'assigner un plafond précis à la vie humaine".

actual data on the mortality of long-lived people: on the assumptions which are made, a temporary decline should be observed in the force of mortality after age 91, reaching 21% by age 96, and then mortality should start to increase (Gavrilov et al., 1984). In fact, the force of mortality in this age interval grows slowly but steadily (Gavrilov et al., 1984), and by age 96 there is an increase of 45% compared with the level at age 91 (Gavrilov et al., 1984). Therefore, even if we postulate the desired heterogeneity of the population, the notion of an absolute life span limit still does not agree with the actual data. Moreover, it must be noted that arguments about an "admixture of long-lived mutants" do not correspond with the facts cited earlier about the exceptionally low heritability of life span at ages greater than 70 years (see Gavrilov et al., 1984, p.33). Finally, this explanation is completely untenable in the case of genetically uniform populations of laboratory animals, in which the same regularities are observed to hold as for human beings (Gavrilov, 1980; Economos, 1980; 1983).

To complete this discussion of the problem of the species-specific life span limit, we quote the conclusion of the well-known ecologist Graeme Caughley (1977), who summarises his observations in the book "Analysis of vertebrate populations". It is significant that the problem of the species longevity limit is allotted a whole section entitled "Misinterpretation of statistics". In particular, in this section, Caughley observes: "... The most accurate life tables available for man and captive mammals provide not the slightest evidence of either a strict potential longevity or a mean potential longevity surrounded by a variance.

"'Potential longevity' is not an appropriate general concept. It must be replaced by a probability statement. An animal dies before the age of infinity not because it cannot pass bounding age but because the probability of its riding out the ever present risk of death for that long is infinitesimal. This statement is not advanced as convenient mathematical sophistry. It is likely to be a closer approximation to biological reality than a model utilizing a cut-off. Cut-off effects are rare in biological systems and no evidence dictates their necessity in a model of mortality" (Caughley, 1977, p.103).

The importance of this conclusion for the biology of life span is obvious, and we will return to it on several occasions as we discuss its consequences. Moreover, the significance of the conclusion that there is no absolute limit to the duration of life extends beyond the biology of life span and even biology as a whole. Indeed, since this issue concerns the way people view the world, it can have philosophical and general scientific importance, occupying a worthy position amongst such fundamental problems as, for example, the question of the finiteness and boundlessness of the Universe.

On the other hand, it seems that the issue has not only theoretical, but also practical importance, for example in health policy. Thus, starting from the species-specific limit hypothesis, the conclusion was once drawn that as the mean life span approached the maximal life span, there would be a "compression of morbidity", as a result of which people would remain healthy for the greater part of their lives, and then in a relatively short time span become senile and die (Fries, 1980). Therefore it was predicted that there would be a reduction in the proportion of chronically ill people in the total population (Fries. 1980). Several well-known scientists have argued against such "seductive predictions", pointing out that reducing expenditure on research into chronic diseases on the basis of such predictions would be a great mistake, given the actually observed trends (Schneider and Brody, 1983). It has also been noted that the notion of a species-specific limit to life on which this prediction is based has no foundation, and is in contradiction with well-known facts (Schneider and Brody, 1983). This example shows what serious practical consequences the solution of this seemingly abstract scientific problem may have.

## 4.3. The concept of species-specific life span invariants

As we have seen, discussing and testing the species longevity limit hypothesis turns out to be so fascinating that it is possible to forget why we embarked on it in the first place. It makes sense, therefore, to return to the beginning of the present chapter and briefly reformulate the original problem.

In order to clarify the causes of inter-species differences in lifetimes with the aid of rigorous quantitative methods, it is essential to have a measure which characterises the life span of individuals of the given species. Since there are significant individual differences in lifetimes, even within the bounds of a single population, the problem arises as to how this value should be selected. If we additionally take into account the fact that life span distributions vary widely from population to population within the same species, our situation seems worse than the one faced by Buridan's ass: he only had two possibilities to choose from, while we are faced with a whole continuum of possible values.

The way out of this impasse might be to find an age which has special properties possessed by no other age. In particular, if there really existed an absolute upper limit to longevity which was a property of individuals of the given species, it would be impossible to invent a better measure of the species-specific life span. However, as we have already ascertained, the species-specific life span limit is like a mirage which escapes us whenever we try to get closer to it by investigating the mortality of long-lived people. It is therefore necessary to seek other approaches to the determination of the species-specific life span.

It is well-known that searches are greatly simplified when the searcher knows what he is looking for. If we are looking for a quantity which characterises the life span of individuals of a given species, it must be the same for all populations of the given species, independently of their genetic characteristics and living conditions. In other words, the most important property of the species-specific life span must be its *invariance* relative to environmental influences and genetic characteristics within the limits of a single species. This desired property of invariance in the species-specific life span was constantly presented as a fact, and attributed to the absolute life span limit. As we have noted on several occasions, it has been claimed of the species-specific limit to the human life span that it "has remained unchanged across time, races and civilisations" (Economos, 1985). Although this assertion itself is without foundation, it concisely encapsulates the desired properties which the species-specific life span must possess. Indeed, if we were to succeed in finding a quantity which characterises the duration of human life and is the same "across time, races and civilisations", the age in question could be called the species-specific life span even if it had no special properties other than invariance. Thus, invariance is the most important criterion for a quantity which represents the species-specific life span.

One compromise which can be used in determining the species-specific life span is to select an age up to which only a small section of the population survives (Dubina and Razumovich, 1975). Thus, the maximal life span can be determined not as the absolute upper limit, but as the age up to which only 1%, 0.1%, 0.01% or 0.001% of the initial number survive. The estimates obtained in this way, in distinction to longevity records, will not grow unrestrictedly with an increase in the size of the sample. Moreover, theoretically, the values of such estimates ought not to depend at all on the number of observations. For such estimates, it is possible to determine the variance and set confidence intervals. On the other hand, a construct of this kind ought fully to satisfy the supporters of the species-specific limit hypothesis, who might use the quantity in question as the "practical life span limit". In this case, to universal satisfaction, this quantity could be named the species-specific life span if it were invariant within the limits of a single species. Unfortunately, this is not what is observed. The value of the maximal life span does not even satisfy the criterion of historical invariance: it changes with time for the population of a single country. For example, for the population of the USA, the age corresponding to the survival of 0.001% of new-born children was 104.8 years for men and 105.4% for women at the beginning of the century (1900). In 1980, this age had increased to 111.4 years for men and 113.9 years for women (Social Security Administration, 1982). There is therefore no basis for asserting that the value of the maximal life span "has remained unchanged across time", when even in the last 80 years, during which some element of testing has been possible, it has increased by 6.6 years in men and 8.5 years in women. There is, moreover, nothing surprising in this increase in the maximal life span, since even a reduction in infant mortality ought to have led to such an increase. Thus, quantities of this kind (corresponding to an age up to which a certain proportion of the population survives) cannot in principle be species-specific characterisations of life span.

Another way of seeking the species-specific life span is not to use the probability of surviving to a given age ($l_x$), which depends on mortality at all ages, including childhood, but to study the probability of dying in the course of a year ($q_x$), which reflects mortality only in the age group under investigation. In the scientific literature, there are attempts to determine the species-specific life span limit by extrapolating the age dependence of the probability of death up to the theoretically maximal value of unity (Depoid, 1973; Khalyavkin, 1983), although in fact, when the definition of the probability of death is taken into account, this "limit" ought to be one year greater (Depoid, 1973). Most often, this kind of extrapolation is carried out on the hypothesis that the probability of death grows exponentially with age. However, as was demonstrated above, this hypothesis does not fit the data (see Fig.20). If an exponential growth hypothesis is introduced, it ought to be introduced not for the probability, but for the force of mortality. Then the quantity calculated by the supporters of the species-specific limit is not the absolute life span limit, but simply the age at which the force of mortality is close to unity. The only special feature of this age is that the corresponding number of survivors ($l_x$) decreases over the year by approximately "e" times (approximately 2.7 times). Clearly, this "special feature" is totally insufficient for us to consider the age in question a species-specific characteristic of life span. Nevertheless, if the value of this age were identical for all human populations, it could be called a species-specific characteristic of life span. In this case, it would also be possible to make a sensible compromise with the supporters of the longevity limit hypothesis, since the quantity in question might be called the "theoretically maximal age" for a hypothetical population with an exponential increase in the probability of death. However, this compromise likewise turns out to be impossible, since such a "theoretically maximal age" does not satisfy the criterion of historical invariance. Thus, the magnitude of this limit, calculated simultaneously for both sexes on the basis of statistical data relating to the pre-war period (the 1930's) in France, the Netherlands, Sweden and Switzerland, was 110.69 years. The same quantity, calculated by the same author for similar data relating to the post-war period (20-25 years later), was equal to 117.28 years for men and 119.28 years for women (Depoid, 1973). F.Depoid, having discovered this phenomenon of an increase in the "theoretically maximal age", considers it "all the more surprising, since it apparently had no precedent in history" (see Depoid, 1973). We, however, would consider the absence of a historical precedent significantly more remarkable if it were a proven fact that this "theoretically maximal age" is historically invariant. In fact, it would be surprising if this indicator remained invariant despite the reduction in the mortality of older people.

It must also be noted that the value of this theoretically maximal age is different for the populations of different countries. Indeed, it is not difficult to show that the theoretically maximal age (i.e. the age at which the extrapolated probability of death is unity), following the model of exponential growth for one-year values of the probability of death, is given by:

$$T_W = 1 + \frac{1}{\alpha} \ln\left(\frac{1}{R}\right), \qquad (50)$$

where $R$ and $\alpha$ are respectively the pre-exponential multiplier and the exponential index in the model of exponential growth for the probability of death. Since, as we have already observed, the values of the probability of death and the force of mortality are numerically close to each other in the region in which mortality values are small, the values of the parameters $R$ and $\alpha$ practically coincide for the Gompertz model and the model of exponential growth for one-year values of the probability of death. Therefore we can use the parameters of the Gompertz formula cited in table 5 to calculate values of the theoretically maximal age for men and women in different countries. When these calculations are carried out, it is not difficult to convince oneself that the values of the theoretically maximal age are not equal for different human populations. Thus, since this indicator does not satisfy the criterion of invariance, it likewise fails as a measure of the species-specific life span.

So, amongst the attempts to estimate maximal longevity, we have not succeeded in finding a species-specific life span invariant. We must therefore abandon the search for a compromise with the adherents of the species-specific limit hypothesis, and turn to other ways of characterising life span. This decision is all the more justified since extreme values are not the only or by any means the best way of characterising the distributions under investigation. This fact was pointed out, in particular, by B.Ts.Urlanis, who noted that the probability of achieving a very high age is negligible, and "therefore it is totally false to consider a very high age to be the real human lifetime" (see Urlanis, 1978, p.255). In a study of longevity, A.A.Zubov and V.I.Kozlov also express the opinion "that exceptional longevity is probably just as poor a characterisation of the biological laws of aging as are individual sporting records of the average physical capabilities of human beings" (see Zubov and Kozlov, 1982, p.131).

In fact, if we wish to establish the magnitude of the species-specific life span and use it in further research, we must make sure that it agrees with values of other species-specific characteristics. Let us clarify this statement with an example. At the present time, there is intensive research on links between the magnitude of the species-specific life span and various anatomical (body weight, brain weight, etc.) (Korchagin et al., 1973; Sacher, 1978; Hofman, 1983; Cutler, 1984), physiological (for example, total diurnal duration of systole, length of sleep, number of offspring in the litter, etc.) (Korchagin et al., 1973; Allison and Cicchetti, 1976; Mineev and Gershkovich, 1979), ontogenetic (length of pregnancy, rate of growth, age of sexual maturity, length of reproductive period) (Korchagin et al., 1973; Allison and Cicchetti, 1976), biochemical (activity of various enzymes, rate of macromolecular metabolism, content of various metabolites, etc. (Kato et al., 1980; Cutler, 1984a; 1984b; 1984c; 1984d; 1984e; Spector, 1984), and molecular-genetic (DNA repair rate, linking of carcinogens with DNA) (Hart and Setlow, 1974; Schwartz and Moore, 1979) characteristics of the species under comparison. Of all the dozens of indicators that have been used, not a single one characterises the maximal values of the features under investigation. Why then should we make an exception for life span? This incongruity which lies at the heart of the majority of inter-species investigations of life span strongly reduces the validity of the obtained results, despite the undoubted relevance of the research itself.

In fact, it is well-known that typical values of the properties under investigation are not at all characterised by maximal, but by average quantities (arithmetic mean, median, mode. etc.). Therefore, if we are interested in typical values for the life span of a particular species, we must turn to average rather than maximal quantities. If we do this, however, the problem of invariance again arises, since all the average values turn out to be non-identical for different populations of the same species.

One of the ways of solving this problem is to split the observed life span distribution into its components. Under this approach, it is hoped to isolate an invariant component of the life span distribution which can be characterised by the average.

This is the technique used in 1978 by Urlanis to determine the magnitude of the species-specific life span of human beings: 86 years for men and 88 years for women (Urlanis, 1978). Urlanis's method of determining the species-specific life span was based on the separation of the observed distribution of the number of deaths ($d_x$) into three subtotals: deaths as a result of injury and other unfavourable environmental factors, deaths as a result of "normal social aging", and lastly, deaths at times corresponding to "purely biological characteristics of the human organism" (see Urlanis, 1978, pp. 245-248). It is for this third component of the distribution of the number of deaths that the value of the mode was determined and called the species-specific life span. It must be noted that B.Ts.Urlanis based his estimates on the hypothesis, formulated as early as 1877 by W.Lexis, of a normal distribution of biological lifetimes, and therefore the values of the mode, median, and mean (arithmetic) life span would coincide for the third subtotal of deaths.

The results obtained by Urlanis can be interpreted in the following way: when all unfavourable factors ("social frictions", to use his definition) are eliminated, the distribution of people according to their lifetimes is described by the normal law with a mode equal to 86 years for men and 88 years for women. It is interesting to note that two years later the American scientist James Fries used a strikingly similar approach, apparently without knowledge of Urlanis's work (Fries, 1980). J.Fries likewise started with the assumption that the lifetime distribution ought to follow the normal law, and that the observed divergences are connected with eliminable cases of premature death. Using this hypothesis and statistical data on the mortality of the US population, Fries established that "under ideal societal conditions mean age at death is not far from 85 years" (see Fries, 1980, p.131). This value, calculated for both sexes together, is exceptionally close to Urlanis's estimates, despite the approximate character of the calculations (see Urlanis, 1978, p.251) and differences in the statistical data used.

Of course, the coincidence of results obtained by different investigators is a definite argument in their favour, but is no proof of their correctness. Moreover, there are some doubts as to the validity of the calculations.

Indeed, the estimates given by B.Ts.Urlanis and J.F.Fries are based on the hypothesis of a normal lifetime distribution "under ideal societal conditions". However the authors provide no justification for this hypothesis, apart from the statement that "all spheres of life are governed by some *distribution*, and in the majority of cases this is subject to the normal law" (see Urlanis, 1978, p.240). The validity of this statement is somewhat controversial, since there is not a single work which proves the wide applicability of the normal law as opposed to other distribution laws. Moreover, actually observed distributions are as a rule asymmetrical and characterised by a non-zero kurtosis. Despite these at times significant deviations from a normal distribution, data are in reality often analysed on the hypothesis that a normal distribution exists. Such a practice, however, does not reflect the validity of the normal law, but only the desire of the investigators to use it since it is in this case possible to use standard statistical techniques. It must, however, be noted that in recent years a growing number of investigators have recognised the necessity for a shift to techniques of non-parametric statistics which do not require the assumption that the features under investigation have a normal distribution.

There is also no theoretical basis for preferring the normal law to a host of other distribution laws known in probability theory and mathematical statistics. What is more, it is possible by a simple argument to show that nature cannot in principle be constructed so that all features under investigation follow the normal distribution law. For example, if we are studying the distribution of particles according to size, three quantities can be used: linear measurements (the effective radius or diameter), surface area, and finally volume (or weight). It can easily be observed that if one of the quantities has a normal distribution, the two others cannot in principle have the same distribution, since the three quantities are connected with each other in a non-linear way. Thus the distribution law even depends on the choice of units of measurement. Therefore, even if some fundamental characteristics of the survivorship of organisms are distributed normally, it does not follow that the distribution of observed lifetimes must also be normal. Finally, there is direct experimental evidence for the inapplicability of the normal distribution law to a description of variation in lifetimes. As is well-known, the distribution of people according to lifetimes is characterised by a marked left-sided (negative) asymmetry. This pronounced deviation from normal of the observed distributions is attributed by B.Ts.Urlanis and J.F.Fries to the existence of premature eliminable aging caused by "social frictions". The correctness of this hypothesis can easily be tested by looking at life tables for laboratory animals kept under standard conditions, where there can be no consideration of social frictions. As was shown in ch.2, the life span distribution of laboratory animals is distinctly non-normal, and is generally characterised by the same left-sided (negative) asymmetry as is observed in human populations (Semyonova, 1983). Moreover, when laboratory animals are kept under especially favourable conditions, their life span distributions have a marked positive kurtosis, i.e. the distributions are significantly more "leptokurtic" than the normal distribution curve (Semyonova, 1983). The problem of life span distributions is discussed in greater detail in ch.2, where it was shown that the normal distribution law is untenable, and that the actual data are in quite good correspondence with the Gompertz-Makeham distribution.

Thus, these examples show once again that the problem of individual differences in lifetimes is indeed a key problem in the biology of life span. Whatever problems we have tried to solve, including the problem of determining the species-specific life span, we have again and again returned to the problem of the variability of organisms with respect to their lifetimes.

In fact, the search for species-specific life span invariants is significantly simplified if the process is split into several stages. During the first stage, when we are determining the parameters of the life span distribution for the population under study, we are in practice determining quantities which are independent of age. For example, having calculated the parameters of the Gompertz-Makeham equation, we can always calculate the values of the force of mortality at any age we are interested in. The parameters themselves, however, do not depend on age. Subsequently, comparing these quantities for the same population at different moments in time, we can establish the parameters which are invariant with respect to time. In our case, this property of historical stability belongs to the age-dependent component of mortality, and the parameters which determine it ($R$ and $\alpha$). These characteristics, which are constant in time, nevertheless display significant regional variation and depend on sex. Therefore, the next natural step in the investigation is to compare the values of these parameters for different populations of the same species in the hope of establishing yet more general (species-specific) invariants. This method of "sequentially extracting invariance" from the actual data may turn out to be significantly more productive than attempts to discover worldwide constants directly from raw, unprocessed material. If we determine invariants in this staged, sequential manner, we may discover an entire hierarchy of invariants of different degrees of generality, nested inside each other like the well-known Russian dolls. This kind of approach to the search for species-specific characteristics of life span requires a comparison of the parameters $R$ and $\alpha$ for different populations of the same species. We now turn to a discussion of the results of such an investigation.

## 4.4. The Strehler-Mildvan correlation

In 1960, the American researchers Bernard L.Strehler and Albert S.Mildvan published an article entitled "General Theory of Mortality and Aging" in the leading scientific journal Science (Strehler and Mildvan, 1960). In this article, it was noted in particular that there is an inverse relationship between the parameters of the Gompertz law: in those countries where the values of the pre-exponential multiplier (designated as $R_0$) were high, the values of the exponential index ($\alpha$) were reduced. If these values, obtained for a number of countries, are plotted on a single graph, a near-linear inverse relationship is observed between $\alpha$ and the logarithm of $R_0$ (Strehler and Mildvan, 1960). Subsequently, this observation became known as the Strehler-Mildvan correlation, and it acquired the status of a fundamental law describing the survivorship of organisms (Strehler, 1962; 1977; Khalyavkin, 1983, 1985).

It can easily be seen that this phenomenon, if it really exists, is of great importance for the determination of the species-specific characteristics of life span. Indeed, the parameters of the Strehler-Mildvan correlation, linking the quantities $R_0$ and $\alpha$, will be species-specific invariants as a consequence of the very principle which has been used to calculate them. Therefore, the first question to arise must be how reliable this important law is.

Strehler and Mildvan (1960) describe their research method as follows: "The age specific mortality rates of all countries in the UN Demographic Yearbook 1955 were plotted on semi-log paper. The curves were smooth for nearly all countries. However, in those few instances where great scatter was exhibited, the data were not further analyzed. the best straight lines through points from age 50 to 70 and, in most cases, from age 35 to 80 were drawn visually, and the values of $R_0$ and $\alpha$ were extracted by graphical extrapolation and measurement, respectively" (Strehler and Mildvan, 1960, p.21).

Three comments arise in connection with the above-mentioned methodology. First of all, the authors have analysed not the force of mortality, but the probability of death. However, as has already been noted, exponential growth is characteristic of the force of mortality rather than the probability of death. Nevertheless, this comment is not very important, since for ages 35-80 the numerical values of the force of mortality and the probability of death (in the course of the year) practically coincide.

The second comment concerns the quality of the

statistical processing of the data. In fact, the data were analysed by subjective methods ("straight lines ... were drawn visually"), for different age intervals (sometimes for ages 50-70, sometimes for ages 35-80), and with no estimate of the accuracy of the determined quantities (absence of confidence intervals). This method of processing the data is without doubt not the best method available even at that time. However, this comment too can be considered unimportant in comparison with the third and final comment.

In their investigation, Strehler and Mildvan have completely ignored the existence of the background component of mortality, although this quantity should not have been neglected, especially in the case of the following tables used by the authors: Algeria (1948), Argentina (1947), Brazil (1950), Costa Rica (1950), Egypt (1947), etc. Of course, when the background component of mortality is zero, the analysis of the data is greatly simplified, reducing to the determination of the parameters of the linear dependency between the logarithm of the force of mortality and age. However, in this particular case, this simplification is completely unfounded, as can be seen even in the graphs presented by the authors. Thus, Fig.1 of the article in question contains a graph of the logarithm of the mortality rate against age for Egyptian men in 1947. This graph, which has a pronounced concave shape bearing witness to a high background mortality, cannot, even with a feat of the imagination, be called linear (see Strehler and Mildvan, 1960). Despite the fact that the authors cite the Gompertz-Makeham formula at the beginning of the article, the entire subsequent analysis is based on the Gompertz formula. As will be demonstrated below, this "simplification" alone is sufficient to generate a false correlation which coincides with the Strehler-Mildvan correlation.

Fig.21 shows, in semi-logarithmic coordinates, the graph of the logarithm of the force of mortality against age *for the same values* of parameters $R$ and $\alpha$, but different values of parameter $A$. It can be observed that, as the background component of mortality ($A$) is reduced, the graph becomes steeper and steeper, corresponding to an increased estimate for parameter $\alpha$ as determined graphically from a measurement of the slope. It can also be seen that at the same time there is a reduction in the value of the point at which the straight line drawn by Strehler and Mildvan intersects the y-axis. This corresponds to a reduction in the estimate for parameter $R_0$, likewise determined graphically. Consequently, the method used by Strehler and Mildvan to determine the parameters leads to biassed estimates for $R_0$ and $\alpha$, and what is more, when the

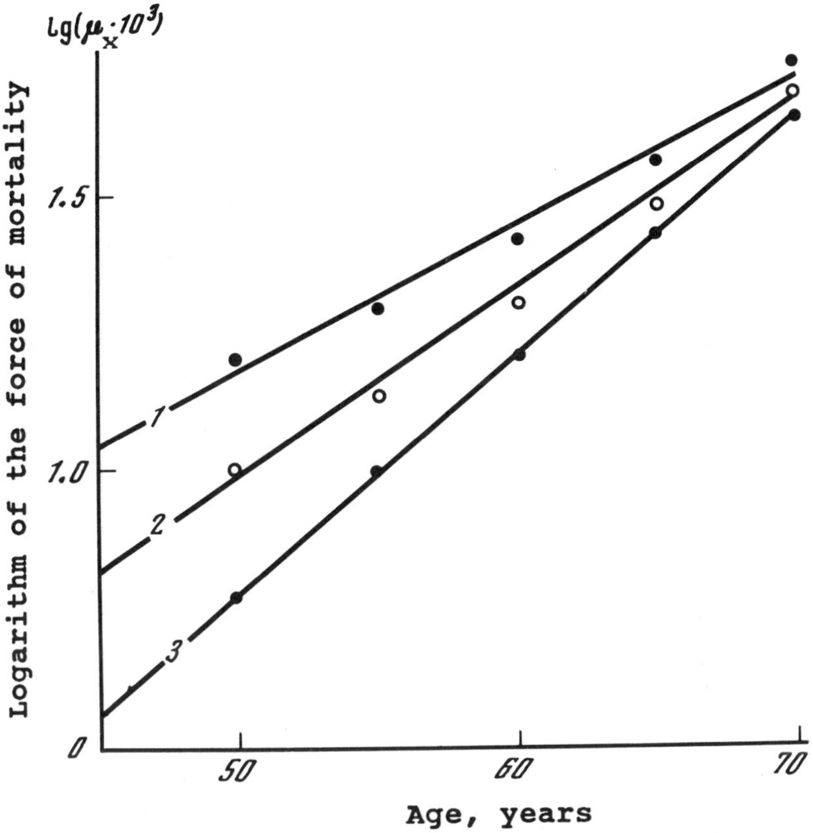


**Figure 21** Biassing influence of the background component of mortality on estimates for parameters $R$ and $\alpha$ as determined by the Strehler and Mildvan method.

1 - the dependence of the logarithm of the force of mortality on human age for $A$ = 0.010 $\text{year}^{-1}$

2 - the dependence of the logarithm of the force of mortality on human age for $A$ = 0.004 $\text{year}^{-1}$

3 - the dependence of the logarithm of the force of mortality on human age for $A$ = 0.000 $\text{year}^{-1}$

The real values of $\alpha$ and $\lg(R \cdot 10^5)$ are 0.1 $\text{year}^{-1}$ and 0.60 respectively.

background component of mortality (*A*) is altered, these estimates are biassed in opposite directions, generating a false correlation.

Naturally, the question which arises is whether this false correlation, connected with the incorrect processing of data, coincides with the Strehler-Mildvan correlation. In order to answer this question, we carried out a numerical experiment using typical values for the parameters of the Gompertz-Makeham formula: $\alpha = 0.1$ year$^{-1}$, $R = 4 \cdot 10^{-5}$ year$^{-1}$, $A$ = 0.0-0.01 year$^{-1}$. With these constant values of parameters $R$ and $\alpha$, dependencies between the force of mortality and age were modelled for a whole range of values of parameter $A$. Then the data were processed in conformity with the methodology used by Strehler and Mildvan. Since these authors did not indicate the cases in which they used the age interval 35-80 for the linearisation, and the cases in which they used 50-70, one and the same interval (50-70 years) was used for the analysis of the data. In conformity with the methodology used by Strehler and Mildvan, the intervals between the points were 5 years, so that each dependency contained 5 points corresponding to ages 50, 55, 60, 65 and 70. The only essential divergence from the methodology of Strehler and Mildvan was that the "linear" dependency was determined not visually, but by the least squares method, in order to eliminate any subjectivity in estimating the parameters of the "straight" line, and in order to determine their accuracy. The results of these calculations are presented in table 7.

It can be seen that, depending on the level of the background component of mortality, a significant bias is observed in estimates of the parameters $R_0$ and $\alpha$. If these results are compared with Strehler and Mildvan's data, there is a striking coincidence (see Fig.22). Thus, the Strehler-Mildvan correlation is not distinct from the false correlation created by the incorrect method used by Strehler and Mildvan.

This example shows that a discussion of what appear to be purely methodological questions can lead to exceptionally important conclusions. Indeed, the correlation claimed by Strehler and Mildvan is widely cited without any serious critical analysis (Strehler, 1962; 1977; Ďoubal, 1982; Koltover, 1983; Khalyavkin, 1983; 1985), and is even considered to be a fundamental law which every theory of aging, mortality and life span must obey (Strehler and Mildvan, 1960; Strehler, 1962; 1977). However, attempts to use the Strehler and Mildvan correlation in constructing mathematical models of aging lead to absurd results. Indeed, from the data presented in the article by Strehler and

**Table 7** Biassing influence of the background component of mortality on estimates of parameters $R_0$ and $\alpha$

| Magnitude of parameter $A$, year$^{-1}$ | Estimate of parameter | |
|---|---|---|
| | $\lg(R_0 \cdot 10^5) \pm \sigma_{\lg R_0}$ | $(\alpha \pm \sigma_\alpha) \cdot 10^2$, year$^{-1}$ |
| 0.000 | 0.60 ± 0.00 | 10.0 ± 0.0 |
| 0.001 | 0.80 ± 0.02 | 9.3 ± 0.1 |
| 0.002 | 0.98 ± 0.05 | 8.8 ± 0.2 |
| 0.003 | 1.13 ± 0.06 | 8.3 ± 0.2 |
| 0.004 | 1.27 ± 0.07 | 7.9 ± 0.3 |
| 0.005 | 1.39 ± 0.08 | 7.5 ± 0.3 |
| 0.006 | 1.50 ± 0.09 | 7.2 ± 0.3 |
| 0.007 | 1.60 ± 0.09 | 6.9 ± 0.4 |
| 0.008 | 1.69 ± 0.10 | 6.6 ± 0.4 |
| 0.009 | 1.77 ± 0.10 | 6.3 ± 0.4 |
| 0.010 | 1.85 ± 0.10 | 6.1 ± 0.4 |

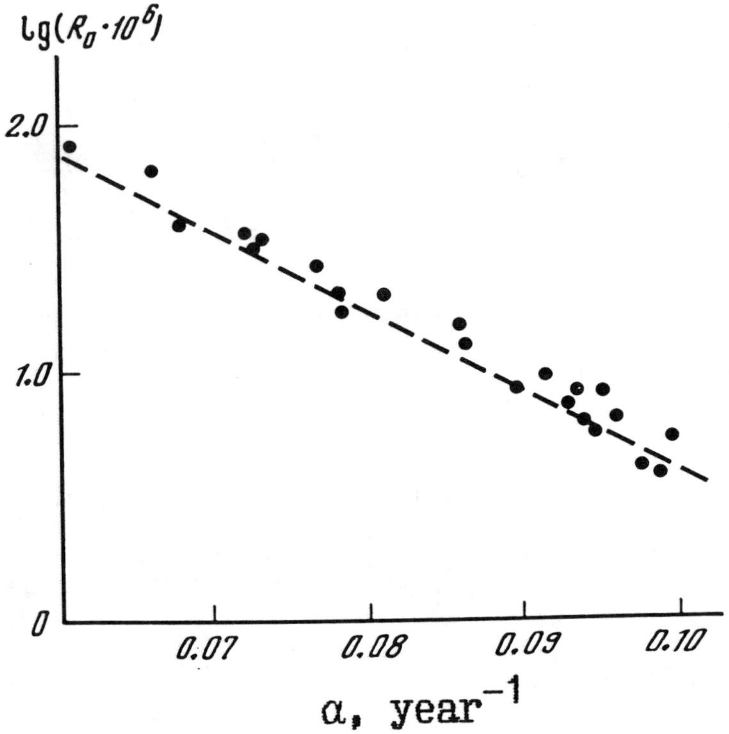


**Figure 22** Coincidence of the false correlation between *R* and α with the Strehler-Mildvan correlation.

The false correlation, constructed on the basis of table 7, is shown by a dashed line. The Strehler-Mildvan correlation is constructed on the basis of data published in Strehler and Mildvan (1960).

Mildvan (1960), it follows that the angular coefficient of the linear regression of ln $R_0$ with α is only 68.5 years. However, in the framework of the "General theory of mortality and aging" by Strehler and Mildvan (1960), this quantity ought to correspond to the age at which so-called vitality, "the capacity of an individual organism to stay alive" (p.15), becomes zero! The absurd result that this parameter, equal to 68.5 years, corresponds to an absolute limit to the human life span, follows from another mathematical model of aging and mortality (Koltover, 1983). The incongruity of these consequences of the Strehler-Mildvan correlation was pointed out by the famous French demographer Herve Le Bras (Le Bras, 1976). Indeed, the magnitude of the species-specific limit to the human life span turns out to be significantly less than average life span of the population of many developed countries! It would seem that this artefact alone ought to have put researchers working in this area on their guard. However, unfortunately, this did not happen (Strehler, 1962; 1977; Ďoubal, 1982; Koltover, 1983; Khalyavkin, 1983; 1985), although the existence of a contradiction is obvious and has been pointed out on several occasions (Le Bras, 1976; Gavrilov et al., 1978; Gavrilov, 1984b).

Since the data in Strehler and Mildvan's article are processed using the subjective visual method, we attempted in 1978 to arrive at a new estimate of the angular coefficient of the given dependency, in order to exclude the possibility of error at this point also (Gavrilov et al., 1978). The values of the parameters $R_0$ and $\alpha$ were taken from the same article by Strehler and Mildvan (from the table which presents 32 paired values of the parameters for the male population of different countries in the world), and the least square method was used to determine the value of the angular coefficient of the linear regression of $\ln R_0$ with $\alpha$. The value of this coefficient and the corresponding standard deviation turned out to be 74±3 years (Gavrilov et al., 1978). Although the calculated value is somewhat larger than the estimate obtained by Strehler and Mildvan (68.5 years), it is nevertheless significantly less than the required value for a species-specific limit to the human life span.

Of course, the contradiction which has been exposed may be connected with a whole range of causes: the artefactual nature of the Strehler-Mildvan correlation, the incorrectness of the proposed mathematical models of mortality, and the erroneous nature of the very notion of an absolute species-specific life span limit. It is possible that all three of these factors are present in this case. Naturally, the authors themselves (B.L.Strehler and A.S.Mildvan) could not help but notice the lack of correspondence between their theory of mortality and the estimate of the species-specific life span limit which follows from the correlation they discovered. However, they did not place any stress on this disagreement, and quickly turned to other methods of estimating the magnitude of the species-specific limit which are based on additional assumptions (Strehler and Mildvan, 1960). For instance, on the assumption that the life span limit corresponds to the age at which the exponentially increasing mortality rate becomes unity, it was calculated that this limit is 103 years (Strehler and Mildvan, 1960). It can be seen that even this estimate is significantly lower than the vast number of documented cases of exceptional longevity. In addition, as can be calculated from Strehler and Mildvan's (1960) data, this value is not a species-specific invariant and varies widely from population to population, oscillating within the range 90-116 years. Nevertheless, in their subsequent publications, the authors did not return to check the reliability of the described correlation, referring to it as to an established fact (Strehler, 1962; 1977). Their example was followed by other investigators (Ďoubal, 1982; Koltover, 1983; Khalyavkin, 1983; 1985). This situation may seem rather strange, but it has to be admitted that the canonisation of artefacts through lengthy uncritical citation is often encountered in modern science.

Thus, although the Strehler and Mildvan correlation is widely cited in the scientific literature, it cannot be considered a proven scientific fact. What is more, the failure of Strehler and Mildvan to take into account the background component of mortality inevitably generates an artefactual dependency which, as we have shown, coincides with the dependency desribed by Strehler and Mildvan. Therefore, the Strehler and Mildvan correlation is unsuitable for determining species-specific characteristics of life span. Nevertheless, this does not mean that the very idea of searching for species-specific invariants by comparing the parameters of the Gompertz-Makeham formula is flawed. On the contrary, as will be shown below, this idea can be brought to fruition if only we correctly estimate the parameters of the Gompertz-Makeham formula.

## 4.5. The compensation effect of mortality

In 1978, the so-called compensation effect of mortality was discovered (Gavrilov et al., 1978), and its existence was confirmed in subsequent research (Gavrilov, 1984b; Gavrilov and Gavrilova, 1979a). It turns out that, within the limits of a single species, the values of the age-dependent component of mortality are correlated in such a way that, when extrapolated, they meet at a single point (see Fig.23). This surprising phenomenon is not only observed in man, but also in the fruit fly *Drosophila melanogaster* (see Fig.24). It can be seen that this intersection at a single point is connected with the fact that the reduction in the level of the age-dependent component of mortality in the transition to more fortunate populations is *compensated* by an increase in the relative rates at which it grows with age, hence the name compensation effect of mortality (Gavrilov et al., 1978; Gavrilov, 1984b).

It is clear that the coordinates of the point of intersection, because of the very principle used to calculate them, are invariant relative to the living conditions and genetic characteristics of the populations under comparison. In other words, they reflect the most general (species-specific) characteristics of the life span distribution for populations of the same biological species. On these grounds, the coordinate corresponding to the age at which all the dependencies intersect has been called the species-specific life span (Gavrilov and Gavrilova, 1979b). For *Homo sapiens*, this quantity is 95±2 years (Gavrilov, 1980; 1984a; 1984b).

Naturally, in a discussion of this interesting phenomenon, a whole number of questions arise. The first of these is the following: what does intersection at a single point actually mean, and what is the use of knowing such species-specific constants? It is rather easy to answer this question. In fact, if we know the coordinates of the point at which all the dependencies meet, it suffices to know just one value of the age-dependent component of mortality at an arbitrary age in order to be able to draw a straight line through the two points and thereby restore the value of the age-dependent component of mortality for other age groups. In practice, this means that from one or two values of the force of mortality it is possible to restore an entire life table in the age interval 20-80 years.* A direct test of this method, using the population of Sweden as an example, confirmed that it really does allow the reasonably accurate restoration of a life table using incomplete data (Gavrilov,

* One value of the force of mortality is sufficient when the value of the background component of mortality can be neglected (contemporary situation in the developed countries). In the general case, it is essential to know the value of mortality in two age groups so as to take account of the existence of background mortality.

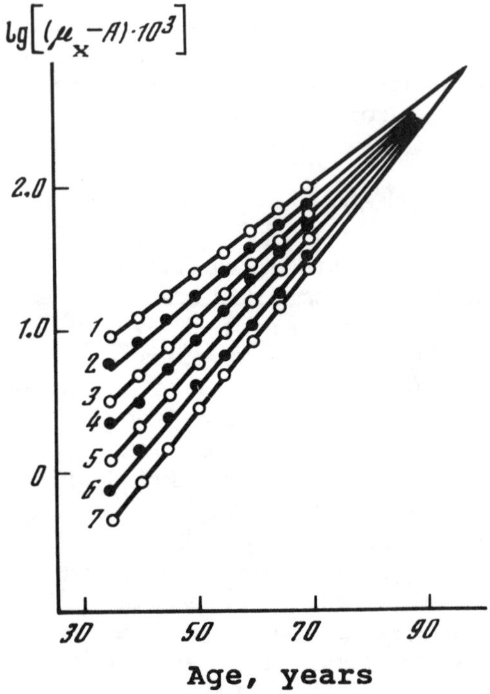


**Figure 23** The phenomenon of the compensation effect of mortality in *Homo sapiens*. The dependence of the logarithm of the age-dependent component of mortality on age.

1 - India, 1941-1950, men, $A$ = 0.00787 year$^{-1}$
2 - Turkey, 1950-1951, men, $A$ = 0.00316 year$^{-1}$
3 - Kenya, 1969, men, $A$ = 0.00531 year$^{-1}$
4 - Northern Ireland, 1950-1952, men, $A$ = 0.00043 year$^{-1}$
5 - England and Wales, 1930-1932, women, $A$ = 0.00238 year$^{-1}$
6 - Austria, 1959-1961, women, $A$ = 0.00068 year$^{-1}$
7 - Norway, 1956-1960, women, $A$ = 0.00055 year$^{-1}$

Calculated and constructed on the basis of abridged life tables published by the UN. Values of the probability of dying in the course of one year were used as an approximation to the total force of mortality. Taken from Gavrilov and Gavrilova (1979b).

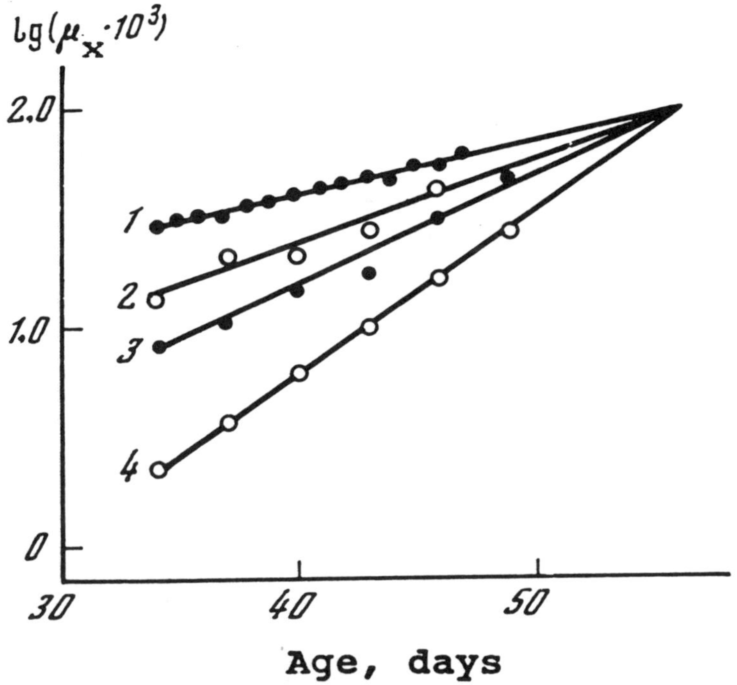


**Figure 24** The phenomenon of the compensation effect of mortality in laboratory drosophila.

1 - drosophila of the Old Falmouth, New Falmouth, Sepia and Eagle Point strains (1000 virgin females)
2 - drosophila of the Canton-S strain (1200 males)
3 - drosophila of the Canton-S strain (1200 females)
4 - drosophila of the Canton-S strain (2400 virgin females)

In calculating the force of mortality, a 6-day age interval was selected. Source of the data: Pearl and Parker (1921); Hall (1969).

1980). This example also shows that the practical significance of intersection at a single point does not depend on whether this point really exists or is a result of extrapolation (analogously, in optics, the parameters of the virtual focus are the most important characteristic of a diverging lens).

The second important question concerns the reliability of the phenomenon. Indeed, the figures 23-24 presented above, although very striking, must be treated merely as an illustration, rather than as a proof of the existence of the compensation effect of mortality. In order to prove the existence of the phenomenon, it must be demonstrated that for populations of a single species the Gompertz-Makeham law can be written in the following way:

$$\mu_x = A + M\cdot\exp(\alpha(x - B)), \tag{51}$$

where $B$ is the age common to all populations at which all values of the age-dependent component of mortality become equal to $M$. Thus, $B$ and $M$ are respectively the x-coordinate and y-coordinate of the point of intersection of the dependencies between the age-dependent component of mortality and age. As has already been noted, the quantity $B$ can be called the species-specific life span, given its properties and dimension. Analogous considerations lead to the quantity $M$ being called the species-specific force of mortality (Gavrilova, 1982; Gavrilov, 1984a).

It is not difficult to show that intersection at a single point is possible if and only if the following relationship holds between the pre-exponential multiplier ($R$) and the exponential index ($\alpha$):

$$R = M\cdot\exp(-B\alpha). \tag{52}$$

Indeed, if this relationship is substituted into the Gompertz-Makeham law, we arrive at the above-cited form of this law.

Thus, the proof of the validity of the compensation effect of mortality reduces to the proof of relationship (52). The problem can be simplified if we take the logarithm of this relationship:

$$\ln R = \ln M - B\alpha. \tag{53}$$

Thus, the proof of the compensation effect of mortality reduces to the proof that there exists an inverse linear relationship between ln$R$ and $\alpha$, i.e. to the problem which Strehler and Mildvan were trying to solve. However, as was shown in the preceding section, this problem must be solved anew, taking into account the background component of mortality.

Table 8 presents the results of such an investigation, based on 209 abridged human life tables (Gavrilov, 1980; 1984b):

**Table 8** Characteristics of the compensation effect of mortality for *Homo sapiens*

| | Men | Women | Both sexes |
|---|---|---|---|
| $\ln M \pm \sigma_{\ln M}$ (per year) | -0.74 ± 0.28 | -0.67 ± 0.44 | -0.67 ± 0.21 |
| $B \pm \sigma$ (years) | 94 ± 3 | 96 ± 4 | 95 ± 2 |
| Correlation coefficient between ln$R$ and $\alpha$ | -0.946 | -0.918 | -0.949 |
| Number of pairs ln$R$ and $\alpha$ | 106 | 103 | 209 |

It can be seen that between the quantities ln$R$ and $\alpha$ there is a clear, near-linear inverse relationship. This is demonstrated by the values of the correlation coefficients, which range from -0.92 to -0.95. It is interesting to note that the magnitude of the species-specific life span calculated for men (94±3 years) and women (96±4 years) turns out to be practically identical. There are therefore grounds for uniting these figures and calculating a species-specific life span for human beings (95±2 years), common to men and women. On the same grounds it is possible to calculate a single value, common to men and women, for the logarithm of the species-specific force of mortality (-0.67±0.21), corresponding to a hypothetical value for the age-dependent component of the force of mortality common to all human populations at age 95 and equal to 0.51 year$^{-1}$. The question as to whether this point of intersection really exists, and what the dynamics of mortality look like after the species-specific age, remains open for the time being, given the lack of accurate data concerning very high ages. Some hypotheses and observations concerning this question can be found in special publications (Gavrilov, 1980; Gavrilova and Gavrilov, 1982). It is essential however to emphasise that, independently of whether this intersection really exists, the very fact that the majority of dependencies meet in a single region is nevertheless of great practical importance (see above), and needs its own theoretical explanation.

One of the possible explanations for the compensation effect of mortality, based on the principles of the theory of reliability, is that the rate of the primary processes which destroy the organism in aging is a species-specific invariant (Gavrilov, 1978; Gavrilov et al., 1978). This at first sight strange assertion should not be identified with the programmed death hypothesis, which presupposes the existence of a kind of time bomb (Kogan et al., 1977). Simply, the existence of a powerful homeostasis has the consequence that the majority of the organism's vital parameters, on which the rate of its destruction also depends, are indeed invariant with respect to many external influences and genetic characteristics. For example, the human body temperature is approximately 36.7°C and practically independent of sex, racial characteristics and climate. It is therefore natural that many destructive processes which are dependent on temperature and other similar factors are so invariant that some of them (for example, the racemisation of L-aminoacids in the organism) can even be used to estimate the age of the organism (Man et al., 1983). If this explanation for the compensation effect of mortality is correct, we may observe significant deviations from the discovered phenomenon in species with a less powerful homeostasis than man (for example, in poikilothermal animals).

Indeed, in laboratory drosophila, the compensation effect of mortality is more often the exception than the rule. For example, in first generation hybrids the age-dependent component of mortality is characterised by a parallel shift of the mortality curve in relation to the corresponding curve in the parent lines (Sacher, 1977; Gavrilov, 1980), therefore in this case there can be no discussion of intersection at a single point. Nevertheless, as can be seen from Fig.24 and table 9, this effect is still sometimes observed in drosophila.

**Table 9** Characteristics of the compensation effect of mortality for *Drosophila melanogaster*

| | Males | Females | Both sexes |
|---|---|---|---|
| $\ln M \pm \sigma_{\ln M}$ (per 6 days) | -1.17 ± 0.33 | -0.83 ± 0.09 | -0.81 ± 0.19 |
| $B \pm \sigma$ (days) | 50 ± 5 | 57 ± 1 | 57 ± 2 |
| Correlation coefficient between $\ln(R)$ and $\alpha$ | -0.992 | -0.9997 | -0.995 |
| Number of pairs $\ln R$ and $\alpha$ | 4 | 4 | 8 |

**Table 10** Characteristics of the compensation effect of mortality for rats of the Wistar strain

| | Males | Females | Both sexes |
|---|---|---|---|
| $B \pm \sigma$ (days) | 660 ± 5 | 580 ± 50 | 610 ± 50 |
| Correlation coefficient between $\ln(R)$ and $\alpha$ | -0.987 | -0.968 | -0.941 |
| Number of pairs $\ln R$ and $\alpha$ | 10 | 10 | 20 |

Attempts to check the existence of the compensation effect of mortality in other biological species are hampered by the absence of data of the required accuracy, and their lack of diversity. Indeed, in order to discern an intersection at a single point and determine its coordinates, it is essential to have data on the mortality of organisms which are not only accurate, but also sufficiently diverse. Therefore, the results obtained for laboratory rats (see table 10) must be considered as preliminary, although they do provide evidence in favour of the existence of the compensation effect of mortality.

Summarising these attempts to determine the species-specific life span, we have to admit that the problem has turned out to be more complicated than previously thought. Although for the majority of species it is still not possible to quote well-founded estimates for the species-specific life span, the fact that species-specific invariants exist can no longer be in doubt. It cannot be ruled out that the method of estimating species-specific characteristics of life span based on the Gompertz-Makeham law and the compensation effect of mortality will turn out not to be the only possible method, or even the best. However, the idea of the "sequential extraction of invariance" from raw data through examining different types of dependency and correlation in sequential stages and isolating parameters of varying degrees of generality will, it seems, form the basis of all future attempts to determine the species-specific life span.

## CHAPTER 5. THE SEARCH FOR THE MECHANISMS WHICH DETERMINE LIFE SPAN

In moving to a discussion of the mechanisms which determine the duration of life, we must immediately admit that these mechanisms are still not understood. It is not even known how similar the mechanisms are which determine the life span of different biological species. What is more, it cannot be ruled out that at different stages in life different mechanisms act to terminate it.

On the other hand, enough information is presently available to test the validity of a wide range of hypotheses, to determine the basic properties of the mechanisms under investigation, and to chart a course towards their decipherment. In particular, the facts known today are sufficient to answer one of the most important questions of the biology of life span: is the life span of organisms programmed?

### 5.1. A self-destruction program or wear and tear?

When discussing the hypothesis that life span is programmed, its suporters usually use as their basic argument the fact that there are large inter-species differences in lifetimes (see Comfort, 1964; 1979). This at first sight convincing argument in fact proves nothing, demonstrating only the necessity of clarifying the meaning of the concept "program". Indeed, different technical devices can also have totally different characteristic "lifetimes" (compare, for example, the working life of a battery and a photoelectric cell). Huge differences can also be observed in the lifetimes of the atoms of different chemical elements and their isotopes. It would, however, be precipitate on the basis of these facts to draw the conclusion that inside each atom there is a program which limits its lifetime. It seems more sensible that the lifetimes of systems (including biological systems) are simply determined by aspects of their construction and function. Thus, the fact that there are large inter-species differences in life span is evidence only that the magnitude of life span (just like the magnitude of all other characteristics) is determined by the material nature of the objects being studied.

Another argument habitually presented by the supporters of the programmed life span hypothesis is connected with the assertion that the heritability of lifetimes is high (see Gavrilov, 1985). This argument immediately contains two errors. Firstly, as we convinced ourselves on the basis of the data presented above (see section 2.3, p.31), the heritability of lifetimes (in the sense of a similarity between blood relatives) is unusually low. The notion that the heritability of life span is high is partly connected with a failure to distinguish between two concepts: the authenticity of an influence and the strength of that influence. As a result, having established the incontrovertible validity of the fact that the life spans of parents influence the life span of their offspring, people draw the erroneous conclusion that life span must be genetically determined (see Gavrilov, 1985). Secondly, a high heritability of life span is a necessary, but not a sufficient condition for considering life span genetically determined. For example, the high gastronomic qualities of sturgeon are without doubt genetically programmed, but it does not follow from this that their genome contains a special program calculated to satisfy the tastes of human beings. Thus, we arrive at the necessity of defining the concept "program" more precisely, despite its apparent simplicity.

A consideration of the meaning of the concept "program" leads us to the conclusion that what is presupposed is the existence of a definite *purpose* which the program serves to achieve. In this sense, it is possible to assert that many useful characteristics of agricultural animals and plants are genetically programmed, since they have arisen as a result of a long period of *purposeful* selection. In the case of natural living conditions, the choice of purpose is determined by the struggle for existence and the selection of those characteristics, following a particular direction, which aid survival in the given concrete circumstances. Thus, we can speak of programmed characteristics only when the concept of purpose is involved, and it is possible to answer who (man) or what (evolution) shaped the program. Consequently, the question whether life span is programmed can be answered by asking what its significance is for biological evolution.

If the question whether life span is programmed is

approached from the evolutionary point of view, it becomes obvious that special mechanisms for the termination of life could hardly help the individual to fight successfully for its survival. On the contrary, those individuals in which the action of such a program of self-destruction had been impaired by a random mutation would quickly displace all the remaining individuals, since in their longer life span they would produce more offspring.

It is therefore all the more surprising that belief in the existence of a special genetic program for the self-destruction of the organism is wide-spread and is being actively propagandised. For example, in one textbook on biological cybernetics for students, a special section is devoted to the exposition of such ideas, and we even find an italicised assertion that there exists "*an active additional regulatory mechanism which at a particular stage in ontogenesis enters into battle with the mechanisms which support life and interrupts its course*" (Kogan et al., 1977, p.65). Finally, even in such a respectable edition as Encyclopaedia Britannica (1989), one can read the following: "Locked within the code of the genetic material are instructions that specify the age beyond which a species cannot live given even the most favourable conditions" (The New Encyclopaedia Britannica, 1989, vol.20, p.471). Not one of these assertions is proven, and moreover, as we have already noted, propositions of this type are even theoretically unsound. However, bearing in mind the popularity of such ideas, we must subject them to especially thorough testing, not restricting ourselves to comments about absence of proof or theoretical unsoundness, but undertaking the analysis of real data.

At first sight, it might seem self-evident that the genetically programmed death hypothesis can only be tested with the aid of direct molecular-genetic, biochemical, cytological and physiological experiments. In this case, data on the life span distributions of organisms in different populations would appear to be of no use whatsoever. However, if we analyse the problem more deeply, at least two different ways can be proposed of using life tables to carry out a reasonably reliable test of the programmed death hypothesis.

The first way of testing the programmed death hypothesis is based on a comparison of life tables for individuals of a single species in natural (wild) and artificial (laboratory, domestic, civilised) surroundings. If the hypothesis under test is correct, such life tables should not show very large differences in the lifetimes of adult individuals. Indeed, for a self-destruction program to arise, take hold and be maintained in the course of evolution, it must at least have some opportunity, however small, of *coming into operation* in natural (wild) conditions. Consequently, the age at which such a program is switched on cannot be too high, otherwise, because of the high mortality in the wild from predators, hunger, infections and harsh natural conditions, no-one will live to the fateful age and the self-destruction mechanism will not even be able to come into operation. It follows from this that life spans in even the most favourable conditions cannot significantly exceed the ages reached by the most hardy section of the population, if, of course, the concept being tested is correct.

Now, when we start to analyse the actual data, we see a picture diametrically opposed to what would be expected from the concept being tested: the life spans of organisms in conditions conducive to longevity greatly exceed the life spans observed in natural (wild) conditions. In view of the abundant evidence, we shall restrict ourselves to just a few examples.

It is well-known that in favourable conditions the chaffinch (*Fringilla coelebs*) can live for 29 years (see Comfort, 1964). However, in the wild this is practically impossible, since around a half of all birds perish in the course of a year from hunger, cold, diseases, and attacks by predators, and the mean life span is only 1.4-1.5 years (Paevsky, 1985). As a result of this high mortality, only 0.1% of the initial number of chaffinches survive to age 11 (Paevsky, 1985).

Similar results have been obtained for voles (*Microtus arvalis* Pall). In laboratory conditions, the average life span of voles is 7-8 months, while individual specimens survive to 25 months (see Fenyuk and Sheikina, 1940). On the other hand, in the wild, the average life span of voles is only 1.2 months, and only 0.1% of the original number of voles survive to 10 months (Fenyuk and Sheikina, 1940). Thus, if we attempt to estimate the age of programmed death on the basis of life spans in laboratory conditions, it becomes clear that no self-destruction program could arise or be maintained in evolution, if only because it would not be able to come into operation in natural conditions, where practically no individual lives to the requisite age.

The same conclusion is reached from an analysis of data on the human life span. At present, the mean expectation of life in developed countries is 70-80 years, while the documented record for longevity is 120 years. If we take these figures as an estimate of the age at which the self-destruction program is switched on, we are forced to admit that such a program could not have arisen in human evolution, since, accoding to palaeodemographic data, virtually nobody survived to such an age. For example, according to the data of Acsádi and Nemeskéri (1970), only a half of those born in the Late Palaeolithic reached 8-9 years, and only a half of those born in the Neolithic reached 26 years. Moreover, even in the Middle Ages (9-12th centuries), life expectancy at birth was no greater than 27-29 years. Investigations of the skeletons of American Indians have shown that only 2 centuries ago only 4% of the population survived to age 50 (Owsley and Bass, 1979). By way of comparison, we note that the probability of surviving to this age in the developed countries is 94-96%. If these facts are compared, it is difficult to refrain from the following question: can the guaranteed destruction of a few old people, who are chance survivors and doomed in the wild, be a sufficient evolutionary basis for the formation and preservation of a special self-destruction program in the human genome? Viewed in this light, the absurdity of the programmed death hypothesis becomes obvious.

The facts just cited, despite their fundamental significance, may nevertheless seem unconvincing to those who are sceptical of the theory of evolution and inclined to see a divine intention in a limited life span. Since our life spans are neglible compared with the characteristic periods of biological evolution, arguments based on evolutionary considerations may seem to such people too abstract and hypothetical. We have also had occasion to encounter the following standpoint: "All this occurred so long ago that nobody can now know what really happened". It therefore makes sense to look at a second approach to testing the programmed death hypothesis, one which is in no way linked with evolutionary historical ideas.

The second way of testing the programmed death hypothesis is to study the pattern (shape) of the dependence of mortality on age (Gavrilov, 1984a; 1984b; Gavrilov and Gavrilova, 1986). If the hypothesis being tested is correct, the moment at which the self-destruction program is switched on ought to be accompanied by a sharp jump in mortality. A thorough analysis of this prediction has already been carried out in section 4.2, with the result that no catastrophic growth in mortality occurs in the region of the proposed life span limit. This fact refutes the idea that "locked within the code of the genetic material are instructions that specify the age beyond which a species cannot live given even the most favourable conditions." (The New Encyclopaedia Britannica, 1989, vol.20, p.471).

A study of the dependence of mortality on age also enables us to test more complicated versions of the programmed death hypothesis. For example, there is the idea that the program "enters into battle with the mechanisms which support life" as early as age 40-50, but "interrupts its course" only after 20-40 years of exhausting struggle (otherwise the hypothesis under scrutiny cannot accommodate the universally known fact that the average life span of people in developed countries is now 70-80 years). In order to illustrate this notion, we cite a characteristic assertion in which the post-reproductive self-destruction hypothesis is presented as a universally accepted fact: "It is well-known that in human life the period of aging begins after the period of highest sexual activity. After 40-50 years, accumulating signs of approaching senescence gradually appear in a person's appearance and behaviour. All the research in gerontology points to the same result: this is the age when destructive changes begin to develop in the organs at the physico-chemical, cell and tissue levels, the efficiency of the visceral functions begins to drop, and a person's physical strength and intellectual capabilities are reduced. "Rejuvenation" effects, achieved with the aid of sex hormones ..., cytotoxic serum ..., undernutrition ..., apart from their general stimulatory action, are perhaps to some extent linked with the postponed development of the self-destruction function" (see Kogan et al., 1977, p.67).

If such assertions are correct, we should expect that, at the moment when the self-destruction program is switched on (40-50 years), we should observe a *discontinuity* in the graph of risk of death against age, marking the beginning of a new stage of accelerated growth in mortality (see Fig.25A). If no self-destruction program exists, and the growth of mortality with age is caused simply by an accumulation of diverse defects and injuries (wear and tear in the wide sense of the phrase), we should expect the monotonic growth of mortality with age without any discontinuities (see Fig.25B). In other words, if the program hypothesis is correct, there should be a *qualitative change* at age 40-50 in the dependence of mortality on age, whereas the wear and tear hypothesis allows for the possibility that mortality at older ages may be calculated by simple extrapolation of the mortality trends observed at younger ages (up to 40-50 years).

Turning to the actual data (Fig.26), it is easily seen that no discontinuity can be observed in the graph of mortality against age, and what is more, mortality at older ages can be easily estimated by simple extrapolation of the trends observed at age 30-40. Reduced viability, manifesting itself as an increase in the risk of death with age, becomes

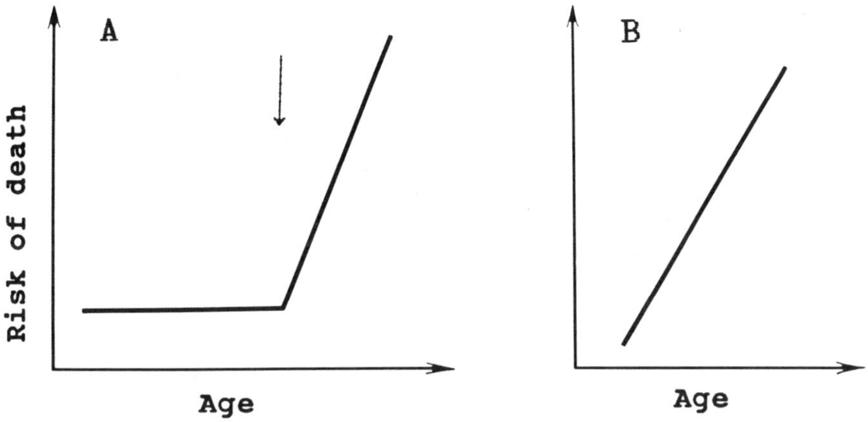


**Figure 25** Two possible graphs of the dependence of mortality on age.

A - the dependence of mortality on age which would be expected if the hypothesis of the actively programmed self-destruction of the organism were valid. An arrow marks the point at which the self-destruction program is switched on.

B - the dependence of mortality on age which would be expected if the hypothesis of the gradual accumulation of defects and wear and tear were valid.

obvious not at 40-50 years, but much earlier, at 10-15 years (see Fig.26). It is at age 10-15 that we see the beginning of the age-dependent growth in total mortality and mortality from a number of "endogenous" causes, and the first signs of degenerative change appear (for example, atrophy of the thymus). However, it cannot be ruled out that more careful investigation will reveal an even earlier appearance of unfavourable changes.

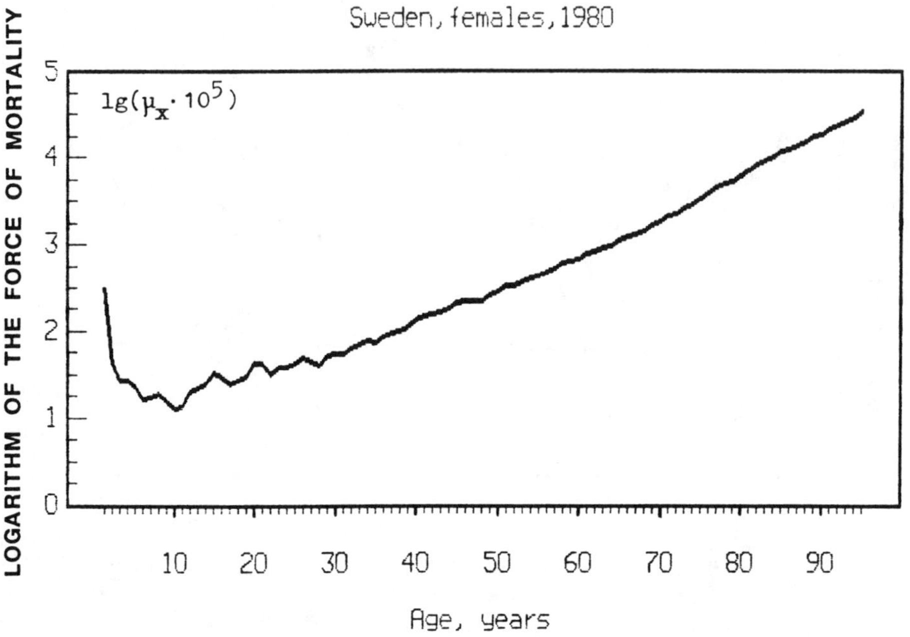


**Figure 26** The actually observed graph of mortality against age. The dependence of the force of mortality on age for Swedish women in 1980. Source of the data: Befolkningsförändringar, 1980.

It must be noted that evidence against the programmed life span hypothesis is provided by the very shape of the life span distribution described by the Gompertz-Makeham law. In fact, this distribution is well-known in the theory of reliability as the first asymptotic function of the distribution of least values (Gumbel, 1958). As has already been stated, this distribution describes the destruction of technical devices and phenomena of electrical break-down (Kase, 1953; Gumbel, 1958). Thus, the graph of the mortality of organisms coincides with the graph of the destruction of materials, supporting the wear and tear hypothesis, in the widest sense of this phrase.

Opponents of the wear and tear hypothesis adduce the following argument. Technical systems, they say, in distinction to biological systems, have no life span limit. This point of view is held, in particular, by the well-known gerontologist A.Comfort: "The constancy of the specific age in forms which senesce is a strong argument against the primacy of 'mechanical' ageing" (Comfort, 1979, p.42). However, as we have already convinced ourselves, no species-specific limit to life span seems to exist at all. Therefore, this argument of the opponents of the wear and tear hypothesis cannot be considered convincing.

Thus, an analysis of the survivorship of organisms allows us confidently to reject the programmed life span hypothesis and conclude that searches for a special "death hormone" will not be successful. At the same time, however, it is necessary to make some notes. Firstly, wear and tear in one of an organism's systems may lead to increased loading and accelerated deterioration of its other systems. In this way, a cascade of dependent failures may occur in the organism, simulating its self-destruction. Secondly, when far-reaching destruction has taken place in the organism, crisis regulation mechanisms may be switched on, bringing about a change in the structure of the system accompanied by a partial failure to carry out its basic function (Gubler, 1978). In this situation, the crisis regulation reactions may be superficially very similar in their manifestations to processes of self-destruction. Thirdly and finally, we should not yet exclude in principle the possibility that a program for the active self-destruction of organisms might exist in some exotic species, formed as a result of particularities in these species' ecology. Life is so diverse in its manifestations that exceptions are possible even to the most general laws.

## 5.2. Analysis of inter-species differences in the duration of life.

One of the possible ways of analysing the mechanisms which determine life span is to study the huge differences in life span which exist between species. It must be noted that comparing different species is in general an exceptionally interesting procedure which helps to clarify a number of fundamental problems in biology (Schmidt-Nielsen, 1984). Inter-species comparisons of life span goes back into the depths of history: Aristotle noticed that there is a definite link between the life span of organisms and their growth period, namely that the life span of a species is about 5-6 times greater than its growth period. Subsequently, in the 18th century, this law attracted the attention of the famous French naturalist Buffon. In 1908, M.Rubner discovered an inverse relationship between the life span of organisms and the intensity of their metabolism (Rubner, 1908). In 1910, Friedenthal noticed that life span is higher in those species which have a greater relative brain weight (Friedenthal, 1910). The history of the study of inter-species differences in life span is described in greater detail in the book by Lamb (1977), as well as in the works of Sacher (1959; 1977), who made an important contribution to the development of this line of research. In this section, therefore, our attention will be basically focussed on the state of contemporary research, which has recently begun to develop especially quickly due to the efforts of Richard Cutler (1983; 1984a; 1984b; 1984c; 1984d; 1984e; 1985a; 1985b; 1986).

At the present time, the number of features which have been studied is so large that a simple listing occupies three tables (see Appendix 2). Some of the features seem to be positively correlated with life span (Appendix 2, table *A*), others show a negative correlation (Appendix 2, table *B*), and finally, many of the features investigated show practically no correlation with life span (Appendix 2, table *C*). The extent of the accumulated material, as well as the great accuracy and complexity of the calculations (Sacher, 1977; Hofman, 1983) create the impression that scientific research in this area is well-founded and well-developed.

At the same time, when this line of investigation is studied more closely, it becomes clear that its development is blocked by three fundamental problems which must be solved: *the problem of the reliability of the original data*, *the problem of the correct statistical processing of the data*, and *the problem of how to interpret the regularities which emerge*. Below we carry out a brief discussion of these problems and the ways in which they might be solved.

*The problem of the reliability of the original data*. The importance of this problem arises from the fact that deficiencies in the original data cannot in any way be compensated by the complexity and accuracy of the subsequent computational procedures. Accordingly, the accuracy and reliability of any results which are obtained is illusory whenever the accuracy and reliability of the original data is unknown. When comparing species, features can be divided into several groups according to their suitability for this kind of analysis. Some features, like, for example, the number of chromosomes or the quantity of DNA contained in the nucleus of a single cell, are ideally suited to inter-species comparisons. Other features, like body weight or brain weight, vary significantly amongst the individuals of a species, and their use entails a wide range of problems. However, it is possible to establish typical values and estimate the dispersion of these features, and to use this information in subsequent investigations. In the case at hand, the saving factor is the fact that inter-species differences with respect to the feature greatly exceed the dispersion of the values of the feature amongst the individuals of a single species. There are some features, however, for which the dispersion is so great that they are suited least of all to inter-species comparisons. One of these is life span.

There are several ways of estimating the life spans of the species to be compared. We can use the mean life span of organisms under natural (wild) conditions, or use record values of longevity under the same conditions. Analogous life span characteristics can be recorded under laboratory conditions, under zoo conditions, in domesticated and farm animals, and finally, in the case of man, under the conditions of modern civilisation. It is clear that under such diverse conditions, and with different choices of life span characteristic (mean, mode, median, age to which 10% of the population survives, record values of longevity), we will obtain highly divergent values for "the species-specific life span". In different publications, depending on the goals and tastes of the investigator, different life span indexes are used, although these often bear similar or identical names. It is neverthless clear that, since essentially different quantities are involved, the results of such investigations are practically incomparable.

In many cases, in an attempt to standardise the data and avoid the effects of external conditions, use is made of maximal life span indexes. However, this method not only fails to solve the problem at hand, but also introduces a whole range of new problems. As has already been noted (see chapter 4, table 6), even for such a well-studied species as man, estimates of the biological life span limit, despite the abundance of statistical data, vary from 60 to 200 years! It is even the case that one and the same investigator, in one and the same publication (Cutler, 1984a), uses 10 different values for the human life span, ranging from 86 to 115 years. Despite the inaccuracy of the original data, heated discussions nevertheless take place in which the argument hinges on the existence of a small difference between two compared coefficients of inter-species correlation (Sullivan, 1982; Cutler, 1983). It is clear, however, that the solution of such problems requires special attention to be paid to the quality of the original data, rather than the differences in the results of calculations.

Thus, the use of record life span values in no way solves the problem of the reliability of the original data. What is more, the use of record values adds two other fundamental problems. The first problem is connected with the fact that longevity records depend on the number of observations. Therefore, as we have noted before in this book, the documented record for the human life span, given the existence of abundant and long-accumulated statistical data, is bound to be significantly greater than the analogous record for a species with the same longevity, but represented by a dozen individuals in a zoo. In other words, information on record life spans cannot be compared because it has been obtained using different numbers of observations, and no correction has been made for these differences. Indeed, data on the human life span show a strong tendency, in many correlations, towards large values (Sullivan, 1982). Sometimes, far reaching conclusions are based on this fact (see Cutler, 1983). However, it cannot be ruled out that such features of the data are more probably artefacts caused by the number of observations.

The second problem connected with the use of record life span values is the fact that like is not being compared with like. Indeed, all the remaining quantities used in inter-species comparisons are not records, but rather typical values of the features under investigation. It is therefore incomprehensible why we should make an exception for life span, and why we should believe that record values characterise the longevity of the species better than typical values (arithmetic mean, modal life span, median life span, etc.).

The switch from record life span values to average values removes the two problems listed above, but cannot radically solve the problem of how to standardise the data under comparison. The problem is that average life span indexes, even within the limits of a single species, strongly depend on the living conditions and genetic characteristics of the compared populations. For example, the life spans of mice belonging to different genotypes, kept under identical, rigorously controlled conditions, differ by a factor of almost two (Sacher and Duffy, 1979; Gelman et al., 1988), even when individuals of the same sex are compared. If we compare the life span of mice from the long-living line C57BL/6J, living under "standard" laboratory conditions, it turns out that the average life span of these mice has increased 1.5-2 times over the period 1948 to 1971 (Kunstýř and Leuenberger, 1975). It must be noted that these large differences are observed when living conditions have been partially standardised. In general, values for the average life span of individuals belonging to the same species can vary by very large factors depending on the living conditions and the genotype (see Lints, 1985).

Several ways can be suggested of solving the problem of standardising the quantities uner comparison. The first way is direct standardisation of the living conditions of the compared species. This technique is suitable for closely related species living in similar ecological conditions. An example of this direct standardisation is the comparative study of the life span of the house mouse (*Mus musculus*) and the white footed mouse (*Peromyscus leucopus*) (Sacher and Hart, 1978). However, the method of direct standardisation of living conditions, while acceptable for the comparison of such species as, for example, the mouse and the rat, becomes much more difficult when comparing, for example, the rat and the chicken, and totally impossible when comparing the elephant and the whale. Therefore, for the purposes of wide-scale inter-species comparisons, it is essential to search for other, more general methods of standardising life span indexes.

One of the most promising approaches to this kind of standardisation is not to compare the life span characteristics of species living under identical conditions (which cannot ensure equality for animals adapted to different ecological environments), but under conditions in which the shape of the life span distribution is identical. In order to convert this principle into practice, it is necessary to have a sufficiently large collection of life tables for each species, so that those tables can be selected for which the shapes of the life span distribution curves coincide. In the event that a whole series of coincident distributions emerges, preference should be given to the most symmetric distribution, in which the differences between the mean, modal and median life span are minimal. The advantage of a distribution with this shape is that the question of which life span characteristic to use (mean, mode or median) then has no significance. However, since this approach is quite complicated, it is worth considering some simpler methods of standardising life span values.

One of the simple methods of standardising life span indexes is to select some way of averaging the mean life span values obtained for different populations of the same species. This averaging process can be a calculation of the arithmetic mean of the mean life span values for different populations. It can be a calculation of the most frequently encountered (modal) mean life span value, median or mode (modal mean, modal median, modal mode). Finally, it is possible to calculate the median value of the life span characteristics of each population (mean, mode and median). It can easily be seen that if three indexes (arithmetic mean, mode and median) are used to characterise the centre of each life span distribution, and each of these can be averaged in three different ways at the inter-population level, we end up with nine possible indexes for the species-specific life span. Although this approach might seem primitive, it is nevertheless more rational than the current practice of using an arbitarily selected life span value with the fine name of "expert estimate".

Finally, a radical solution to the problem of the reliability of the original data can be achieved if the calculations are carried out using species-specific life span values of the type described in chapter 4. However, no universal methods of estimating species-specific life spans yet exist, and actual species-specific life span values have only been determined for human beings and the fruit fly (*Drosophila melanogaster*). Therefore, strictly speaking, it is premature to carry out inter-species comparisons, since no solutions are yet available to the scientific problems which the development of such comparisons will necessarily involve. Despite the abundance of publications and results, it has to be admitted that they all possess an exceptionally preliminary character. In order for the results of such investigations to be considered reliable, the following conditions must be fulfilled:

(1) The quantity selected to characterise the life span of the given species must be the same not only in the work of each individual author, but also in the publications of all the authors carrying out inter-species comparisons.

(2) Such a congruence of values ought to follow not from consensus or agreement between the investigators, but from the very procedure used to determine the species-specific life span.

(3) The computational procedures used should be based on information about the life span distribution of individuals of the given species for varying living conditions and genotypes, and should be a generalisation of this information.

(4) The values obtained for species-specific life spans should be accompanied by appropriate estimates of confidence intervals and measures of the reliability of the results.

It must be admitted that not one of these conditions has yet been fulfilled, and the reliability of the data used is such that it cannot produce reliable results. At the same time, it must be noted that the problem is not a totally insoluble one, and possible solutions have already been presented above. If these solutions were put into practice, research into inter-species differences in life span could be placed on a rigorous scientific footing.

*The problem of the correct processing of the data.* The analysis of inter-species differences is characterised by two factors: the inaccuracy and unreliability of the original data, and variation in the vast number of interdependent characteristics that different species have. As has been noted before, the accuracy of currently used species-specific life span indexes is illusory. Therefore, at the present time, the only result that can be calculated with sufficient reliability is the ranking of each species according to the longevity of the individuals which belong to it. For instance, it is possible to state confidently that human beings live longer than dogs, dogs live longer than guinea pigs, guinea pigs live longer than rats, and rats live longer than mice. Thus, only the ranking is reliable, or put more simply, the position which each species occupies in an ordered sequence of life span values. Therefore, if data of this kind are to be processed correctly, what is needed is precisely a calculation of coefficients of rank correlation (Spearman and Kendall correlation coefficients), rather than the traditional Pearson correlation coefficients used in practically all current investigations. As a result of this practice, the unreliability of the original data leads to unreliability in the final results of the calculations, whereas the use of coefficients of rank correlation would lead to more stable results which were less dependent on the actual numerical life span values. Another argument in favour of replacing the traditional Pearson correlation coefficients with coefficients of rank correlation is the presence of dispersion in the analysed data. It is well-known that the use of the Pearson correlation coefficient and the estimation of confidence intervals for this coefficient are justified only when the joint distribution of the characteristics under investigation can be considered bivariately normal (Aivazyan et al., 1985). In the case of inter-species comparisons, this condition is practically never satisfied, and what is more, a very typical situation is for one or two points to lie widely outside the general run of points. Amongst these discrepant points, the point relating to man is most frequently found (see, for example, Fig.25 in the article by Cutler, 1984a). We have already discussed the reasons for such dispersions, noting their possible artefactual nature. The following is also important: widely discrepant points can, if processed in the traditional manner, create the illusion of a strong connection between the variables under investigation, since very high values are obtained for the Pearson correlation coefficients. Besides, it is well-known that high values for correlation coefficients in such cases prove nothing, and that analysing the data in this way is a typical example of incorrect procedure (see Aivazyan et al., 1985). It is often sufficient to exclude one of these widely discrepant points from consideration for the "supposed" connection between the variables totally to disappear. However, the more correct method is not to exclude this data from analysis, but to use coefficients of rank correlation, which are stable in relation to the magnitude of divergence of discrepant points.

Another deficiency in current methods of analysing inters-species differences is the fact that the analysis is usually limited only to the paired comparison of variables, i.e. to the calculation of paired correlation coefficients. This is the method used in the majority of investigations whose results are presented in Appendix 2. However, it is a feature of inter-species differences that the compared species differ according to a huge number of characteristics which are at the same time quite closely correlated. Therefore, the connection between life span and the chracteristic in question may not be direct, but indirect, i.e. caused by the effect of a third variable, omitted from the analysis, on the two chracteristics under investigation. For this reason, in order correctly to investigate the effect on life span of the characteristic in question, it is necessary to calculate not paired correlation coefficients, but partial correlation coefficients in which the effect on life span of other confounding factors has been eliminated. An example of this kind of investigation is the work of Harvey and Zammuto (1985), in which it was shown that there is a strong positive relationship between age at first reproduction and life span in the wild (the correlation coefficient was 0.976). Since body weight also positively correlates with life span and age at first reproduction, the partial correlation coefficient was calculated between life span and age at first reproduction, eliminating the effect of body weight. This correlation coefficient was 0.892, showing that the connection in question between the two variables exists even when the effect of body weight is excluded. Unfortunately, in this particular investigation, the authors used the traditional Pearson correlation coefficient, although partial correlation coefficients can be calculated using the Kendall coefficient of rank correlation (Kendall, 1962).

Thus, in order to determine the genuine connections between life span and a particular characteristic, it is essential to calculate the partial correlation coefficient, excluding the effects of all other characteristics. This approach can be effectuated using the Kendall coefficient of rank correlation. Another approach to the same problem is to solve a multiple regression equation including the entire set of possible characteristics. This approach can be found in the works of Sacher (1959; 1977; 1978), although it has to be noted that the results of these kinds of calculations may strongly depend on the actual numerical life span values. However, no matter how the calculations are carried out, we inevitably come to the third problem in the analysis of the data: the calculation of the partial effects on life span of the characteristic in question, eliminating the effect of all other confounding factors, requires the measurement of a huge number of characteristics. On the other hand, a huge number of analysed characteristics can only be used in the analysis when the number of analysed species is even greater, since the number of cases must be greater than the number of variables used. As a result, we arrive at the necessity of studying a large number of species whose life spans may turn out to be so close to each other that the problem of accurately determining the species-specific life span again comes to the fore.

Thus, the correct analysis of inter-species differences in life span is possible if the following three conditions are satisfied:

(1) Changing from the calculation of traditional Pearson correlation coefficients to coefficients of rank correlation would lead to results which were stable in relation to the large indeterminacy of species-specific life span indexes and the presence of widely divergent values for the variables.

(2) Changing from paired correlation coefficients to partial Kendall coefficients of rank correlation would permit the effect on life span of each characteristic under

investigation to be taken into account, excluding the effect of all other characteristics.

(3) A calculation of the partial effect of a characteristic, eliminating the effect of all other confounding variables, requires the analysis of a large number of charateristics and an even greater number of species. Naturally, as the number of compared species increases, there will also be an increase in the number of doubtful cases concerning the ranking of species according to life span. Therefore, the limited reliability of data on species-specific life spans becomes a fundamental limitation which cannot be circumvented by improving statistical methods.

It must be admitted that the majority of investigations into inter-species differences do not satisfy the conditions set out above. Therefore, the results of inter-species comparisons are highly preliminary in nature. However, as has already been shown, it is comparatively easy to improve the reliability and depth of analysis of the accumulated data, and it is to be expected that this will be done in the near future. At the same time, it must again be emphasised that, without a radical improvement in the quality of the original data, the possibilities of statistical analysis are very limited.

*The problem of how to interpret the results of inter-species comparisons of life span.* It is well-known that the existence of a correlation between two variables in no way implies the existence between them of a causal link (the two variables may be determined by a third, unknown variable). Therefore, the problem of how to interpret the results which emerge from the correlational analysis of inter-species differences in lifetimes is quite a complicated one. The problem is made more complex by the fact that the results of inter-species comparisons do not always agree with the results of comparisons within a species. For example, it has been firmly established that the life span of species increases with the growth in their body weight (see Appendix 2, table *A*). At the same time, it is well-known that large values of body weight not only do not lead to large life spans in human beings, but are even a risk factor which is important enough to be frequently taken into account in life insurance (Manson et al., 1987). In this particular case, it is clear that it would be highly precipitate to attribute body weight with a direct positive effect on life span. However, situations of this kind often arise in the study of other characteristics, for which the absurdity of such arguments is not so obvious. For example, it has been established that superoxide dismutase activity correlates positively with life span in a variety of species (see Appendix 2, table *A*). On the basis of this inter-species correlation, it was concluded that there is a causal link between superoxide dismutase activity and life span (Koltover, 1983). This conclusion was drawn despite the fact that no intra-species comparisons of superoxide dismutase activity and life span were carried out. What is more, it was recently discovered that heightened superoxide dismutase (SOD) activity in man, caused by microduplication of the small segment of chromosome 21 which contains the SOD-1 gene, leads to the appearance of all the features of Down's syndrome, a typical disease of premature aging, accompanied by a sharp drop in life expectancy (Huret et al., 1987).

These examples show that reaching any conclusions on the basis of inter-species comparisons is a high-risk activity. Does it follow from this that it is pointless to carry out inter-species comparisons? It seems to us that the correct answer to this question is as follows: inter-species comparisons are totally useless as a form of proof, but are an excellent method of formulating hypotheses and submitting them to a preliminary test. In particular, an analysis of the data in Appendix 2 leads to the formulation of the following hypothesis. It turns out that life span is greater in those species which have more powerful systems of self-preservation and defence against harmful influences (see Appendix 2, table *A*). At the same time, short-lived species are characterised by a high metabolic rate and damage to their own structure (including DNA) by the intermediate and side products of metabolism (see Appendix 2, table *B*). Therefore, an analysis of inter-species differences likewise leads us towards the wear and tear (accumulation of defects) hypothesis as a possible mechanism limiting the life span of organisms. In a more concrete form, taking into account the biochemical nature of life span markers, this hypothesis might be called the hypothesis of the biochemical amortisation of the organism.

## 5.3. Analysis of intra-species differences in the duration of life

An analysis of intra-species differences includes the following three lines of research:

(1) An analysis of differences in life span between different strains of the same species. In this section, a concrete example of this approach will be discussed.

(2) An analysis of sex differences in lifetimes. In view of the magnitude of this problem, as well as its distinctive nature, we will devote a special section to it (section 5.4).

(3) An analysis of individual differences in lifetimes, the search for predictors and markers of individual life span. This line of research is so extensive that it deserves a special monograph, Some light was shed on the problem in the second chapter of the present book.

Despite the fact that in all the cases listed the differences in lifetimes are much less than in inter-species comparisons, many of the problems cited in section 5.2 concerning the comparability of the data nevertheless practically disappear. We give below a concrete example illustrating the great potential of the method of intra-species comparisons.

In studies of the mechanisms which determine life span, incomparably greater attention is paid to the characteristics of the organism itself than to the analogous characteristics of its parents. It is considered obvious that any connection between the characteristics of the parents and the life span of their offspring, even if it existed, would be indirect rather than direct in character, thanks to the inheritance by the offspring of these characteristics. Therefore, given the existence of "direct" information concerning the characteristics of the organism itself, "indirect" information concerning the characteristics of the parents is considered useless for predicting life span, and is usually not used at all (Sacher and Duffy, 1979).

A diametrically opposite view follows from the reliability theory of life span proposed in 1978 (Gavrilov, 1978; Gavrilov et al., 1978). According to this theory, the life span of organisms has a great deal in common with the durability and strength of materials (Gavrilov and Gavrilova, 1986): in both cases a crucial role is played by the original distribution of defects, since the successive, avalanche-like multiplication of defects is what determines the process of subsequent destruction (Gavrilov, 1987). From this point of view, parental characteristics determining the quality of the gametes, and especially maternal characteristics determining the accuracy of the early stages of development, would be expected to have much greater influence on the life span of the offspring than these same properties of the offspring themselves. In other words, the reliability concept leads us to a paradoxical prediction: the best predictors of life span should be sought not amongst the characteristics of the organism itself, but amongst the characteristics of its parents.

We have tested and completely confirmed this paradoxical prediction of the reliability theory of life span. A previously unknown phenomenon has been revealed, namely the priority of parental characteristics over the organism's own characteristics in the prediction of life span. This new regularity radically changes traditional ideas concerning the mechanisms which determine life span, and opens up a new direction for research in the biology of life span.

In view of the fundamental importance of this issue, special attention was paid to the quality of the original data. To this end, we used for analysis the highly reliable data obtained at the Argonne National Laboratory (USA) under the leadership of the well-known specialist George Sacher (Sacher and Duffy, 1979). It must be emphasised that these data were obtained independently of the hypothesis under test, and were also published in full (Sacher and Duffy, 1979), permitting any investigator to carry out an independent check of the correctness of our results.

In Sacher and Duffy (1979), the following parameters were measured in male mice with various genotypes: mean life span; body weight; metabolic rate according to the consumption of oxygen per unit weight at rest; and on average throughout the day. These last three parameters were measured twice, at young (6-8 months) and old (24-34) ages, resulting in 6 different variables. Out of the 21 genotypes under investigation, 5 belonged to the following inbred mouse strains: A/JAnl, BALB/cJAnl, C57BL/6JAnl, C3Hf/JAnl and DBA/1J. The remaining 16 genotypes belonged to $F_1$-hybrids, obtained by crossing the above-mentioned inbred strains in various combinations and directions.

The analysis of the data in Sacher and Duffy (1979) was carried out in the following way: the mean life span of the $F_1$-hybrids was treated as a dependent variable (the life span of the offspring) whose value was to be predicted on the basis of 18 other variables: 6 characteristics of the $F_1$-hybrids themselves, 6 characteristics of the maternal strains, and 6 characteristics of the paternal strains. Since the number of predictor variables under consideration (18) was greater than the number of observations (16), the method used was stepwise regression, successively including in the equation only those variables which were the best predictors of life span (program *2R* in the well-known *BMDP* package; forward stepping procedure).

The results of this analysis were striking and exceeded even our boldest expectations. The characteristics of the organism itself turn out to be totally useless for the prediction of life span, whereas the parental characteristics allow life span to be predicted with great accuracy. The "best" correlation between life span and a characteristic of the organism itself was insignificant even at the 90% confidence level ($r$ = -0.39; $F(1.14)$ = 2.51; $P$ > 0.1), whereas the 4 parental characteristics were much more closely linked with the life span of the offspring. Table 11 contains the results of successively including the best predictors of life span in the multiple regression equation. All 18 variables were treated equally and competed with each other for a place in the equation on the basis of their predictive value. It turned out that characteristics of the maternal and paternal strains were included in the equation in approximately equal proportions, whereas the characteristics of the organism itself cannot compete with the parental chracteristics and are not included in the equation at all (see table 11).

Table 12 contains the results of analysing the predictive value of 4 different sets of variables: characteristics of the organism itself (6); characteristics of the maternal strains (6); characteristics of the paternal strains (6); characteristics of the parental strains (12 characteristics: 6 characteristics of the maternal and 6 characteristics of the paternal strains). It turned out that all 6 characteristics of the organism itself account for less than 25% of the observed variation in life span, and not one set of variables representing the characteristics of the organism resulted in the statistical significance of the regression even at the 90% confidence level (see table 12). By way of comparison, we note that the first characteristic of the maternal strains to be included (minimum rate of oxygen consumption at a young age) accounts for 42% of the dispersion of life spans between the offspring populations with a confidence level of greater than 99% ($r$ = - 0.65; $F(1.14)$ = 10.14; $P$ < 0.01). If the 6 parental characteristics are used as a basis for prediction, they account not for 25% (as in the case of the characteristics of the organism itself), but 95% of the variation in life span (see table 12). Thus, life span can be accurately calculated from the values for the parental characteristics, while the characteristics of the organism itself turn out to be practically useless in this respect.

**Table 11** The priority of parental characteristics over the characteristics of organisms themselves in the prediction of life span.

| Step number (in iter-ation)[a] | Quality of life span predic-tion[b] | Number of variables included in the equation: | | |
|---|---|---|---|---|
| | | organisms' own character-istics | character-istics of maternal strains | character-istics of paternal strains |
| 1 | 0.38 | 0 | 1 | 0 |
| 2 | 0.61 | 0 | 1 | 1 |
| 3 | 0.65 | 0 | 2 | 1 |
| 4 | 0.75 | 0 | 2 | 2 |
| 5 | 0.84 | 0 | 2 | 3 |
| 6[c] | 0.91 | 0 | 3 | 3 |

[a] At each step one new variable with the best predictive value was included in the regression, so the step number coincides with the total number of variables in the equation and equals the sum of the figures in the three columns on the right.

[b] The quality of the prediction was measured by the the square of the adjusted correlation coefficient, i.e. with account taken of the reduction in the number of degrees of freedom as the number of variables was increased.

[c] After the 6th step the inclusion of further variables did not lead to a significant improvement in the quality of prediction, and the value of the *F*-criterion for the new variables did not exceed 1.1. Therefore there was no point in including further variables.

**Table 12** Comparison of the predictive value of organisms' own and parental characteristics.

| Number of variables in the equation | Percentage of explained life span variation $(100 \cdot r^2)$ [a] | | | |
|---|---|---|---|---|
| | organisms' own characteristics | characteristics of paternal strains | characteristics of maternal strains | characteristics of paternal and maternal strains |
| 1 | 15.2[c] | 20.0 | 42.0 | 42.0 |
| 2 | 19.2[c] | 35.2 | 50.9 | 66.5 |
| 3 | 22.3[c] | 45.1 | 51.3 | 72.0 |
| 4 | 23.1[c] | 52.2 | 51.3 | 81.3 |
| 5 | 24.8[c] | 66.9 | -"-[b] | 89.3 |
| 6 | 24.9[c] | -"-[b] | -"-[b] | 94.7 |

[a] Since at each step the best predictor variable was included, the percentage of explained variation is one of the highest possible with the given number and type of variables.

[b] New variables were not included in the equation since their coefficient of multiple correlation with the other variables already present in the equation was close to unity and no improvement in prediction could be expected. In other words, the tolerance of the new variables was so close to zero that further calulations were pointless.

[c] The regression is insignificant even at the 90% confidence level, as established by the value of the *F*-criterion.

The newly-discovered paradoxical phenomenon of the priority of parental characteristics is in good agreement with the notion that life span, mortality and morbidity are largely determined by events which occur long before birth, and possibly even before the formation of the zygote (otherwise it is difficult to explain the priority of paternal characteristics over the characteristics of the organism itself). One of the possible mechanisms which might determine the life span of the offspring is the accumulation of damage in the DNA of the gametes, caused by active side products of the oxygen metabolism of the parents. It has recently been discovered that the rate of such oxidative damage to nuclear DNA, and especially mitochondrial DNA, caused by the side products of oxygen metabolism, is very high (Richter et al., 1988). In the light of this data, it becomes clear why, of all the characteristics we have investigated, the parental respiratory rate indexes turn out to be the best predictors of the life span of the offspring, and why the mean life span of the offspring is lower when the respiratory rate of the parents at rest is higher.

Accordingly, we carried out a study of the dependence of the mean life span of the offspring on the respiratory rate of the parents. To calculate the respiratory rate of the parents, we divided by two the sum of the oxygen consumption rates of young sleeping mice from the paternal and maternal lines. The mean life span of the offspring was seen to decrease sharply with an increase in the respiratory rate of the parents (correlation coefficient $r = -0.81$; confidence level $P < 0.01$). Extrapolating this dependence to a zero parental respiratory rate leads to a mean life span for the offspring of 1662 days. In the framework of the hypothesis under consideration, this means that if the gametes of the parents are fully protected from oxidative damage, the life span of the offspring can be increased on average by a factor of 2! It is striking that the respiratory rate of the sleeping young offspring, in distinction to the repiratory rate of their parents, has no effect on life span (the correlation coefficient is only -0.1, and the relationship is not statistically significant). This fact likewise agrees well with the notion that life span is determined exceptionally early.

A consequence of these ideas is the fundamental possibility that life span might be significantly increased and morbidity reduced by preventing damage to the gametes and disturbances in the early stages of ontogenesis. Evidence in support of this notion is the fact that the life span of the offspring is increased when antioxidants preventing oxidative damage are added to the maternal diet

(Harman and Eddy, 1979). It seems that this fact should be treated as merely the first step in a promising direction involving the "superprimary" prevention of aging and disease. As far as fundamental research in the biology of life span is concerned, it is obviously essential that a new line of investigation be developed with the aim of explaining the role of parental characteristics in the determination of life span.

## 5.4. Analysis of sex differences in lifetimes

One of the promising approaches to an understanding of the mechanisms which determine the life span of organisms is without doubt the study of sex differences in lifetimes. Early works refer to the longer life span of females. Darwin, for example, considered the shorter life span of males as "a natural and constitutional peculiarity due to sex alone" (Darwin, 1847). A.Comfort, whose analysis is based on modern data, also came to the conclusion that, for the majority of species which had been investigated, the female had the longer life span, although he noted that there were a number of exceptions (Comfort, 1964; 1979). Figures 27 and 28 contain some examples in support of the notion that females have a longer life span compared to males.

**Figure 27** (opposite page) Relationship between the mean life span of men and women.

Each point corresponds to data on a particular country at a particular time. The mean life span of males is plotted along the x-axis and the mean life span of females is plotted along the y-axis. Points which lie above the bisecting line at 45° correspond to the case in which the life span of females exceeds the life span of males. Points which lie below the line correspond to the opposite case. Constructed on the basis of data published by Bruk (1981).

abscissa - mean life span of males, years

ordinate - mean life span of females, years

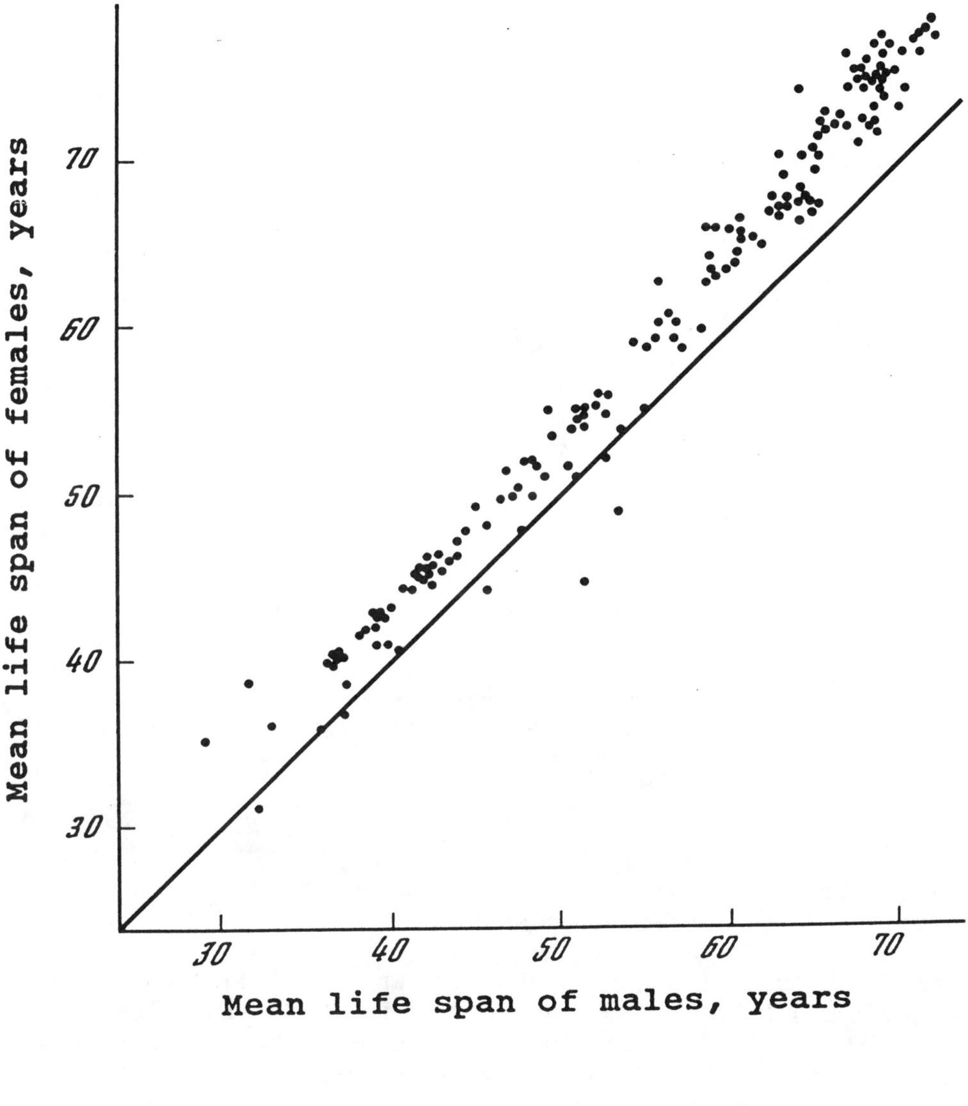
Mean life span of females, years
70
60
50
40
30
30
40
50
60
70
Mean life span of males, years

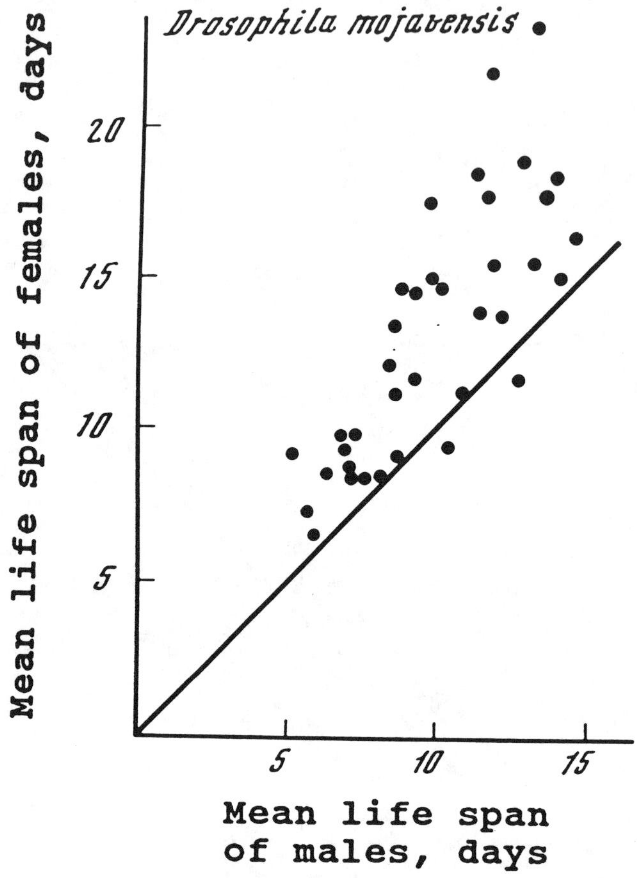


**Figure 28** The relationship between the mean life span of *Drosophila mojavensis* males and females.

The graph contains 39 points, of which 36 lie above the bisecting line and 3 lie below the bisecting line. The mean life span of females is significantly more frequently greater than the mean life span of males ($P > 0.99$). Source of the data: Starmer et al. (1977).

abscissa - mean life span of males, days

ordinate - mean life span of females, days

Thus, it seems as if the fact that females have a longer life span cannot be doubted. On the other hand, the genetic, biochemical and physiological characteristics of each sex are now quite well researched. It will therefore be possible, through a comparison of these characteristics with the observed differences in lifetimes, to formulate and test a variety of hypotheses concerning the mechanisms which determine life span.

The first thought which crosses the mind when making such a comparison is that women have two X-chromosomes, while men have only one. Genetic defects in the X-chromosome are therefore fully manifested in males, reducing their ability to survive, but are practically innocuous to the cells of the female organism. This explanation, which was proposed at the beginning of the century (see Comfort, 1964), turns out however to be oversimplified, in that it has become clear that only one X-chromosome is active in women (Schneider, 1978). Nevertheless, this hypothesis has been preserved in a somewhat modified and unusual form. D.Harman proposes that the greater life span of females is determined in the first 48 hours of the zygote's development (Harman, 1981). During this early period, both X-chromosomes are functional, and therefore the cells of the female organism show greater redundancy than those of the male organism. Harman draws attention to the fact that the X-chromosome contains the gene encoding glucose-6-phosphate dehydrogenase, a key enzyme in the formation of NADPH. NADPH, in its turn, is essential for the maintenance of glutathione in a reduced form. Glutathione is a powerful inhibitor of free-radical oxidation processes, as well as a substrate of glutathione peroxidase, which reduces hydrogen peroxide to water and organic hydroperoxides to the corresponding alcohols (Harman, 1981). The hypothesis proposed by Harman can also explain why women may be less sensitive to a deficiency of selenium, which is a component of glutathione peroxidase (this problem is discussed above, see section 3.4). According to Harman's hypothesis, it is to be expected that the protective effect of antioxidants in inhibiting free-radical processes will be more pronounced in the less protected males. Indeed, it has been demonstrated that the addition of antioxidants to the food of pregnant mice leads to an increase in the life span of the offspring, and that the life span of the male offspring is increased to a significantly greater degree than that of the female offspring (Harman and Eddy, 1979). At the same time, it has to be admitted that this interesting hypothesis requires further verification and more direct proof.

One important property of the inactivation of one of the

X-chromosomes during the early stages of the development of the female organism is the fact that this inactivation is a random event. Consequently, the female organism is an amazing mosaic of cells, half containing an active paternal X-chromosome and half containing an active maternal X-chromosome (Schneider, 1978). In the event of a defect in one of the X-chromosomes, a system of this kind ought in principle to be more stable because half the cells will be normal. Thus, the inactivation of one of the X-chromosomes leads to the appearance of redundancy not at the intracellular level, but at the level of the organism's population of cells. It must be noted that, in addition to its obvious advantages, this redundancy may also have certain disadvantages. For example, it turns out that the antigen diversity of the cells of the female organism is significantly greater than in males. Because of this, women might be expected to suffer more frequently from autoimmune disorders, and this is indeed observed (Hazzard, 1986).

The explanation of sex differences in lifetimes which has been outlined above suffers from one significant deficiency. It is well-known that in a whole range of organisms the females are heterogametic, and not the males (for example, butterflies and chickens). Nevertheless, in these species the males do not live longer than the females, in clear contradiction with the hypothesis (see Comfort, 1964). Therefore, this hypothesis cannot serve as a general explanation for the greater life span of females.

In the Soviet Union, the problem of sex differences in general, and lifetime differences in particular, has received a large amount of attention in the works of V.A.Geodakian (1982; 1983). V.A.Geodakian has developed an original concept which appears to explain a series of differences between males and females. He proposes that the greater life span of females is a general biological phenomenon, and therefore any explanation must also be a general biological one, based on evolutionary principles. According to this hypothesis, it is the male sex whose characteristics show the greater variability, and therefore the main burden of natural selection always falls on the male portion of the population, leading to its increased mortality. Thus, it emerges that the shorter life span of males indeed has a general biological character, connected with the fact that males are the evolutionary avant-garde of the species on which nature performs its experiments.

V.A.Geodakian's hypothesis seems at first sight very attractive. When it is examined more carefully, two drawbacks can be observed. The first is the fact that the hypothesis does not reveal a concrete molecular-genetic mechanism resulting in a greater female life span. This drawback should not however be considered significant, since it can in principle be eliminated as the hypothesis is developed and made more concrete.

A much more significant drawback in the hypothesis is the problem that it conflicts with a number of facts. Firstly, Geodakian's hypothesis predicts that amongst long-lived people there should be more males (Geodakian, 1982; 1983). In fact, the direct opposite is observed (see Gavrilov, 1985). Secondly, according to the hypothesis, as living conditions are improved and mortality is reduced, the differences in male and female life spans ought to disappear, and under especially favourable external circumstances the mean life span of males ought to be even higher than that of females (Geodakian, 1982; 1983). However, at the present time the directly opposite process can be observed. As life expectancy increases, there is also an increase in the differences between male and female life spans (Bednyi, 1979; Lopez, 1983). A special study of this process has shown that in recent years in the developed countries what is observed is an accelerated reduction in the mortality of old-aged women compared to men (Gavrilov and Nosov, 1985). Thirdly and finally, it must be noted that the greater life span of females is not at all a general biological law, despite the wide-spread opinion to the contrary. In view of the importance of this point, we shall consider it in more detail.

In fact, when we turn to the works dealing specifically with sex differences in lifetimes, we note that the conclusion that females have a longer life span is illustrated by only a few examples (Comfort, 1964; Geodakian, 1983). Since there is a mass of contradictory data in the scientific literature, there is a danger of arriving at one or other conclusion through the intentional or unintentional choice of examples. An objective resolution of the issue requires an analysis not only of the works which deal specifically with the problem of the greater life span of females, but also of all the remaining works where this is not the aim. When a scrupulous investigation of this kind is carried out, it turns out that there is no significant tendency for females to have a greater life span in the majority of species.

In this regard, an especially revealing example is the analysis of the sex differences in the life spans of laboratory drosophila. It is believed to be firmly established that the females of this species live longer than the males (Comfort, 1964). Indeed, there are many case studies which justify this assertion. However, if the literature dealing with the drosophila life span is examined in toto, it turns out that there are just as many examples in which the direct opposite is observed (see Fig.29). Out of 279 cases, the mean life span of females was greater than mean life span of males in 141 cases, while the opposite was observed in 135 cases. A statistical analysis of these results shows that there is no tendency for either sex to have a greater life span. This result confirms the conclusion of Lints et al. (1983) that there is no tendency for female drosophila to have a greater life span compared to males. The same conclusion has been reached using other, more refined methods of analysing life span data. We processed the data of Gonzales (1923) on the life span of the Old Falmouth strain of drosophila, as well as of vestigial, arc, speck, black and purple mutants, including mutants bearing different combinations of these mutations (Gavrilov, 1980). The processing of the data involved calculating a multiple linear regression equation for the life span of drosophila depending on the presence of one or other mutation. When a particular mutation was present, the argument was assigned the value +1. When the mutation was absent, the argument was assigned the value -1. Since only five mutations and their various combinations were studied, the number of elementary variables was five. As the sixth independent variable, the sex characteristic was used, assigned the value -1 for males and +1 for females. The result of analysing the data in this way was the following linear regression equation (only the most significant arguments are included):

The mean life span of drosophila (days) =

$$25.3 \text{ days} - 6.0x_1 + 3.2x_2x_3 - 2.3x_2 + 2.2x_1x_4 + \ldots *$$

where $x_1$ is the variable reflecting the presence or absence of the vestigial mutation, $x_2$ the purple mutation, $x_3$ the arc mutation, $x_4$ the variable representing male ($x_4 = -1$) or female ($x_4 = +1$) sex.

When analysing these data, it must be borne in mind that the combinations of variables are placed in order of decreasing effect. Thus, the vestigial mutation ($x_1$) has the most important effect on the drosophila life span: the mean life span is sharply decreased, as reflected in the negative multiplying factor. As far as sex ($x_4$) is concerned, its effect is so weak that of the four above-mentioned variables it occupies the last position. However, what is especially

---

* The end of this equation may be found in Gavrilov (1980).

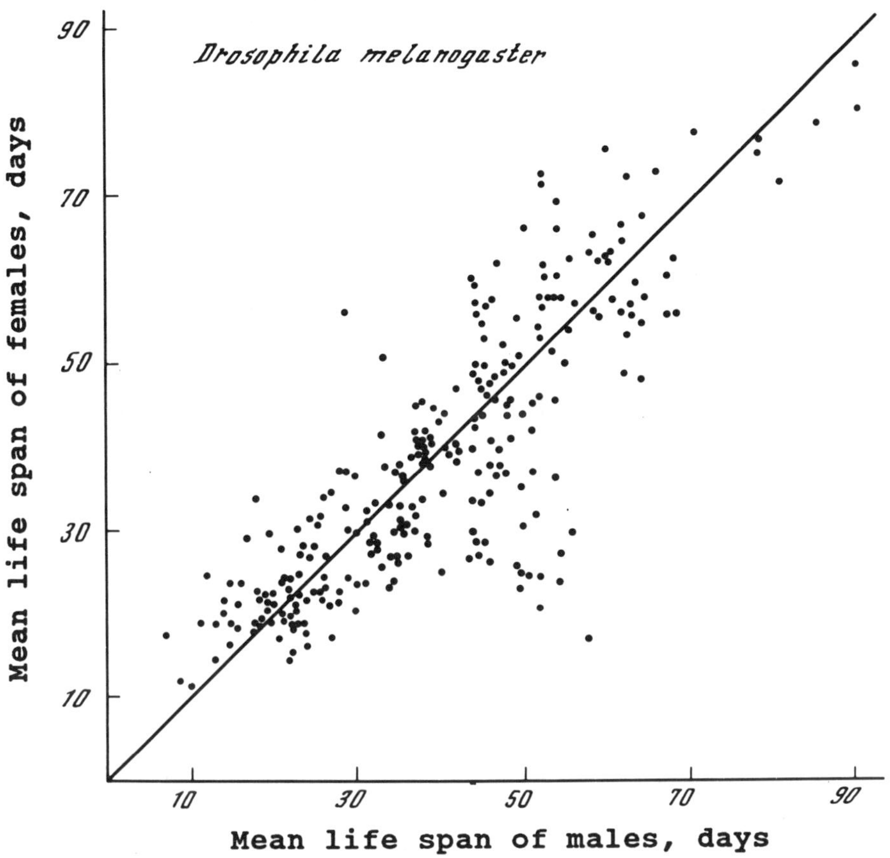


**Figure 29** The relationship between the mean life span of *Drosophila melanogaster* males and females.

The graph contains 279 points, of which 141 lie above the bisecting line and 135 lie below the bisecting line. Differences between the mean life span of males and females are statistically insignificant. Sources of the data: Gonzales (1923); Pearl and Parker (1924); Steinfeld (1929); Lüers (1935); Hall (1969); Van Herrewege and David (1970); Allemand et al. (1973); Gould and Clark (1977); Bileva et al. (1978, 1981); Bozcuk (1978); Lints et al. (1979); Unlü and Bozcuk (1979); Giess (1980); Shakhbazov et al. (1980); Nekrasova and Shakhbazov (1981).

abscissa - mean life span of males, days

ordinate - mean life span of females, days

important is that sex does not affect life span independently of the mutations: there is a strong interaction with the vestigial mutation. When this mutation is present, females live longer than males, and when it is absent, the opposite is observed. Thus, depending on the genetic background, the life span of females can be either greater or smaller than that of males. This example shows how dangerous it is to make wide generalisations on the basis of isolated, even though statistically reliable, cases of greater female life spans.

Figures 30-35 contain some more examples demonstrating either the absence of any tendency for females to have a greater life span, or even the presence of the opposite tendency. These examples show that the conclusion that the greater life span of females is a general biological law has insufficient foundation.

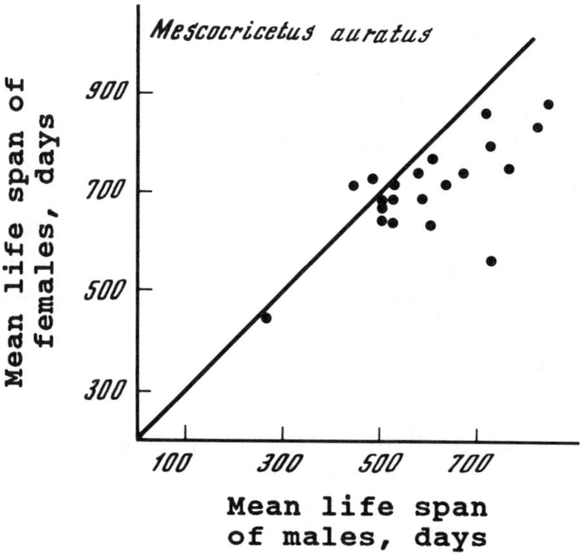


**Figure 30** The relationship between the mean life span of male and female Syrian hamsters.

The graph consists of 25 points, of which 2 lie above the bisecting line and 23 lie below the bisecting line. The mean life span of males is significantly more frequently greater than the mean life span of females ($P > 0.99$). Source of the data: Kirkman and Yau (1972).

abscissa - mean life span of males, days

ordinate - mean life span of females, days

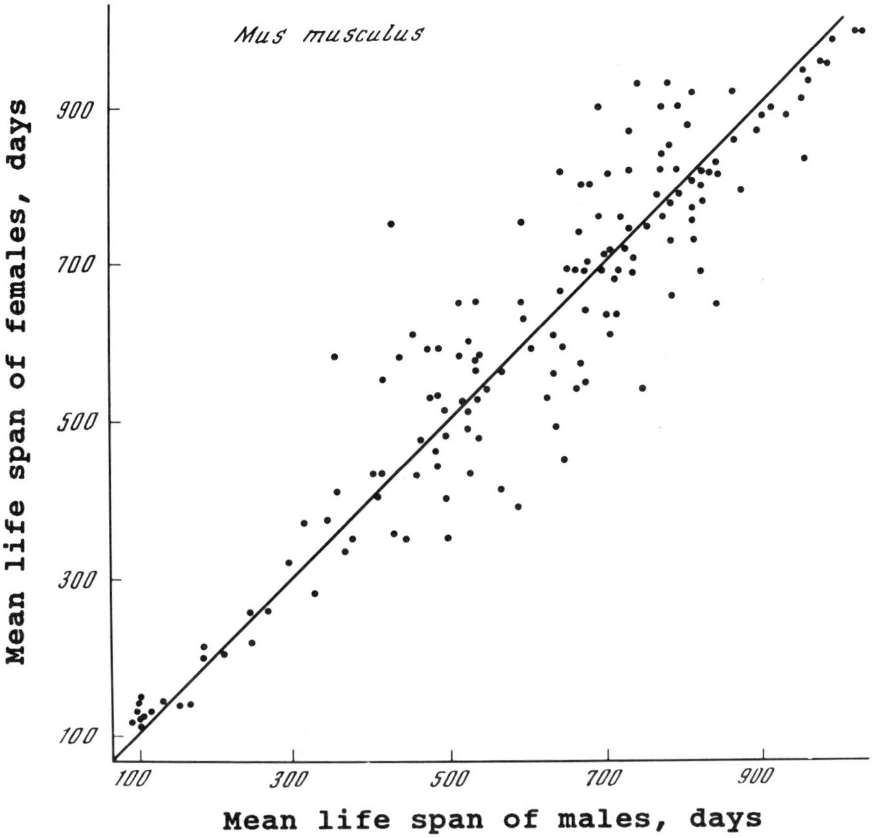


**Figure 31** The relationship between the mean life span of male and female laboratory mice.

The graph consists of 153 points, of which 69 lie above the bisecting line and 83 lie below the bisecting line. Differences between the mean life span of males and females are statistically insignificant. Sources of the data: Lorenz et al. (1955); Neary et al. (1957); Nowell and Cole (1959); Kohn and Guttman (1963); Gordon et al. (1966); Sacher (1966); Hollicroft et al. (1967); Goodrick (1974); Fernandes et al. (1976); Eklund and Bradford (1977); Holliday and Stevens (1978); Clapp et al. (1979); Storer (1979); Cheney et al. (1980); Thomson et al. (1981).

abscissa - mean life span of males, days

ordinate - mean life span of females, days

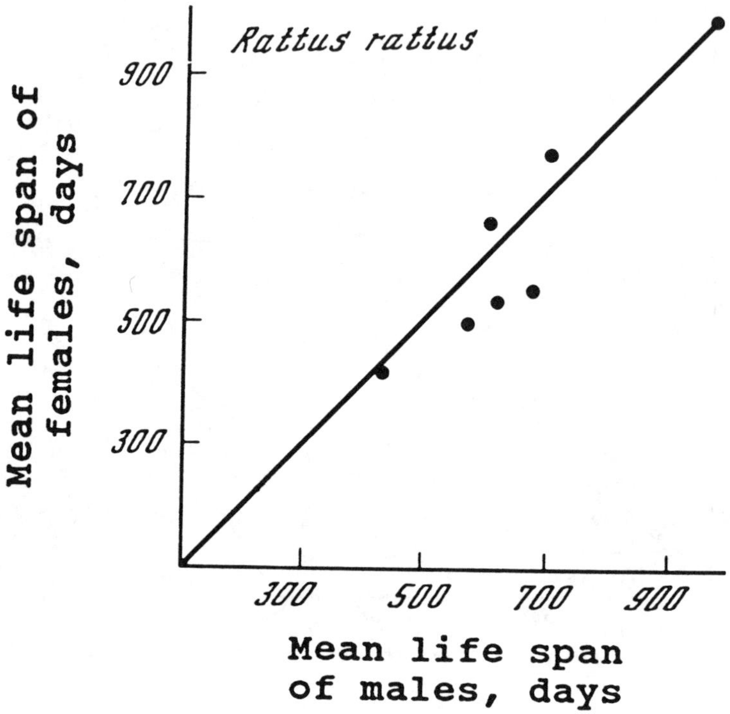


**Figure 32** The relationship between the mean life span of male and female laboratory rats.

The graph consists of 7 points, of which 2 lie above the bisecting line and 4 lie below the bisecting line. Differences between the mean life span of males and females are statistically insignificant. Source of the data: Hursch et al., 1955.

abscissa - mean life span of males, days

ordinate - mean life span of females, days

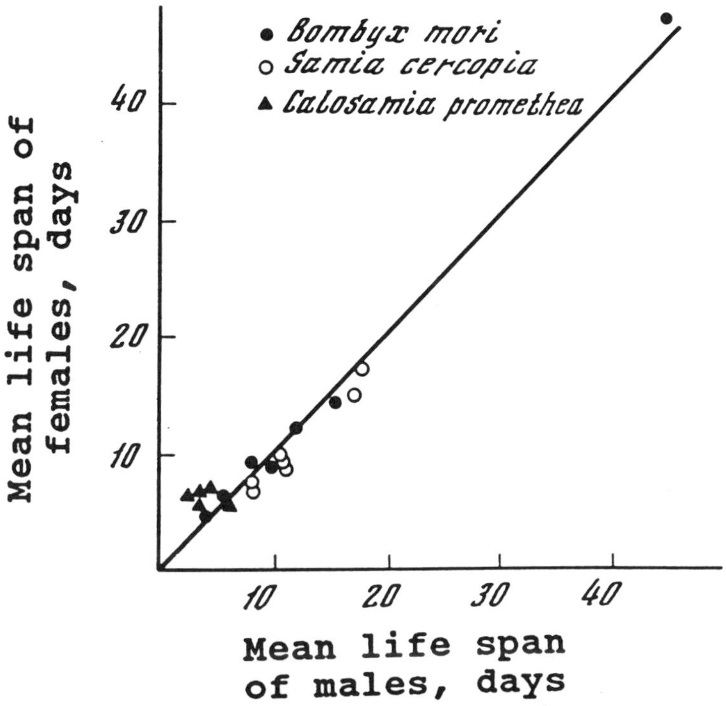


**Figure 33** The relationship between the mean life span of male and female silkworm moths.

The graph consists of 19 points, of which 8 lie above the bisecting line and 11 lie below the bisecting line. Differences between the mean life span of males and females are statistically insignificant both for all silkworm moths, and for each species taken individually. Sources of the data: Alpatov and Gordeenko (1932); Osanai (1978).

abscissa - mean life span of males, days

ordinate - mean life span of females, days

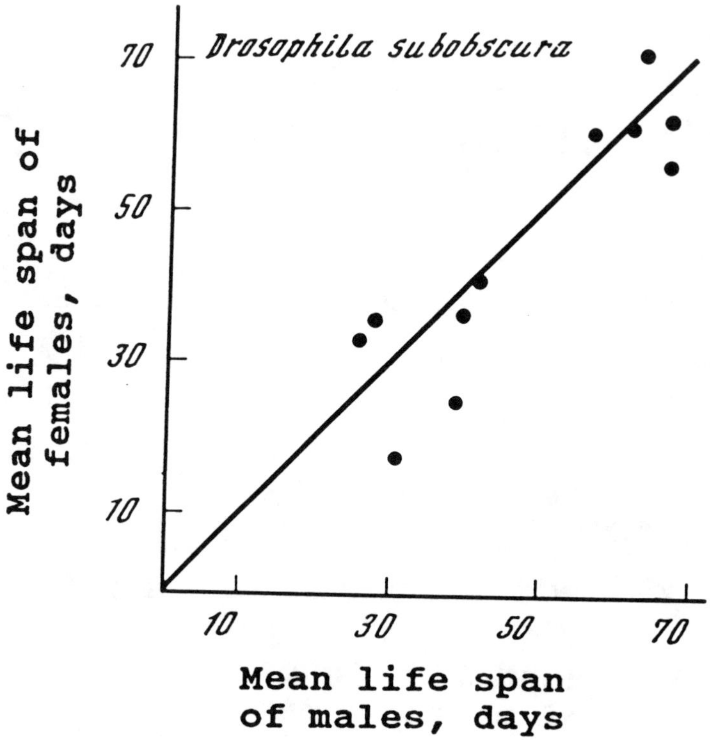


**Figure 34** The relationship between the mean life span of *Drosophila subobscura* males and females.

The graph consists of 11 points, of which 4 lie above the bisecting line and 7 lie below the bisecting line. Differences between the mean life span of males and females are statistically insignificant. Source of the data: Clark and Rockstein (1964).

abscissa - mean life span of males, days

ordinate - mean life span of females, days

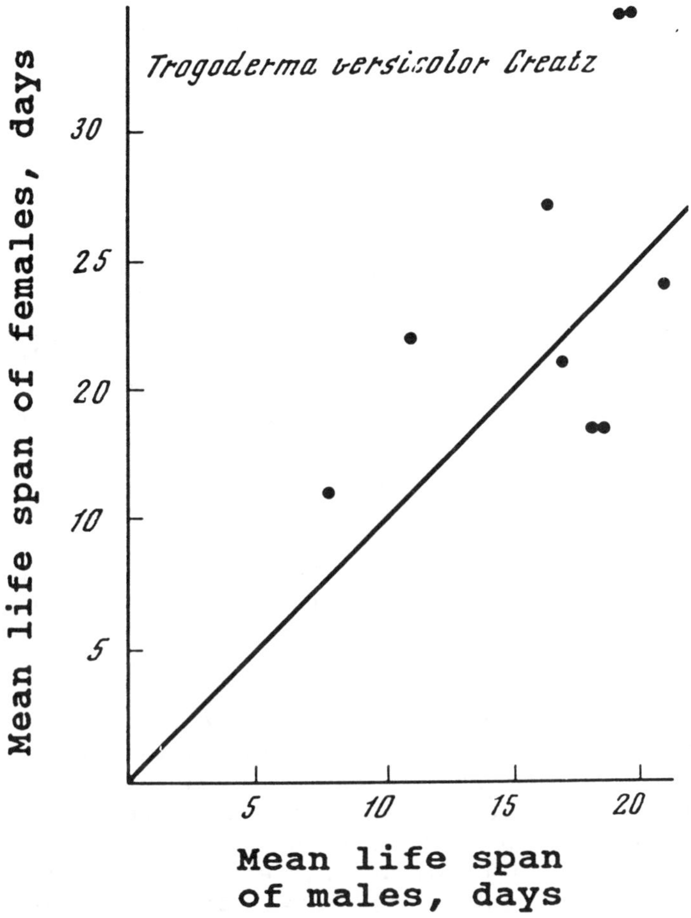


**Figure 35** The relationship between the mean life span of male and female hide beetles *Trogoderma versicolor C.*

The graph consists of 9 points, of which 5 lie above the bisecting line and 4 lie below the bisecting line. Differences between the mean life span of males and females are statistically insignificant. Source of the data: Davis (1945).

abscissa - mean life span of males, days

ordinate - mean life span of females, days

Therefore V.A.Geodakian's hypothesis, which starts from general biological notions of the greater life span of females, turns out to be inapplicable in this case. The examples cited prove that it is necessary to seek not universal, but specific explanations of sex differences in lifetimes for each individual species.

In conformity with this conclusion, it makes sense to return to a discussion of the sex differences in the human life span. In this particular case, there does indeed exist a tendency for women to have a greater life span (see Fig.27).

Of course, sex differences in human mortality may be partially caused by social factors. However, an explanation of this kind cannot be exhausting. For example, the increased mortality of representatives of the male sex is observed in all age groups and even before birth, when there can be no talk of social differences between the sexes. According to the data of a number of investigators, there are significantly more male than female fetuses amongst those which are spontaneously aborted (Byrne and Warburton, 1987). Honoré, 1988), and during the early stages of pregnancy (3 months), the male/female ratio reaches 3.8 (McMillen, 1979). It is clear that in order to explain these sex differences it is necessary to turn to medico-biological, rather than socio-cultural hypotheses.

It was noted above that sex differences in human lifetimes are connected with the biological component of mortality and are differently expressed in different human populations (see section 3.4). However, even within this single species, the mechanism of sex differences in mortality can be different at different ages. While for adults a change in living conditions leads to approximately equal changes in male and female mortality (see Fig.36), in infancy (0-1 year) boys are significantly more sensitive to a change in external conditions (see Fig.37). Therefore, the differences in the mortality of the two sexes may even on purely phenomenological grounds be connected with two different factors: a mortality factor which is independent of external conditions, but dependent on age (corresponding to the difference between the age-dependent components of male and female mortality), and a factor related to greater sensitivity to the damaging effects of the environment. It can be seen from the figures that the first of these sex difference mechanisms is characteristic of adults and the second of infants. Thus, we arrive at the necessity of further subdividing the problem of sex differences, even within a single species, by taking into account different age groups.

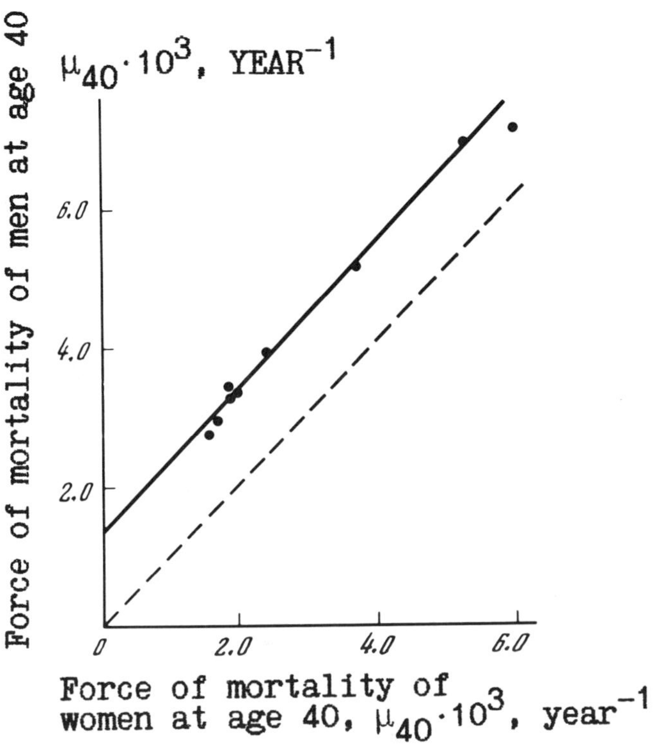


**Figure 36** The linear relationship between the force of mortality of men and of women.

The x-coordinate and y-coordinate of each point correspond respectively to the force of mortality of women and the force of mortality of men at age 40. The data are taken from life tables for the white population of the USA in 1920-1929, 1929-1931, 1939-1941, 1949-1951, 1959-1961, 1964, 1970, 1975, and 1977. Source of the data: see Gavrilova et al. (1983)

abscissa - force of mortality of 40-year-old women; $\mu_{40} \cdot 10^3$, year$^{-1}$

ordinate - force of mortality of 40-year-old men; $\mu_{40} \cdot 10^3$, year$^{-1}$

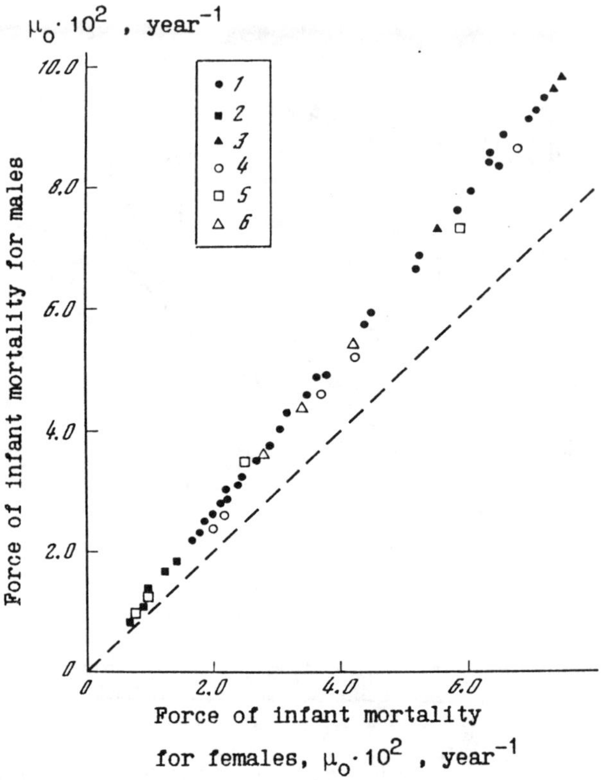


**Figure 37** The linear relationship between the force of infant mortality for boys and girls.

The x-coordinate and y-coordinate of each point correspond respectively to the force of mortality of girls and the force of mortality of boys at age 0-1. The data are taken from life tables for the following countries: 1 - France, data for 1931-1937, 1947-1965, 1967, 1969; 2 - Sweden, data for 1956-1960, 1961-1965, 1966-1970, 1971-1975, 1975-1979; 3 - Denmark, data for 1921-1925, 1926-1930, 1936-1940; 4 - USA (coloured population), data for 1939-1941, 1949-1951, 1964, 1975, 1977; 5 - Norway, data for 1911-1920, 1946-1950, 1971-1975, 1976-1980; 6 - Switzerland, data for 1933-1937, 1941-1950, 1948-1953. Source of the data: see Gavrilova et al. (1983).

abscissa - force of infant mortality for girls; $\mu_0 \cdot 10^2$, year$^{-1}$

ordinate - force of infant mortality for boys; $\mu_0 \cdot 10^2$, year$^{-1}$

If we carry out a more detailed analysis of life span differences between the two sexes in human beings, and investigate the different causes of mortality, some new and unexpected regularities are revealed. As a result of the research we have carried out, it has been discovered that causes of death differ strongly in the extent to which there is an interdependence between the male and female mortality rates. To exemplify this, table 13 contains values for the sexual concordance coefficient (the correlation coefficient between the rates of male and female mortality) for some of the 84 studied causes of death.

**Table 13** Values for sexually concordant causes of death*

| Cause of death | Sexual concordance coefficient $r$ | Slope coefficient of regression $\alpha$ | Free term of regression $b \cdot 10^6$, $year^{-1}$ | Relative excess of male mortality (REMM) |
|---|---|---|---|---|
| Malignant neoplasm of stomach | 0.963 | 2.06±0.04 | 60 ± 14 | 2.27 |
| Influenza | 0.950 | 1.50±0.04 | 4 ± 2 | 1.63 |
| Cirrhosis of liver | 0.947 | 3.18±0.08 | -44 ± 19 | 2.96 |
| Diseases of the digestive system | 0.926 | 2.75±0.08 | -183 ± 40 | 2.33 |
| Cerebrovascular disease | 0.919 | 1.32±0.04 | 148 ± 51 | 1.46 |

* The calculations are based on the statistical data of the WHO for 26 developed countries over the period 1965-1987 in the age group 55-64 years (179 cases in total).

Table 13 (continued)

| Cause of death | Sexual concordance coefficient $r$ | Slope coefficient of regression $\alpha$ | Free term of regression $b \cdot 10^6$, $year^{-1}$ | Relative excess of male mortality (REMM) |
|---|---|---|---|---|
| Pneumonia | 0.902 | 1,46±0.05 | 73 ± 10 | 1.94 |
| Hypertensive disease | 0.886 | 1.10±0.04 | 47 ± 8 | 1.40 |
| Venous thrombosis and embolism | 0.891 | 1.10±0.04 | 19 ± 5 | 1.31 |
| Ischaemic heart disease | 0.885 | 2.66±0.11 | 1209 ± 183 | 3.45 |
| Tuberculosis of respiratory system | 0.874 | 3.87±0.16 | 21 ± 8 | 4.45 |
| Infections of kidney | 0.871 | 0.74±0.03 | 0 ± 3 | 0.74 |
| Acute respiratory infections | 0.866 | 1.64±0.07 | 2 ± 1 | 1.83 |

It can be noted that for a number of diseases (cancer of the stomach, cirrhosis of the liver, strokes), the values of the male and female mortality rates are in such close agreement that the mortality rate of one sex can be determined from the mortality rate of the opposite sex. However, for some other diseases (cancer of the lung, oesophagus, buccal cavity, larynx), the sexual concordance coefficient turns out to be surprisingly small (less than 0.5) (see Figs.38 and 39).

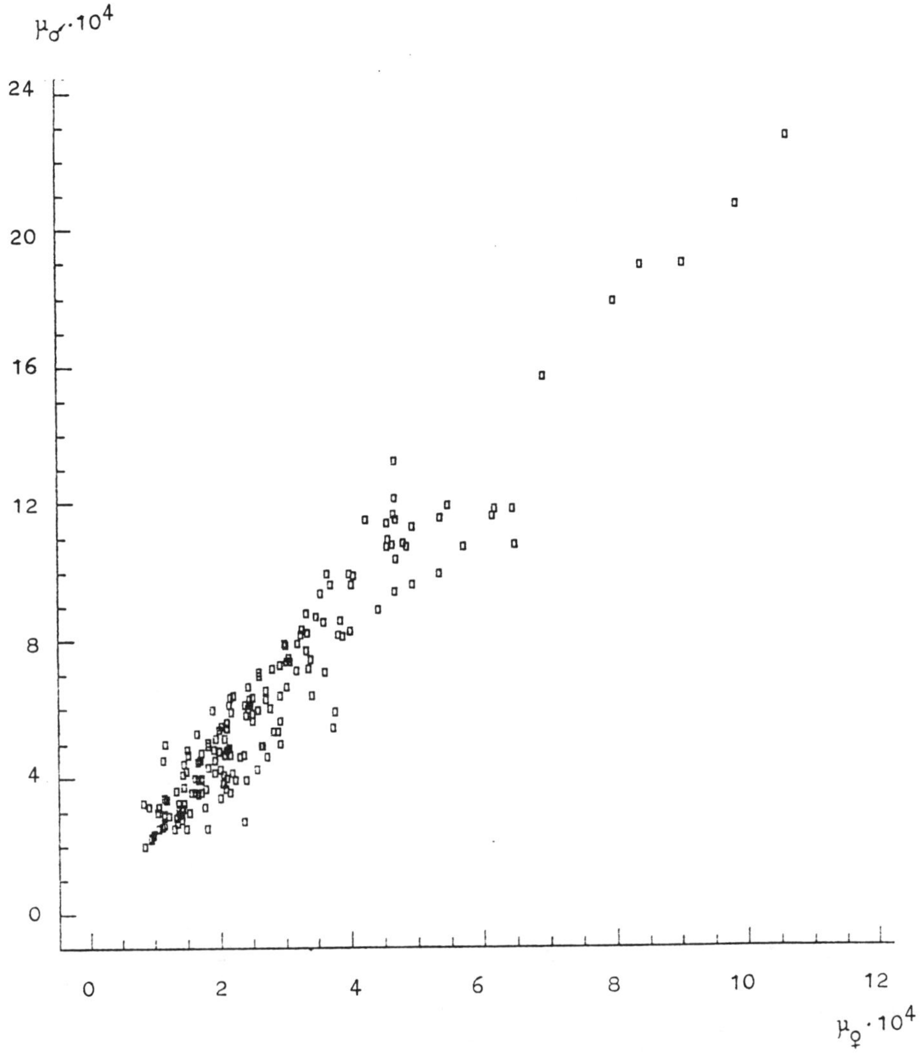


**Figure 38** Example of a sexually concordant cause of death. The correlation between male and female mortality at age 55-64 from malignant neoplasms of the stomach. Correlation coefficient $r = 0.963$.

Constructed on the basis of data for 26 countries (including 179 cases).

abscissa - female mortality from stomach cancer; $\mu \cdot 10^4$

ordinate - male mortality from stomach cancer; $\mu \cdot 10^4$

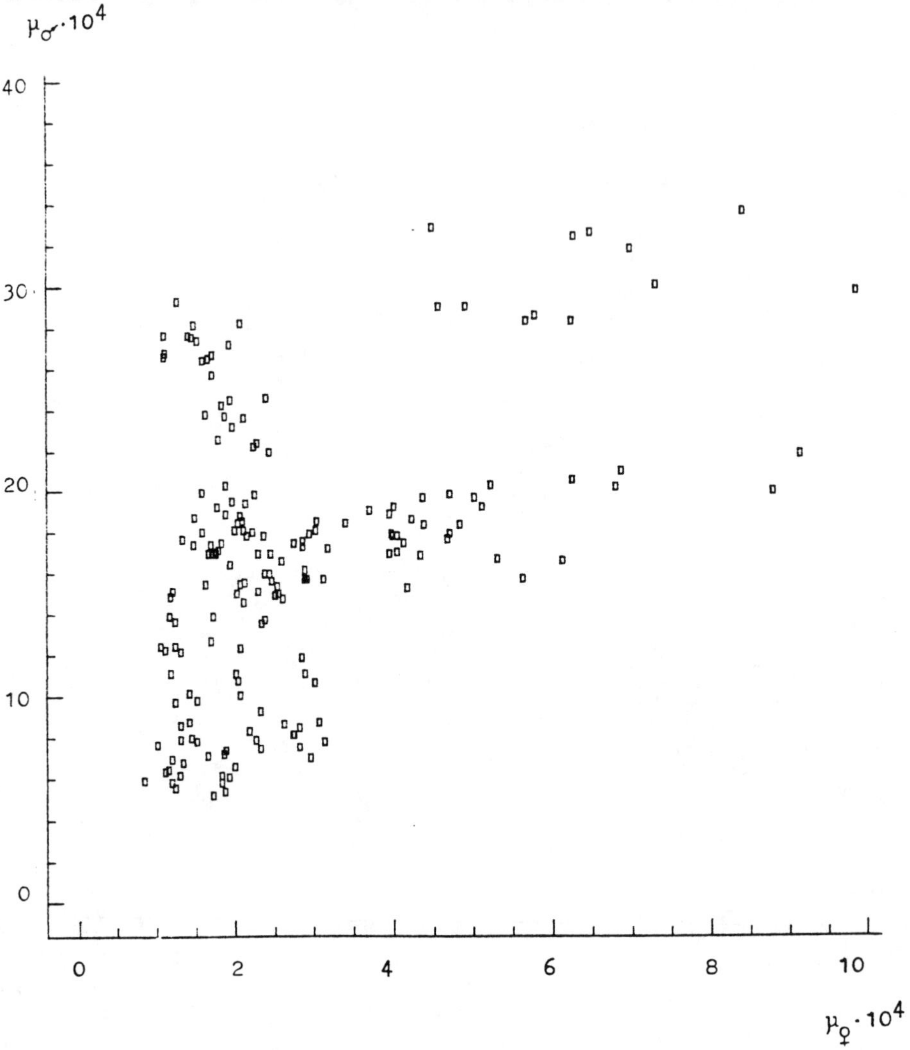


**Figure 39** Example of a sexually inconcordant cause of death. The correlation between male and female mortality at age 55-64 from malignant neoplasms of the lung. Correlation coefficient $r = 0.432$.

Constructed on the basis of data for 26 countries (including 179 cases).

abscissa - female mortality from lung cancer; $\mu \cdot 10^4$

ordinate - male mortality from lung cancer; $\mu \cdot 10^4$

In order to clarify the mechanism which lies behind the observed differences in the sexual concordance coefficient, we carried out a factor analysis of mortality from a sexually concordant cause (stomach cancer), and a sexually inconcordant cause (lung cancer). It became clear that the geographical and temporal variation in mortality from stomach cancer is determined by a single factor common to both men and women. At the same time, mortality from lung cancer is determined by two factors, one of which determines male mortality and the other female mortality (see table 14). Thus, it has been established that sexual concordance of mortality is conditioned by general (non-sex-specific) factors, and sexual inconcordance of mortality is conditioned by sex-specific factors.

If the increased mortality of men is caused by additional risk factors specific to the male sex, the greatest sexual dimorphism in mortality ought to be observed for sexually incongruent causes. Indeed, it turns out that the greatest relative excess of male mortality (REMM) is observed in precisely those causes which have a low value for the sexual concordance coefficient. Thus, for the age group 55-64 years, the greatest REMM was observed in the case of mortality from cancer of the larynx (REMM = 11.8; $r$ = 0.128); accidents caused by firearm missiles (REMM = 11.4; $r$ = 0.395); accidents mainly of industrial type (REMM = 9.5; $r$ = 0.463), accidents caused by means other than motor transport ("other transport accident") (REMM = 6.2, $r$ = 0.593); lung cancer (REMM = 6.1; $r$ = 0.432).

At the same time, it turns out that the contribution of sexually inconcordant causes of death to the observed sexual dimorphism of total mortality is paradoxically small, and does not exceed the contribution of sexually concordant causes of death. One possible explanation for this paradox is that sex differences may be caused by sex-specific factors which, varying little in space and time, do not disturb the sexual concordance. In this case, if we were to calculate the coefficients of the linear regression equation relating male and female mortality, we would expect that the sex differences for sexually concordant causes of death ought to be determined by the magnitude of the free term (intercept) in the regression reflecting the shift in the values of mortality caused by the action of invariant, sex-specific factors. In fact, it turns out that the free term (intercept) in the regression makes a significant contribution to the magnitude of the sexual dimorphism in mortality from sexually concordant causes. For example, the contribution for ischaemic heart disease reaches 33.7% of the observed sex differences. At the same time, it is

**Table 14** The factorial structure of mortality from malignant neoplasms of the stomach, as well as the trachea, bronchi and lung*

| Sex | Disease | Age | Factor I | Factor II | Factor III |
|---|---|---|---|---|---|
| Male | Stomach cancer | 55-64 | 0.994 | | |
| Male | " | 45-54 | 0.991 | | |
| Male | " | 65-74 | 0.968 | | |
| Female | " | 55-64 | 0.967 | | |
| Male | " | 35-44 | 0.965 | | |
| Female | " | 45-54 | 0.958 | | |
| Female | " | 65-74 | 0.943 | | |
| Female | " | 35-44 | 0.916 | | |
| Female | Lung cancer | 55-64 | | 0.972 | |
| Female | " | 65-74 | | 0.964 | |
| Female | " | 45-54 | | 0.912 | |
| Female | " | 35-44 | | 0.736 | |
| Male | " | 45-54 | | | 0.940 |
| Male | " | 35-44 | | | 0.911 |
| Male | " | 55-64 | | | 0.901 |
| Male | " | 65-74 | | | 0.770 |

*Factorial loads of less than |0.25| are omitted from the table

obvious that the free term in the regression is not the main source of sex differences in mortality from sexually concordant causes.

It can be seen from table 13 that the slope coefficient of the linear regression for sexually concordant causes of death is significantly greater than unity, i.e. the male subpopulation turns out to be more sensitive to the action of the very same factors which affect the female subpopulation. As our investigations have shown, this regularity makes the basic contribution to the sexual dimorphism in mortality from sexually concordant causes, and a significant contribution to the sexual dimorphism of total mortality. This regularity can be used in subsequent research in human biology, and can also be used to predict sex differences in human mortality.

Summarising these attempts to clarify the mechanisms determining life span through an investigation of sex differences in lifetimes, we can state the following. The hopes connected with the search for general biological mechanisms underlying these differences seem to be in vain, since, despite the wide-spread opinion to the contrary, the greater life span of females is not in itself a general biological regularity. Even within a single biological species (man), there is justification for the separate analysis of the problem in different age groups. Naturally, these limitations significantly reduce our optimism in investigating the issue. However, they lead to a better understanding of the real processes underlying sex differences in lifetimes.

## 5.5. Experiments in life extension

Another approach to understanding the mechanisms which determine life span is based on attempts to change life span experimentally. From both a practical and a purely scientific, cognitive point of view, experiments in life extension are significantly more interesting than experiments in life curtailment. Indeed, practically any action can lead to life curtailment if its intensity is high enough, and the mechanism destroying the organism may be completely different than under normal circumstances. On the other hand, life extension cannot be achieved without interfering with the processes of life span determination which interest us.

At the present time, experiments in life extension have

become a whole new field of scientific research using a very wide spectrum of agents, ranging from the most diverse pharmacological drugs (Heicklen and Brown, 1987; Richie et al., 1987; Sawada and Carlson, 1987; Anisimov et al., 1988) to electrical fields (Kellog and Yost, 1986). The huge diversity of attempted methods of life extension is quite fully reflected in the extensive review chapters (Cutler, 1981; Nikitin, 1984; Obukhova and Emanuel, 1984) and monographs which deal with the problem of life extension (Kurtzman and Gordon, 1976; Walford, 1983; 1986; Gavrilov, 1985; Frolkis and Muradian, 1986; Gavrilov, 1990).

It is surprising that, out of all the diverse methods of life extension which have been discovered and tried, the most effective methods ironically turn out to be the most simple ones: lowering the body temperature (hypothermia) of cold-blooded (poikilothermal) organisms and reducing the calory intake of hot-blooded (homeothermal) animals. These simple methods can lead to a twofold or threefold increase in life span, and their effectiveness has been demonstrated many times by a wide variety of investigators using a wide variety of biological species (see Cutler, 1981; Frolkis and Muradian, 1988). No other methods of life extension (if we ignore the effect of heterosis, which does not belong to the category of external interventions) can be compared wth these two methods either in the strength of their effect, or the wideness of their applicability. Therefore, these two methods (hypothermia and restricted calory intake) are of special interest to investigators attempting to understand the mechanisms determining life span. Let us consider them in greater detail.

*Hypothermia.* It is widely held that the prolonged life of organisms which results from a lower temperature can be simply attributed to a reduction in the rate of all the processes taking place in the organism. Even if this were really the case, such an approach would be discredited solely as a means of prolonging the active life of people, and not as a means of testing different hypotheses concerning the determination of life span. For example, investigators who believe that there is a special clock in the organism which determines the duration of life are forced to admit that this clock is very poor, since a reduction in body temperature of only 5-10°C slows the hypothetical clock by a factor of 2! Such a twofold increase in the duration of life is observed in a wide variety of organisms when the body temperature is changed by 5-10°C (see Frolkis and Muradian, 1988, pp. 199-200). In this case, however, we have an alternative: instead of considering the clock to be poor, we can consider the hypothesis that it exists to be poor.

For the supporters of the wear-and-tear hypothesis, the increased duration of life which accompanies hypothermia means that the limiting stages which determine life span are not connected with external and internal background radiation, nor with the chain continuation reactions of free-radical destruction, since all these processes are virtually independent of temperature.

Of especial interest are data showing that the effect of hypothermia on life span is not encompassed by the primitive hypothesis that organisms are preserved in the cold like food products in a domestic refrigerator.

Thus, it has become clear that certain organisms (fish, for example) not only live longer when their temperature is reduced, but also grow significantly more quickly, reaching larger sizes (Walford, 1983). Equally, it turns out that an increased life span can be achieved not only when organisms are kept at a low temperature all the time, but also when their temperature is lowered only in the second half of life (see Cutler, 1981). Moreover, in certain cases experiments of this kind even lead to greater life extension than when the organism is kept at a constantly low temperature (see Cutler, 1981). Therefore, the increase in the life span of organisms which occurs when their body temperature is lowered cannot be attributed solely to a non-specific retardation of all life processes, and it is necessary to seek alternative, deeper reasons for the phenomenon.

One possible explanation for life extension by hypothermia is that a reduction in temperature leads to a sharp reduction in the rate of "harmful" non-enzymatic side reactions, but has little effect on the rate of "useful" enzymatic reactions (Gavrilov and Gavrilova, 1986; Frolkis and Muradian, 1988). Indeed, it is well-known that the activation energy of enzymatic reactions is significantly lower than the activation energy of most non-enzymatic chemical reactions (this is after all the basic raison d'être of enzymes, which are catalysts of biochemical reactions). Therefore a reduction in body temperature leads to an increase in the "signal to noise" ratio, i.e. there is an increase in the contribution of controlled processes taking place in the organism. It is not difficult to see that this explanation of the effect of reduced temperatures is in good agreement with the idea of wear-and-tear (accumulation of defects) as the basic reason for the limited life spans of organisms.

*Restricting calory intake.* The second effective method

of prolonging the life of experimental animals is to limit their calory intake while at the same time maintaining a proper diet of vitamins, trace elements and other indispensable substances (Cutler, 1981; Walford, 1983; Nikitin, 1984; Frolkis and Muradian, 1988). In recent years, this line of research has been developed especially intensively, and as a result, a radical revision has taken place in traditional ideas concerning the mechanism of life extension through a restricted calory intake (McCarter et al., 1985; Holehan and Merry, 1986; Masoro, 1988).

For a long time it was considered self-evident that the slow-down in development which occurs with a restricted calory intake is the basic cause of the increase in life span. This kind of explanation was especially attractive to the supporters of the programmed life span hypothesis. From the point of view of the supporters of this hypothesis, restricting the calory intake simply leads to a delay in sexual maturation and in the switching-on of the organism's program of self-destruction. However, it subsequently became clear that a significant increase in life span can also be achieved when the calory intake is restricted after sexual maturation (see Holehan and Merry, 1986; Masoro, 1988).

Another traditional explanation is that restricting the calory intake leads to an increase in life span through a reduction in the metabolic activity, i.e. through a general reduction in the rate of metabolic processes (similar to the mechanism of life extension proposed in hypothermia). Indeed, when the calory intake is restricted, metabolic activity may initially be somewhat reduced. However, further more careful and extensive investigation has revealed that this reduction in metabolic activity is observed only in the intial stage as the organism adapts to the new low-calory diet. When the organism's total body weight and cell mass are reduced to a new stable level, the rate of oxygen consumption per unit of cell mass turns out to be the same as for the control group. If metabolic activity is calculated not for each unit of respiring cell mass, but for each unit of body weight, the metabolic activity of restricted calory intake animals even turns out to be significantly higher (McCarter et al., 1985)!

Another wide-spread point of view is that the extension of life which occurs with a restricted calory intake is brought about by the animal's reversion to a natural feeding regime in which it rarely eats ad libitum, and does not therefore suffer from the overeating now characteristic of laboratory animals and mankind. In other words, it is more correct to speak not of the extension of life through a restricted calory intake, but of the shortening of the lives of the animals in the "control" group through overeating (Cutler, 1981). This original point of view is certainly worthy of attention, but it does not solve the problem of the mechanisms underlying the strong inverse relationship between life span and the calory content of the diet. Attempts have been made to refute this point of view on the grounds that the level of obesity in the "control" laboratory animals does not show an inverse relationship with their life spans, while in the "experimental" animals with a low-calory diet the relationship between obesity and life span can even be positive (Masoro, 1988). It must also be noted that the so-called natural feeding regime is by no means ideal in its balance of vitamins, trace elements and other irreplaceable dietary components, which are strictly controlled when the calory intake is artificially restricted. On the other hand, the harmfulness of overeating is confirmed by a multitude of epidemiological investigations which have revealed a positive correlation between the calory intake of human beings and mortality from a range of diseases including breast cancer (Knox, 1977; Doll and Peto, 1981).

It seems to us that restricting the calory intake in principle leads to an increased life span through the same mechanism as hypothermia: in both cases there is an increase in the "signal-to-noise" ratio, i.e. the ratio between the rates of controlled enzymatic reactions and the rates of harmful non-enzymatic side-processes. The difference is simply that in hypothermia this ratio increases because of differences in the activation energies of enzymatic and non-enzymatic processes (see above), while in the case of a restricted calory diet the same result is achieved through a reduction in the concentration of metabolites and their side-products. A reduction in the concentration of metabolites must inevitably lead to a reduction in the rate of the side-reactions in which they participate (since the rate of these reactions is proportional to the concentration of metabolites). At the same time, metabolic activity can remain unaltered via the heightened concentration of key enzymes and receptors.

In recent years, there has also been some experimental evidence in favour of our hypothesis. For example, it has been shown that the concentration of glucose in the blood plasma is significantly lower in rats which are on a calory-restricted diet (Masoro et al., 1989). As would be expected in this case, a significant reduction is observed in the content of one of the products of the side-reaction of non-enzymatic glycation: glycosylated haemoglobin (Masoro et al., 1989). At the same time, an increased level is observed in a number of key enzymes belonging to the antioxidant system: catalase and glutathione transferase (Yu et al., 1989). Also observed is an increased rate of protein turnover, which in its turn prevents the accumulation of non-active and low-active forms of partially denatured enzymes (Holehan and Merry, 1986). Reducing the calory content of the diet leads to a reduction in the metabolic load on a number of systems and organs, and to a reduction in the organism's intake of a variety of toxins and xenobiotics contained in the food. As a result, it becomes possible for there to be a reduction in the activity of the microsomal xenobiotic oxidation systems, and consequently, a reduction in the rate at which active forms of carcinogens are generated from the metabolised xenobiotics. For instance, it turns out that the activation of polycyclic hydrocarbons into active mutagenic and carcinogenic epoxides which bind with DNA is significantly reduced in restricted-calory mice (Pashko and Schwartz, 1983). Therefore, the increase in life span which occurs when the calory intake is restricted is in good agreement with wear-and-tear (accumulation of defects) hypothesis, although many of the concrete mechanisms behind the destruction of the organism with age remain to be clarified.

Without doubt, life extension experiments deserve more detailed and careful examination than they have received in this section. For us, however, what is important is that the proposed mechanism underlying the most important methods of life extension is in good agreement with the generalised wear-and-tear hypothesis as the basic cause limiting the duration of life. Readers who wish to acquaint themselves more fully with life extension research may consult the specialised literature (Cutler, 1981; Walford, 1983; 1986; Nikitin, 1984; Obukhova and Emanuel, 1984; Gavrilov, 1985; Holehan and Merry, 1986; Frolkis and Muradian, 1988; Gavrilov, 1990).

## 5.6. The limit of cell division: the key to the mechanism which determines life span?

One of the favourite arguments of those who support the hypothesis of a genetically programmed life span limit runs as follows: the length of life is fixed, because an organism's cells can only carry out a limited, strictly determined number of divisions, after which they die (see, for example, Fries, 1980).

It must be noted that this idea has deep historical

roots, and many of the contradictions in the publications on the subject can only be understood if its history is taken into account. Therefore, for clarity of exposition, the arguments will be presented in the form of a brief historical excursus. What is more, in order to guarantee the objectivity and authenticity of this excursus, we will be compelled to make abundant citation of the primary sources. We hope that the reader will approve of the exhaustive documentary style in which the arguments are presented.

More than 100 years ago, the famous German biologist August Weismann (1882; 1884) postulated that the limited life span of organisms is determined by the limited capacity of somatic cells for division. Weismann also suggested that the differences in the longevity of different animal species are caused by the different number of generations that the somatic cells of each species can produce (the cells of long-lived species are capable of completing more divisions). In this way, the supposition arose "that life span is connected with the number of somatic cell generations which follow after each other in the course of an individual life, and that this number, like the life span of individual generations of cells, is already determined in the embryonic cell" (Weismann, 1892; 1914). Weismann tried to explain "the different life span of animals by making it dependent on the number of cell generations which was the norm for each different species" (Weismann, 1892; 1914).

Weismann's view that life span is limited because the capability of cells to multiply becomes exhausted appears to have been extremely popular at the turn of the 20th century, and to have been shared by many famous scientists of the day. For example, the Russian scientist Ivan R.Tarkhanov wrote: "... in an embryonic fertilised egg, the total of cell generations which can develop as a result of the creative forces of the embryo is determined in advance, by the laws of heredity" (Tarkhanov, 1891, pp. 551-552). In Tarkhanov's view, "the number of cell generations which can develop from the embryo in the course of an entire life ... determines longevity, the maximal life span which can be achieved by different organisms" (Tarkhanov, 1891, p.542).

Weismann's concept was dominant for 30 years. For this reason, something of a sensation was caused by the experiments of Alexis Carrel (1912) and Albert Ebeling (1913) on the cultivation of chick heart fibroblasts in vitro, which refuted Weismann's ideas. Carrel and Ebeling reported that under appropriate conditions the cells of a multicellular organism can reproduce themselves practically without restriction, like microorganisms (the cell multiplication experiment lasted all of 34 years, and was terminated only in 1946 after Carrel had himself died). As a result of these experiments, which received general recognition and reached a wide audience, Weismann's concept was abandoned and forgotten, and its place was taken by the diametrically opposed idea of the potential "immortality" of the somatic cells which form a mortal organism. People therefore began to seek the causes of the limited life span of organisms at a supracellular physiological level, in particular at the level of hormonal regulation.

Carrel's concept ruled for more than 40 years, and it seemed to have vanquished finally and irrevocably. Although many investigators repeatedly observed that the capability of cultured cells to multiply became exhausted, it was acceptable to explain these cases as failures of the kind which often accompanied cell cultivation (for example, contamination by viruses, or the toxicity of the particular batch of blood serum used for the culture). The popularity of Carrel's concept was so great that no attention was paid to the many exceptions to the established "rule" (the unlimited growth of animal cell cultures): they were simply considered to be methodological artefacts (see Witkowski, 1987).

H.Earle Swim (1959) seems to have been the first investigator to attempt a radical revision of Carrel's concept. After analysing results from 336 publications, including the results of his own experiments on the serial cultivation of 23 strains of fibroblasts derived from normal tissues of the rabbit and chick embryo (Haff and Swim, 1956), as well as 51 strains of human fibroblasts derived from foreskin, placenta, testicle, uterus and embryonic tissues (Swim and Parker, 1957), Swim came to the following fundamentally important conclusions: "... in most instances where growth occurs the cells eventually undergo nonspecific degeneration" (Swim, 1959, p.145). "The common experience of many investigators indicates that the early cultivation of cells usually follows a characteristic course which can be conveniently divided into three phases. In phase I the cells proliferate rapidly after an initial lag and usually can be transferred serially without difficulty. Phase II is characterized by a decrease in multiplication to a point where it usually ceases and the cells are eventually lost as a result of nonspecific degeneration" (Swim, 1959, p.159). "This was accompanied at first by an increase in the number of granules in the cytoplasm of the cells; later, degenerating cells were observed and their numbers increased progressively until the bottoms of the flasks were covered with a dense layer of cellular debris..." (Haff and Swim, 1956, p.201). The important conclusion was also drawn that when cells stop multiplying, this is not a methodological artefact caused by such factors as inoculum size, toxic media, or their inability to proliferate on glass (Swim and Parker, 1957). Finally, Swim (1957) notes that "infrequently a third stage is recognized by the appearance of actively proliferating cells in phase II cultures." (p.159). "It should be emphasised that phases I and II represent the usual pattern, while phase III is a relatively rare event." (Swim, 1959, p.160). Swim (1959) also noted that cells in phase III often differ from the original cells both in morphology and growth pattern.

Thus the unlimited capability of animal cells to divide turns out to be the exception rather than the rule, and in many cases the cells which divide without limit little resemble the original normal cells either in morphology or growth pattern (Swim, 1959). It might seem that this conclusion provides every ground for reviewing Carrel's concept and reviving interest in Weismann's concept. However, to carry this through and destroy generally accepted ideas which had been established in the public consciousness for decades would have required exceptionally vigorous efforts by Swim to propagandise his own results and conclusions, and a degree of enterprise which he apparently did not possess. However, other investigators emerged who brilliantly performed this function.

In the 60's and 70's, L.Hayflick published an extended series of academic and popular scientific articles describing the serial cultivation of human fibroblasts (see for example Hayflick and Moorhead, 1961; Hayflick, 1965; 1968; 1974). In fact, these articles confirmed Swim's results and conclusions, as well as Weismann's concept. However most readers, being unaware of the works of these undeservedly forgotten authors, took Hayflick's publications to represent a fundamentally new scientific discovery.* Since Hayflick's publications aroused great interest and reached a wide audience, we shall restrict ourselves to a

---

* This reaction to Hayflick's articles was partially caused by the style in which they are written: Weismann's idea that there is a limit to cell division was presented without reference to its originator, and there is only one citation in Hayflick's numerous early papers (Hayflick and Moorhead, 1961), referring to a single work (Swim and Parker, 1957), of the fundamental works of Swim (Haff and Swim, 1956; Swim and Parker, 1957; Swim, 1959). The impression is therefore created that all the results and their interpretation are fundamentally new.

brief exposition of his basic ideas.

In a celebrated article (Hayflick and Moorhead, 1961, p.600), the authors write: "The general history of diploid cell strain may therefore be divided into three distinct phases... Phase I, or early growth phase, constitutes that period when cells have been freed from the intact tissue and are just establishing themselves on glass (primary culture)."* "Phase I ends with the formation of the first confluent sheet, at which time the culture is ready for its first subcultivation and entry into Phase II. Phase II is characterized by rapid cell multiplication... During this phase the diploid strains must be subcultured at least twice a week with split ratios of 2 or 3:1".** "Phase II lasts from 2 to 10 months at the end of which time cell degeneration begins to take place. This degeneration and lessening of mitotic activity heralds the appearance of Phase III or the terminal phase. It is characterized by the appearance of debris..., reduction of mitotic activity and a consequently longer period of time for the development of confluent sheets".*** What is more, the authors note that "an alteration may occur... giving rise to a 'cell line' whose potential life is infinite." (Hayflick and Moorhead, 1961, p.587). It is not difficult to see that this "alteration" corresponds to the transition of the culture into Phase III of Swim's classification (Swim, 1959). Thus, many of the fundamental points in the celebrated and widely cited article of Hayflick and Moorhead (1961) are in fact a repetition of the undeservedly forgotten conclusions of Swim (1959).

On the basis of these results and a number of additional experiments, the conclusion was drawn that human cells are capable of carrying out only a limited, strictly determined number of divisions, after which they perish. According to this cellular version of the "Peau de Chagrin" model, human beings in the course of a lifetime will inevitably exhaust

---

* It can easily be observed that Phase I in Hayflick and Moorhead's classification corresponds to the initial lag in Phase I of Swim's classification (Swim, 1959).

** It is clear that Phase II in Hayflick and Moorhead's classification corresponds to Phase I after initial lag in Swim's classification (Swim, 1959).

*** Phase III in Hayflick and Moorhead's classification exactly coincides with Phase II of Swim's classification (Swim, 1959).

the potential for division present in each cell, and "the timing of this loss may well represent the limit of the human life span" (Hayflick, 1980, p.42). Indeed, "it appears that fibroblasts from human embryos will divide in cell culture 50±10 times, those from persons between birth and the age of about 20 will divide 30±10 times and those from donors over 20 will divide 20±10 times" (Hayflick, 1968, p.35). Thus, after almost 100 years, researchers in fact again returned to the once rejected concept of Weismann (1882), forgetting however who originally invented it.

Weismann's idea that the cells of long-lived species are capable of more divisions (Weismann, 1882) was also confirmed. It turns out that the embryonal fibroblasts of mice are capable of doubling their numbers in vitro only 14-28 times. For chicks the figures are 15-35 times, for human beings 40-60 times, and for tortoises 72-114 times (Hayflick, 1974).

Hayflick's numerous publications in the 60's and 70's were considered by many to be a scientific sensation, opening up a fundamentally new approach to explaining the mechanisms which determine life span. Since practically no reference was made to Swim's works, Hayflick's articles were accepted as the first and decisive refutation of Carrel's myth that cells are capable of unlimited division. Since Weismann's concept of limited cell division as the basis of aging was also fundamentally forgotten, Hayflick's articles were considered to represent an essentially new idea, rather than a return to the old hypothesis. Therefore, the limited capability of cells for division became known as the Hayflick limit (see for example Bremermann, 1982; Walford, 1983; Juckett, 1987), and the history of how this phenomenon was discovered is often presented as starting from the work of Hayflick and Moorhead (1961), as if from the year of Our Lord.

The new (or, more precisely, well-forgotten old) concept that cells have a limited capability for division has by now become so well-known and so fully accepted that it is now even to be read in the New Encyclopaedia Britannica (1989, vol.1, p.148): "Laboratory experiments have demonstrated that cells from complex organisms go through only a limited number of cell divisions before they die off, supporting the idea that cellular events can produce senescence". Hayflick's works, which confirmed this idea, came to be treated as an example of a scientific revolution in biology (see Witkowski, 1987). This might seem to be the right point at which to draw this long story to a happy conclusion. However, there are serious grounds for considering that the story is not yet complete.

In 1978, we carried out a revision of the Weismann-Swim-Hayflick (WSH) concept, resulting in the following conclusions (Gavrilov and Gavrilova, 1978; Gavrilov and Yaguzhinsky, 1978):

(1) The mass death of cells in Swim's Phase II and Hayflick's Phase III is a methodological artefact caused by damage to the cells as a result of the replating procedure (treatment of the cells with trypsin and their subsequent suspending and replanting). If the probability of cell death is evaluated excluding the replating periods, it turns out that *the probability of cell death does not increase with the age* of the culture, i.e. there are no grounds for speaking of aging at the cellular level (Gavrilov and Gavrilova, 1978; 1982; Gavrilov and Yaguzhinsky, 1978).

(2) The basic cause of the Swim-Hayflick phenomenon in vitro is the accumulation in the culture of postmitotic cells which are unimpaired from a metabolic point of view (Gavrilov and Gavrilova, 1978; 1982). The process by which these metabolically unimpaired cells are formed resemble *cell differentiation* rather than damage and aging (Gavrilov and Gavrilova, 1978; 1982; Gavrilov and Yaguzhinsky, 1978). Other investigators simultaneously reached the same conclusion (Bell et al., 1978).

(3) When the whole evolving cell population is taken into account, the accumulation of postmitotic cells in the culture is not accompanied by a reduction in the absolute number of dividing cells. In other words, *what occurs is not the disappearance of dividing cells, but only the dilution of the culture with postmitotic cells.* Consequently, there are no grounds for speaking of the limited capability of *all* the cells in the culture for division. On the contrary, in a culture of embryonal human fibroblasts for example, there are cells which complete around 170 (instead of the generally accepted 50) divisions, and have the properties of stem cells with an unlimited capability for division (Gavrilov and Yaguzhinsky, 1978).

Let us now examine these conclusions in somewhat greater detail, undertaking an analysis of the technique of cell cultivation. The following procedures are involved. A tissue sample (skin, for example) is placed in a flask containing a culture medium and left until a portion of the cells from the primary explant migrate to the inner surface of the flask, where they multiply and form a confluent monolayer covering the base. Nobody knows how many cells migrate from the transplant in this process, or how many divisions they carry out. According to Goldstein et al. (1975), the process corresponds to approximately 10 population doublings. If embryonal tissue is used, it is treated with trypsin, and the suspended cells are strained into a vessel containing a nutrient medium (Hayflick, 1968). When the cell culture forms a monolayer, it is replated by treating it with trypsin, and the required number of cells are suspended and transferred into new vessels containing a nutrient medium. Replating the cells in the ratio 1:2 means that only half the cells which formed the confluent monolayer are transfered into a vessel with the same surface area. When a monolayer is formed in the new vessel as a result of division, doubling of the cell population is considered to have taken place, corresponding in the mean to one division of the cells. This operation is repeated many times (approximately 50 times in the case of embryonal human fibroblasts), until the culture loses the capability of quickly restoring its numbers. We shall now try to estimate the number of divisions which the cells carry out. It is known that one hour after the replating operation only 50-70% of the cells from "young" cultures and only 25% of the cells from "old" cultures remain alive (these latter cells are larger and more easily damaged in suspension) (see Gavrilov and Gavrilova, 1978; 1982).* Thus, 40-60 subcultivations correspond not to 40-60 population doublings, but to 80-120 doublings carried out by the surviving cells (see Gavrilov and Gavrilova, 1978; 1982). Further, one doubling of the cell population corresponds to one cell divison only if all the cells are capable of division. In reality, the proportion of dividing cells drops dramatically during the cultivation process, and by the final subcultivation amounts to only 10-20% (see Gavrilov and Gavrilova, 1978; 1982). It can be shown that if only 10% of cells maintain the ability to divide, they must divide 10 times for the numbers of the whole culture to double (Gavrilov and Yaguzhinsky, 1978; Gavrilov and Gavrilova, 1982). If this fact is taken into account, it turns out that 80-120 doublings of the cell population corresponds to 170±30 divisions (Gavrilov and Yaguzhinsky,

---

* The mass destruction of cells after replating, in conjunction with the low rates at which the numbers of "old" cultures are restored, has given rise to the illusion of natural mass degeneration and death of cells described in a number of works (Swim, 1959; Hayflick and Moorhead, 1961; Hayflick, 1968).

1978). The result is that 50 subcultivations of the cells corresponds not to 50, but to 170±30 divisions (Gavrilov and Yaguzhinsky, 1978; Gavrilov and Gavrilova, 1982).

The most important point, however, is that this number of divisions is not final. In fact, the death of a cell culture is determined in a highly arbitrary way. It seems that a cell culture is called dead only because in an arbitrary time interval (usually from one to four weeks), its numbers *do not grow* to the required size (the usual target is a number 2-4 times greater than the original)!* What is more, in "dead" cultures 10-20% of cells remain capable of division, and the culture can therefore compensate for natural cell death (which, by the way, does not increase), and even slowly grow (see Gavrilov and Gavrilova, 1978; 1982). When the whole evolving cell mass is taken into account, the absolute number of dividing cells turns out not to decrease, but the dividing cells are simply diluted with non-dividing cells (Gavrilov and Yaguzhinsky, 1978; Gavrilov and Gavrilova, 1978; 1982). As a result, when the proportion of dividing cells approaches 10%, the culture cannot indeed grow quickly enough to be considered alive (Gavrilov and Yaguzhinsky, 1978; Gavrilov and Gavrilova, 1982). Thus, the phenomenon of culture "death" not only cannot be equated with cell death, but does not even imply that all the cells in the culture have a limited capability for division (Gavrilov and Gavrilova, 1982; Gavrilov and Yaguzhinsky, 1978). Our conclusion was that cell differentiation in culture proceeds with the formation of non-dividing differentiated cells, some of the cells having properties of stem cells, which can proliferate indefinitely (Gavrilov and Yaguzhinsky, 1978).

Some of the above-mentioned conclusions we have reached were simultaneously and independently confirmed by other authors (Bell et al., 1978, p.1160): "... one of the least documented features of the aging-in-vitro hypothesis is the accepted conclusion that phase 3 cells are moribund and eventually die. No hard data support this conclusion. ... In our experience, replating itself inevitably results in loss of cells. ... We observed the survival of foreskin fibroblasts (strain 1519) in culture for long periods after cell divisions no longer occured. Our first cultures

---

* As an example, we cite the following definition: "Cultures were judged to be terminal when they failed to achieve confluent monolayers after 4 weeks with weekly refeeding" (Goldstein and Singal, 1974, p.360).

of 1519 cells were kept alive in Ham's medium... for 22 months, more than a year to 14 months after cell division terminated. ... Biochemical studies have shown that while phase 3 cells differ in some respects from phase 2 cells, they are essentially normal. ... Reported biochemical differences between phase 2 and phase 3 cells reflect differences of differentiation rather than of age".*

Subsequently Hayflick himself, in conjunction with T.Matsumura and Z.Zerrudo (Matsumura et al., 1979), confirmed the absence of mass cell death in Phase III of his classification. The authors of this work maintained "dead" cultures for 6 months, and some of these cultures *increased* their numbers by 16 times over this period (Matsumura et al., 1979)! No signs of imminent mass cell death could be discovered, although the period the "dead" cultures were observed sometimes even exceeded a year (by way of comparison we note that the entire previous procedure of 50 subcultivations usually takes 6-8 months) (Hayflick, 1968).

As a result of these investigations, Hayflick changed his interpretation of his own experiments. Initially, he presented his results in the following manner (see Fig.40): "We found that fibroblasts taken from four-month-old human embryos doubled in this way about 50 times (the limit ranged between 40 and 60 doublings). After reaching the limit of capacity for division the cell population died" (see Hayflick, 1968, p.35). In the 1979 work, another figure is presented (see Fig.41) and everything is expressed differently: "In early Phase III, the culture still proliferates, although the rate of proliferation is decreasing. In late Phase III, cell proliferation is very low. ... *Neither sign of ultimate death* of a culture nor of spontaneous acquisition of infinite proliferation potential was observed during this period." (see Matsumura et al., 1979, p.332; p.333). It must be emphasised that when an investigator changes his position, it is does honour to him as a scientist. The goal of science is not the defence of dogma, but the search for truth. However, it is regrettable that the newly discovered facts became known only to an extremely narrow circle of specialists, while such leading popular science magazines as Scientific American continued to advertise the old myth that "finally the cells undergo a

---

* Phases 2 and 3 in this article correspond to Hayflick's, rather than Swim's classification. Apparently the authors of the article, like many other investigators, are simply not aware of Swim's works and therefore do not cite any of his articles.

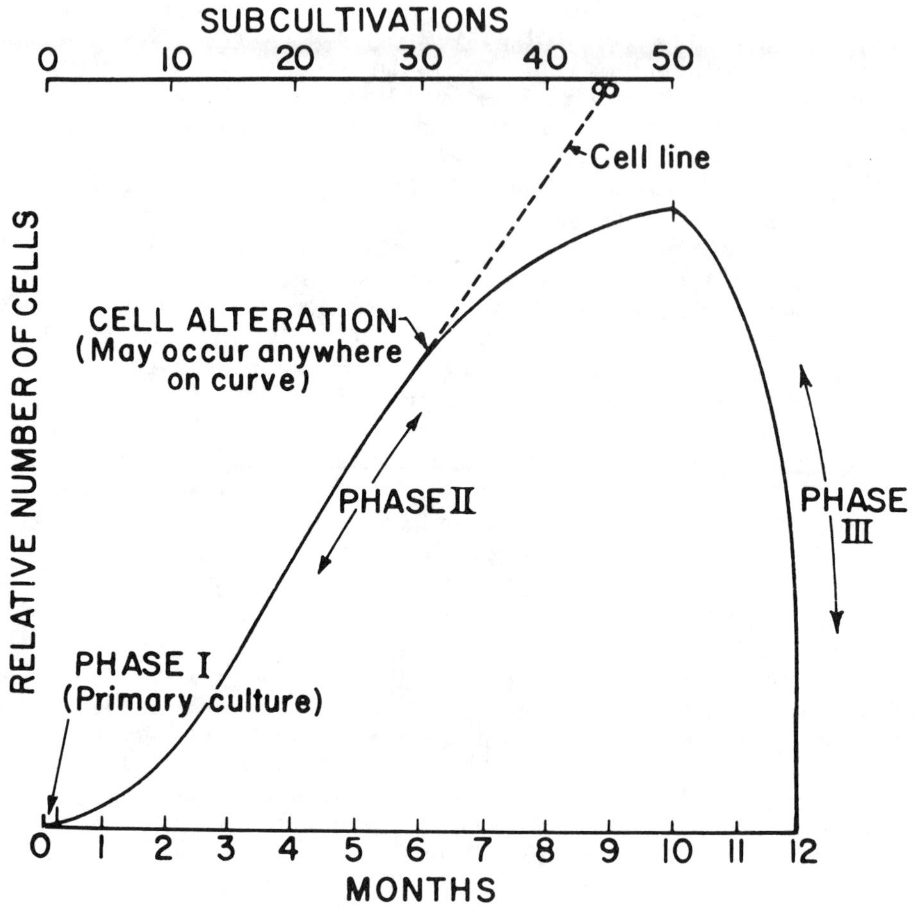


**Figure 40** The initial version of the life history of human diploid fibroblasts in culture according to the work of L.Hayflick and P.S.Moorhead (1961). Reproduced with the permission of the copyright owner (Academic Press).

The figure shows an increase in the number of cells as they are cultivated (Phases I and II), followed by a sharp drop in the number of cells (Phase III). This figure, first published in 1961, has been continually reproduced in both scientific and popular scientific publications (see, for example, Hayflick, 1966; 1968; 1977; 1982), and is even adduced in a textbook on the biology of aging (Lamb, 1977).

ordinate - relative number of cells

abscissa (top) - subcultivations

abscissa (bottom) - months

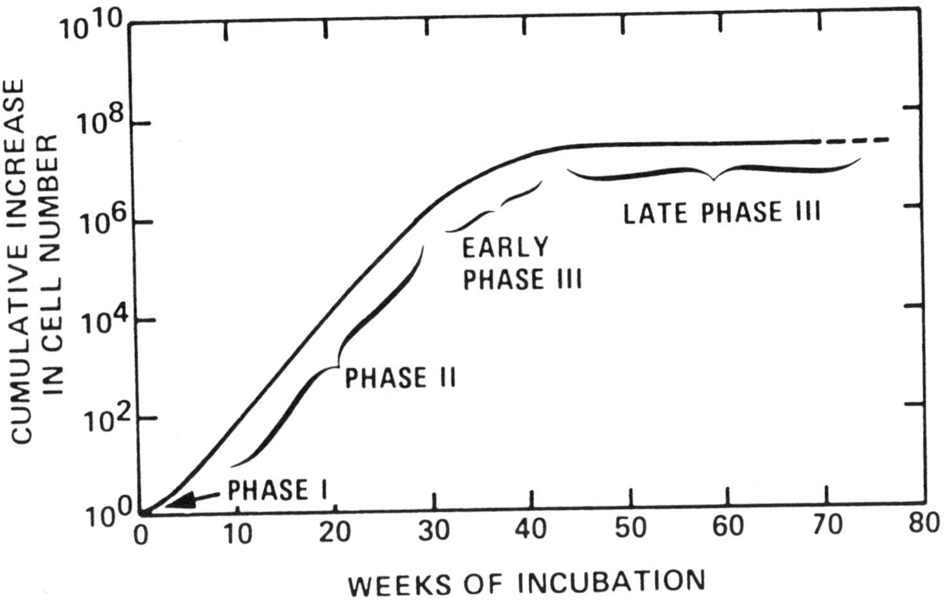


**Figure 41** The new revised version of the life history of human diploid fibroblasts in culture according to Matsumura et al. (1979). Reproduced with the permission of the copyright owner (The Gerontological Society of America).

The figure shows an increase in the number of cells as they are cultivated (Phases I and II), followed by a gradual reduction in the rate of growth (early and late Phase III). In distinction to the initial version (see Fig.40), this more correct schema was not advertised and is known only to a narrow circle of specialists.

ordinate - cumulative increase in cell number

abscissa - weeks of incubation

variety of degenerative changes and die" (Hayflick, 1980, p.42). Of course, it is surprising that the propagation of this myth was carried out by the same author who a year before had so well demonstrated its inadequacy.* This riddle however goes beyond the bounds of the present book. At the same time, it becomes comprehensible how antiquated notions about a limit to cell division can become canonised in the latest volumes of such solid publications as the New Encyclopaedia Britannica (1989).

Bearing in mind the new facts, it is easy to see that the Swim-Hayflick phenomenon is no longer so obviously connected with the issue of the limited life span of organisms. While the mass death of cells might understandably lead to the death of the organism, the limited life span of organisms does not directly follow from a reduction in the rate of growth of the cell population. What is more, reductions in mitotic activity are often a normal physiological process connected with the differentiation of cells. For example, when nerve and muscle tissue is being formed, the cells practically lose all capability of division, and under the above-cited definition the cell populations could be called dead. However it is obvious that this definition is not very constructive, since we are dealing with highly specialised and functionally unimpaired cells. In fact, in many cases the performance of highly specialised functions turns out to be incompatible (at least at the same time) with cell division. It is also known that, when cells differentiate, cell division is often essential for their differentiation and for the switching of their metabolism, the critical point in this shift coinciding with the synthesis of DNA (see Gavrilov and Gavrilova, 1978; 1982). Thus, a clear similarity is observed between the counting of the number of cell divisions which occurs in cell differentiation and the counting of the cell divisions which is observed in cell cultures. The hypothesis naturally arises that the phenomenon discovered by Swim and Hayflick is not a fundamentally new process involving aging at the cellular level, but the long known, although insufficiently studied, process of cell differentiation, accompanied by the counting of cell divisions and a reduction in mitotic activity (Bell et al., 1978; Gavrilov

---

* It is even more surprising that three years later the same author again publishes the out-of-date figure 40 instead of his new amended figure 41, and again writes that "after cessation of mitotic activity, the culture finally undergoes total degeneration" (see Hayflick, 1982, p.64).

and Gavrilova, 1978; 1982; Gavrilov and Yaguzhinsky, 1978). Unfortunately, it is still unclear what particular cell types human fibroblasts differentiate into in vitro. Perhaps a deficiency of certain factors in the culture medium may suppress the expression of the new differentiated phenotype. In principle, it is known that under certain conditions fibroblasts are capable of differentiating in vitro to form non-dividing adipocytes, but the expression of this new phenotype can be held back by deficiencies in a range of factors, for example insulin, biotin, etc. As far as cultured human epidermal keratinocytes are concerned, it has been directly shown that a reduction in the growth rate of these cultures is linked to the differentiation (keratinisation) of the keratinocytes (Rheinwald and Green, 1977).

If we were to accept the differentiation hypothesis, we would not at all expect the number of times the cell population doubles to be a fundamental inherent property of cells, as is supposed by Hayflick, but rather that this number might be relatively easily increased using inhibitors of differentiation (certain growth factors, tumour promotors,* etc.). In fact, it turns out that epidermal growth factor (EGF) increases the number of times the cell population doubles from 50 to 150, with a simultaneous threefold increase in the "lifetime" of the culture (Rheinwald and Green, 1977). At the same time, the increase in the "lifetime" of the culture was accompanied by the delayed appearance of a multitude of signs of terminal keratinocyte differentiation (Rheinwald and Green, 1977).

Another prediction of the differentiation hypothesis is that, contrary to what is claimed by Hayflick, constant active growth of cell cultures may be observed not only in the case of transformed heteroploid cells, but is also possible for normal diploid cells (if these cells do not embark on terminal differentiation). In fact, it turns out that normal diploid mouse embryo cells, which under standard conditions manifest a growth crisis after 7-10 population doublings, may be successfully cultivated without any sign of an approaching growth crisis for at least 200 population doublings. All that is necessary is to change the composition of the culture medium (excluding blood serum and adding a number of ingredients, including the above-mentioned EGF). In this case the cells, which are apparently

---

* In this particular case tumour promotors are used not as the cell transformation agents but as the inhibitors of cell differentiation.

capable of unlimited multiplication, remain diploid and nontumorigenic (Loo et al., 1987).

Certain normal diploid cells show a practically inexhaustible capability for proliferation not only in vitro, but also in vivo. Normal drosophila imaginal disc cells have this property if their differentiation inductors are absent (see Finch, 1976), as do haemopoietic stem cells (Harrison, 1984).

Taking all the above cited facts into account, we are forced to recognise that the Weismann-Swim-Hayflick concept must be revised, especially with respect to the additions put forward by Hayflick (1965; 1968). Although this concept enjoys great popularity, especially amongst investigators who are remote from cell biology (Fries, 1980; Benjamin, 1982), it is at present unclear how the results obtained from cell culture research are related to life span. It cannot be ruled out that the limited capability of certain cells for division is only indirectly linked to the limited life span of organisms. Evidence in favour of this conclusion is provided by the fact that organisms which are basically constructed out of postmitotic cells (laboratory drosophila) essentially have the same life span distributions as human beings. On the other hand, it is clear that the transition of a certain proportion of cells into a postmitotic state is the prelude to a reduction in the organism's capabilities for regeneration, and finally to a reduction in the number of functioning cells, as is indeed observed with increasing age (Lamb, 1977; Walton, 1982). Therefore, research into the kinetics of cell populations at different stages in an organism's life may be of great importance in clarifying the mechanisms which determine life span. However, attempts to justify the existence of a genetically programmed limit to life span by alluding to a limit to cell division cannot be considered convincing. The key to the mechanism which determines life span is not an account of the reasons why proliferation rates decrease, but rather an explanation why cells, as well as other structures of the organism, become damaged and die. It is significant that one of the originators of the WSH concept of a limit to cell division himself nevertheless rejects the genetic programme hypothesis, and supports the accumulation of damage (wear and tear) hypothesis (Hayflick, 1988).

## 5.7. Reliability theory: the methodological foundation for research into the mechanisms which determine life span

As will be clear from the previous discussion, there is a distinct analogy between the processes which limit the life span of organisms and the processes of wear and tear which we observe in inanimate nature. It is paradoxical that this obvious and at first sight even trivial idea meets with great resistance on the part of many gerontological specialists and representatives of other professions. Possibly the reason for this resistance is a belief in the unique perfection of biological systems and the purpose of everything which happens to them. Both to those enthusiasts who would like to solve the problem with a single remedy (the elixir of youth), and to fatalists who consider it pointless and even immoral to try to change the plan predetermined by the creator ("It is the will of God" and "One cannot avoid one's fate"), the idea of a program is significantly more attractive than the idea of wear and tear.

In a materially and atheistically oriented society, one might think that the idea of wear and tear would be treated as self-evident. However, we have been forced on a number of occasions to recognise that this is not the case. Even attempts to draw analogies between the survivability of human beings and laboratory drosophila have encountered accusations of "biologising", i.e. reducing a higher form of the activity of matter (social) to a lower one (biological). The position of the opposition can be summarised as follows: "Man ages and dies in a fundamentally different way than animals. This truth is rather banal..." (see Kron, 1980, p.176). Attempts to draw an analogy with the destruction of non-living systems have been treated as extreme cases of reductionism and mechanicism, and a violation of the Marxist thesis that higher forms of the activity of matter cannot be reduced to lower ones. It must be noted, however, that in this struggle for Marxist purity our opponents seem to be "holier than the Pope", since Marx himself wrote in Das Kapital that "the damage which afflicts individual parts of machines etc. is by nature random, and therefore the repair which it necessitates is also random. However, within the general mass of repairs, it is possible to distinguish two types which have a more or less constant character and belong to different periods in the life of the basic capital: *illnesses of childhood and the incomparably greater number of illnesses belonging to an age beyond the average life span*. For example, no matter how perfectly a machine is constructed before it begins the production process, using it in practice will reveal faults which need extra work to repair. On the other hand, the more it exceeds its average age, and consequently the more it is affected by normal wear and tear and the more the material from which it is made becomes worn and weak with age, the more numerous and more significant become the repairs which are necessary to maintain the machine's existence to the end of its average life span; *in exactly the same way an old man, in order not to die prematurely, must spend more on medical assistance than a man full of youthful vigour*".*

Subsequently it became clear that the form of the relationship between failure rate and age is indeed strikingly similar for living systems (Hutchinson, 1978; Urlanis, 1978) and technical devices (Barlow and Proschan, 1965; 1975), qualitatively corresponding to a bathtub curve. At present, the theory of reliability distinguishes the following three periods in the functioning of systems:

(1) The "working-in" period, also called the period of "burning-out" of defective elements. It is characterised by a high initial failure rate, which decreases rapidly with time. Exactly the same type of mortality rate is observed at the beginning of the life of most organisms, including human beings. In man, this period corresponds to the age of heightened infant mortality (approximately 0-5 years), and is also to a significant extent caused by the preferential death of weakened organisms with inborn defects (Altukhov et al., 1981; Altukhov, 1989).

(2) The normal working period, corresponding to an age of low and approximately constant failure rates. This period corresponds to the period of low and approximately constant force of mortality observed in all living creatures. For human beings, this period is apparently rather short and ends too soon (10-15 years).

(3) The aging period, which involves an inexorable rise in the rate of failure with age. This period corresponds to the period of growth in the force of mortality which is likewise characteristic of most living organisms, and is usually described by the Gompertz-Makeham law. The analysis of mortality in this period has been the theme of the greater part of the present book. For man, the aging period lies

---

* Marx, K. and F.Engels, Works, vol.24, p.196 (our italics - L.G. and N.G.)

aproximately within the interval 20-100 years.

Thus there is a surprising similarity in the distribution of the lifetimes of technical and biological systems. This similarity is reinforced if we take into account the fact that at highly extreme ages there seems to be yet another period which is common to both technical and biological systems. This period of failure saturation corresponds to a very high and approximately constant failure rate (force of mortality) (Gavrilov, 1980; Economos, 1983). The existence of this period is probably due to the fact that under conditions of high mortality the rate at which organisms with defective elements are eliminated becomes comparable with the rate at which they accumulate; the result is the establishment of a stationary distribution of organisms according to the number of defects, and the force of mortality becomes constant. In human beings, this period is observed at ages exceeding 100 years (Gavrilov, 1980; Economos, 1983).

It can easily be seen that the stages which characterise the breakdown of living and non-living systems show a remarkable coincidence. The obvious similarity in the observed periodisation suggests the possible existence of a similarity in the processes which limit the lifetimes of technical and biological systems.

Unfortunately, the scientific investigation of this simple and rather obvious idea was for a long time not pursued. Partly this was because the theory of reliability was itself only developed as a scientific discipline during the second world war in connection with investigations into the functioning of complex radio-electronic systems (Barlow and Proschan, 1965; 1975). Also, it was pointless to compare biological and technical systems since they were simply incomparable from the point of view of their complexity. For this reason, attempts at a comparison were treated as an extreme manifestation of reductionism and mechanicism. The following statement by the well-known gerontologist A.Comfort is very typical: "... the differences between an old cart and an old horse are sufficiently striking to make the extensive acceptance of 'wear' as an explanation of senescence" (see Comfort, 1964, p.45). Thus, the majority of biologists conducting research into aging and life span treated wear and tear as an extreme simplification, instead directing their efforts towards the search for a genetic aging program, the isolation of "age-inducing" proteins and hormones, and the hunt for a biological clock governing the species-specific life span limit (Kogan, 1977).

Now, after several decades of persistent work, the gerontologists themselves have been forced to admit that the research they carried out was not particularly successful (Hayflick, 1988). What is more, as has already been noted, it must be doubted whether there is any justification for the very concept that organisms are genetically programmed for self-destruction.

On the other hand, as technology has advanced, the concept of "machine" has become broader. With the development of cybernetics, machines have begun to be understood as complex systems with a definite purpose. The fading of the boundaries between biological and technical systems has also been aided by the development of the science of bionics, which uses the principles of the functioning of biosystems in the design of technical systems. As a result, the claim that there are fundamental differences between biological and technical systems has acquired a tinge of vitalism, because it has become equivalent to the claim that living things are unfathomable, and that the principles which govern the functioning of living things cannot be exploited in practice (in technology). Moreover, it has become clear that the idea that biological systems are a kind of machine does not refute, but rather confirms their biological specificity, since "machines in principle are not explicable by physical-chemical principles. This is because the understanding of a system requires both a knowledge of laws and of boundary conditions, and in the case of machines the boundary conditions, which are the specification of classes of machines, are not derivable from the laws of physics. Presumably an analogous statement would be that only the functioning, but not the existence of the universe is understandable in the sense of physics. Thus the statement that organisms are machines, so often used to support reductionism, has a quite contrary significance. Life cannot be entirely reduced to chemistry and physics precisely because living things are machines" (see Yčas, 1969, p.286).

Thus, the basic obstacles and misunderstandings which formerly blocked the use of the theory of reliability in biology have now been removed. It is therefore not surprising that there has recently been an upsurge in the number of publications concerning the use of the ideas and techiniques of the theory of reliability in the biology of life span (Gavrilov, 1978; 1980; Gavrilov et al., 1978; Shukailo, 1979; Ďoubal, 1982; Koltover, 1983; Kozlovsky and Gavrilov, 1983).

Indeed, in the mathematical theory of reliability, a wide range of theoretical models of life span distribution have long been developed (including the Gompertz-Makeham distribution), there are efficient techniques for analysing survivorship data, and approaches to the planning of experiments in the field have been developed. The most important virtue of the mathematical theory of reliability is the fact that it is very general in nature and can be applied with equal success to the study of both technical and biological systems. Thus, the mathematical theory of reliability can serve as an excellent methodological foundation for research into the mechanisms which determine the life span of organisms, since it does not impose its own solutions, but merely points to the most efficient research techniques.

In addition to the mathematical theory of reliability, experience accumulated over many years in the use of complex technology may also be of great importance, since it leads to the uncovering of the basic parameters which are essential to the working (vital) capacity of the systems under investigation. Let us enumerate some of these parameters:

(1) Stability - a highly important characteristic of a system, consisting in its ability automatically to restore an established state when this has been suddenly disturbed by some external or internal agent. Static stability is characteristic only of the initial and final state. Dynamic stability determines the length of the transition from the initial to the final state, and the way in which the system evolves during this transitional period. It is well-known that as a system ages there is a significant drop in its dynamic stability with respect to a whole range of perturbatory factors. The classic example of such a drop in stability is the reduced glucose tolerance in aging. When glucose is administered, glucose levels in young organisms quickly fall to their original level, while in old organisms this process takes a significantly longer time (Walford, 1983).

(2) Accuracy in carrying out operations - how the carrying out of an operation in practice approximates in quality to the ideal (theoretical) mode of carrying out this operation. The deviation from the ideal mode characterises the magnitude of error. Data exist concerning the reduced accuracy in carrying out operations which accompanies the aging of an organism. For example, DNA-polymerase isolated from old organisms is more often subject to the error of including non-complementary nucleotides into the DNA (Holliday and Kirkwood, 1981). Another example of the reduced accuracy in carrying out operations which accompanies aging is the reduction in the ability of T-

lymphocytes to recognise their own antigens, leading to an increase in autoimmune diseases in old age (Walford, 1983). The reduced accuracy in carrying out operations which accompanies aging is also observed at the level of the whole organism. For example, the number of errors made by old mice in finding their way through a maze is significantly higher than for young mice, despite preparatory training and the selection of only the well-trained individuals (Ingram et al., 1987).

(3) Speed of operation (nominal and effective) - the mean statistical number of operations carried out in a unit of time. Its inverse is the mean time required for the carrying out of one operation. It is well-known that as an organism ages there is a fall in the speed of operation of many of its systems. As an example, we can cite the fall in the rate of transmittal of nervous impulses which is observed in aging (see Strehler, 1962; Walford, 1983).

(4) Loading characteristic (static and dynamic) - a parameter characterising the relationship between load size and the system's response. As an organism ages, many of its systems show a change in loading characteristic. For example, in response to the action of a hormone, the tissues of an old organism begin to react at lower hormone concentrations, but the magnitude of the maximal response at high hormone concentrations is reduced. Examples of this kind of change are the lowered response of the reproductive system to the action of sex hormones, and the reduced quantity of luteinising hormone and follicle-stimulating hormone secreted in response to the administration of gonadotropin releasing hormone in the aging of rats (Meites, 1988). This means that in aging there is a reduction in the extent to which the system's response to a stimulus can change.

Thus, the use of the theory of reliability allows for a sensible classification of many of the age-related changes in an organism, and draws attention especially towards those which are essential for its ability to survive.

Failure is the term used in the theory of reliability if the values of the listed parameters exceed the permitted limits. A distinction is made between sudden failures, when there is a convulsive change in one or more of the system's basic parameters (for example, cell death), and gradual failures. Gradual failure corresponds to a slow change in one or more of the system's parameters (for example, cell atrophy). In addition, failures are divided into those which are independent and those which are dependent.

Independent failures correspond to a failure in one of the system's elements which is not brought about by damage to other elements (trauma). A dependent failure is a failure which occurs as a result of damage or breakdown in other elements (death from heart failure). Since in biological systems the existence of individual organs totally depends on the normal functioning of other organs, dependent failures are very important for biological systems. Even a small and retrievable failure in one of the organism's systems (for example, the formation of a clot in a blood vessel) can provoke a cascade of dependent and unretrievable failures in all the remaining systems (clinical death). Therefore the majority of systems in an organism are redundant with respect to the number of functioning elements. These redundant elements allow the system to work normally as a whole even when a large group of elements temporarily fail. However, the reduction in the number of elements which accompanies aging (lowering the number of redundant elements) leads to a sharp drop in the system's reliability, and an increase in the probability of death.

The concepts and ideas of the theory of reliability which we have presented above are merely illustrations; a detailed analysis can be found in the specialist literature (Barlow and Proschan, 1965; 1975), and in the last chapter of this book. However, the examples given are fully sufficient for an understanding of the possibilities opened up by the theory of reliability for a rigorous and accurate description of the processes which determine the life span of organisms. Efforts in this direction may lead to the creation of a quantitative theory of the life span of organisms which links the ways in which organisms change in aging to the observed life span distributions. For the time being, we only have a number of mathematical models of these processes, and it is to these that we now turn.

# CHAPTER 6. MATHEMATICAL MODELS OF LIFE SPAN

## 6.1. Introduction

At the present time, the number of mathematical models of the life span of organisms can already be counted in double figures (Strehler and Mildvan, 1960; Sacher and Trucco, 1962; Strehler, 1962; 1978; Brown and Forbes, 1974; Gavrilov, 1978; Skurnick and Kemeny, 1978; Abernethy, 1979), and continues to grow inexorably (Ďoubal, 1982; Kozlovsky and Gavrilov, 1983; Woodbury and Manton, 1983; Sutherland and Bailar, 1984; Witten, 1985; Piantanelli, 1986; Gavrilov, 1987; Guess and Witten, 1988; Hibbs and Walford, 1989).

Models based on the theory of reliability have been developed especially intensively (Gavrilov, 1978; 1987; Gavrilov et al., 1978; Abernethy, 1979). When the fruitfulness of this approach had been demonstrated at the end of the 1970's (Gavrilov, 1978; Gavrilov et al., 1978; Abernethy, 1979; Shukailo, 1979), the circle of investigators engaged in the mathematical modelling of life span using the techniques of reliability theory significantly expanded (Ďoubal, 1982; Koltover, 1983; Witten, 1985).

On the one hand, the abundance of mathematical models of life span reflects the topicality of the problem, but on the other hand, it reflects a crisis connected with the fact that in many cases the correspondence between the model and the actual data is not tested. For example, some models commonly presuppose a catastrophic growth in the force of mortality in the oldest age groups, and the existence of an absolute limit to life span in a region where its existence is blatantly excluded (see section 4.2). However, criteria exist for the adequacy of mathematical models of life span which were formulated as early as 1960 (Strehler and Mildvan, 1960), and which require not an accelerating, but a decelerating growth in the force of mortality in the oldest age groups. These widely known conditions, which any model of life span must satisfy, are frequently ignored, and authors select their own criteria of evaluation in such a way that the model they propose is bound to satisfy the criteria. It is obvious that following this procedure any number of mathematical models of life span can be invented in a single day.

A second problem is connected with the fact that some investigators claim from the outset that they have created a mathematical theory of life span which links the life span distribution of organisms with fundamental biochemical and molecular-genetic processes (Koltover, 1983). Since this noble task requires a high level of competence in many scientific fields, a competence which the investigators generally do not possess, the theories created in this way are often highly speculative in character. Let us cite a concrete example. In one of the models, a calculated coefficient turned out to be close to ten. On the basis of this fact alone, the coefficient was first interpreted as the number of regulatory genes, then the number of releasing factors, and finally as the number of cellular oncogenes, inasmuch as the author knew that the number of these structures is approximately ten (Koltover, 1983). It is clear that the same coefficient could no less justifiably be identified with the number of fingers on two hands, a fact pointed out in a discussion of this particular model by participants in a research seminar on the reliability of biological systems in Chernigov in 1982. Inevitably, all attempts to construct a fundamental mathematical theory of life span without the necessary knowledge only feed the growth of attitudes which are critical towards the mathematical modelling of life span.

These examples lead us to the following question: what do we aim to achieve by modelling life span mathematically, and what conditions must a mathematical model satisfy? It seems that the most sensible answer to this question will be the recognition that what we basically aim to achieve is nevertheless a clarification of the mechanisms which determine the life span of organisms. Starting from this point, mathematical modelling is not a goal in itself, but only one of the means of achieving the intended goal. Therefore, we should pay particular attention not to cumbersome mathematical constructions which claim to be fundamental theories, but rather to comparatively simple heuristic working models which are correspond to the known facts and predict new regularities. For example, starting from a mathematical model based on the theory of reliability, it was possible to predict that there exists an upper limit to the growth of the force of mortality at extremely old ages (Gavrilov, 1978; 1980). This paradoxical prediction stimulated research into the characteristics of the mortality of long-lived people which added to our knowledge of the biology of life span (Gavrilova and Gavrilov, 1982b). Thus, our basic interest is not in the models themselves, but in testing the predictions which they make. Under such an approach, models are not an aim in themselves, but rather a research technique, and they can supplant each other as our knowledge becomes more precise.

It is necessary to note that this view of mathematical modelling is not only justified for the biology of life span, but also for biology in general. According to Yu.G.Antomonov, "the best guarantee of success in applying the technique of mathematical modelling to biosystems is a dynamic change of models. This does not mean that we should never stick to mathematical models once they have been derived. Rather it is necessary to be mindful of the initial restrictions on which both the physiological arguments and the construction of the model are based. A mathematical model should be investigated to see how its capabilities match the aims for which it was created, and once a model has been derived, it should be subjected to criticism and never made into a dogma for any length of time" (see Antomonov, 1977, p.248). Thus, mathematical modelling ought to begin with the construction of simple models which allow simple questions to be studied, and only subsequently should the transition be made to more sophisticated models which generalise the results and make more precise predictions. This inductive research method is likely to lead more quickly to the creation of a mathematical theory of life span than guesswork which relies on luck or flashes of inspiration.

The need for a critical attitude to mathematical models and methods in the biology of life span is reinforced by a dangerous tendency for this research to be obfuscated with unusual terminology which is unfamiliar to most biologists and doctors. As a result, favourable conditions have arisen for the publication of weak speculative models which have only a semblance of scientific rigour, and the elevation of these models to the status of mathematical theories of life span. What is more, the uncritical use of modern statistical techniques and computers has in a number of cases created a thick biometric fog which hides the spuriousness of the published conclusions.

In this chapter, we have attempted to introduce some clarity into questions concerning the mathematical modelling of survival, the aim being to "demystify" and make plain the real capabilities and limitations of this kind of approach. It is essential that this should be done, since the use of mathematical models in the biology of life span is not a goal in itself, but only a single approach to the study of life span which ought to be comprehensible to all investigators.

### 6.2. The need for a critical attitude to mathematical models of life span

The use of mathematical techniques involves the use of special terms, rules and criteria which are often unknown to the inexpert reader. The reader who finds himself in this situation has the option of either refusing to believe the obtained results, trying to fathom out how they have been obtained, or simply accepting them on faith. In the latter case, a decisive argument is whether the work has been published in a respectable edition which guarantees the high scientific level of its articles by careful refereeing.

In what follows, we shall present some concrete examples which demonstrate that accepting results on faith is the most dangerous of the three options described above. The leading international scientific journals do not seem to be able to provide a reliable screen against the publication of mathematical models of life span which contain blatant errors.

Below we cite several instances which clearly demonstrate the need for a highly critical attitude to the published mathematical models of life span, and how dangerous it is to quote the conclusions contained in such works without carefully checking them.

The first example of such a blatant error relates to an article by the well-known American scientist E.A.Murphy, "Genetics of longevity in man", published by the leading international publishing house, Plenum Press, in the widely-known book, The Genetics of Aging (Murphy, 1978).

In this article, the author proposes a simple and original mathematical model of aging which he calls the "Bingo model" (Murphy, 1978). According to this model, an organism is a system consisting of $k$ subsystems such that the breakdown of any one subsystem leads to the organism's death. It is proposed that each subsystem breaks down after $n$ random instances of damage, and therefore the probability that a subsystem will "survive" to age $x$ can be written in the following way:

$$e^{-x} \left[ 1 + x + \frac{x^2}{2} + \ldots + \frac{x^{n-1}}{(n-1)!} \right] \qquad (54)$$

Correspondingly, the probability of the organism surviving to age $x$ is the above expression raised to degree $k$.

Next, the author differentiates the new expression with respect to age and obtains the following formula for the life span probability density, which describes the distribution of organisms according to their lifetimes:

$$f(x) = ke^{-kx}\left(1+x+\frac{x^2}{2}+\ldots+\frac{x^{n-1}}{(n-1)!}\right)^{k-1}\left(1+x+\ldots+\frac{x^{n-2}}{(n-2)!}\right) \qquad (55)$$

It can easily be seen that a typical student error has been committed in this case: only one of the factors of the product has been differentiated, and the exponential factor has been left unaltered as if it were a constant. If the expression is correctly differentiated, it turns into quite another, simpler expression:

$$f(x) = \frac{kx^{n-1}\,e^{-kx}}{(n-1)!}\left[\,1+x+\frac{x^2}{2}+\ldots+\frac{x^{n-1}}{(n-1)!}\right]^{k-1} \qquad (56)$$

In response to our written request, the author of the model under discussion recognised the fact that an error existed, and explained it as a printing error committed by Plenum Press. Since it is rather more important for the reader to know the correct formula than who is responsible for the strange misprint, this example shows the need for a careful independent check of all the calculations in the published mathematical models.

The second example relates to a mathematical model published in the leading international gerontology journal, Mechanisms of Ageing and Development (Skurnick and Kemeny, 1978a). This model is saturated with mathematical symbols and terms, and is capable of inducing a reaction of shock in most biologists and doctors. Indeed, the article uses the rather complicated and special apparatus of the statistics of extremes, examining different types of limiting distributions of extreme values and the conditions for convergence to one or other limiting distribution.

The authors of this model believe that an organism can be likened to a chain whose strength is determined by the strength of its weakest link. Thus, an organism consisting of $n$ links dies when the first link breaks. It is not difficult to discern a certain resemblance between this model and the already cited "Bingo model" (the $n$ links of the organism in this model correspond to the $k$ subsystems of the organism in the previous model). The basic difference of the new model is the fact that the number $n$ is believed to be so large that it possible to make use of limiting distributions of *minimum* values (since the lifetime of an organism is determined by the *smallest* lifetime of its weakest link). The authors of the model believe that the process which destroys the links is a random sequential process involving the accumulation of damage until it reaches a critical number, equal to $\alpha$. On the basis of these propositions, the authors come to the conclusion that in the initial period of life the force of mortality ought to grow with age according to the power law (the Weibull law), and then in the region of old ages it ought to follow the geometric progression law (the Gompertz law). The justification for the first part of this conclusion is well-known to specialists in the statistics of extremes (see Barlow and Proschan, 1975). However, the second part of the conclusion is erroneous, since in the search for the distribution of *least* values the authors of the model mechanically use the criterion for convergence to the distribution of *greatest* values:

$$lim \frac{1 - F(\log t)}{1 - F(\log mt)} = m^{\alpha} \tag{57}$$

$$t \to +\infty \text{ , for all } m > 0 \text{ and } \alpha \neq 0,$$

where $F\ (t)$ is the function describing the distribution of the lifetimes of the organism's links.

This error is all the more surprising since it is well-known (Barlow and Proschan, 1975; Galambos, 1978) and it is completely obvious even from intuitive considerations that the form of the distribution of *least* values should also be sought in the region of *least* values, rather than with $t \to +\infty$. This error demonstrates that the authors of the model, having conscientiously set out the basic propositions of the statistics of extremes, did not fully understand the essence of the theory. This example also shows that even a leading international journal cannot guarantee that works concerning the mathematical modelling of life span will be competently refereed. Readers who get lost in the jungle of mathematical symbols are recommended to place more reliance on intuition and common sense, which are of some assistance in discovering possible errors in mathematical models of aging. The use of the statistics of extremes in modelling life span is examined in detail in section 6.3 of this chapter.

The publication of incorrect mathematical models of life span is also often encountered in the leading Soviet science journals. For example, the prestigious journal Progress in Modern Biology, as well as six other editions, repeatedly published one and the same mathematical model (Koltover, 1983) which turns out to contain a whole assortment of blatant logical contradictions (see Gavrilov, 1987).

The author of this model (Koltover, 1983) proposes that the death of multi-cellular organisms (including man) is always the result of damage to at least one of the $Q$ blocks of genes in the genome. No account is taken of the fact that the genome is contained in almost every cell of the organism! The consequence is that all the calculations in the mathematical model are based on the hidden assumption that the death of a multi-cellular organism occurs *if and only if* one particular genome (cell) in the entire organism is damaged. Damage to any of the other cells in the organism, according to this model, has no effect on its ability to survive! The stupidity of this hidden assumption must be obvious even to non-specialists. Further, the author of the model allocates to each $j$-th block in the special genome a particular quantity $m_j$ corresponding to its so-called "disfunction". It is postulated that the time of "failure-free" operation of the $j$-th block is given by the formula:

$$T_j = b(m_c - m_j), \tag{58}$$

where $m_c$ represents a critical value of "disfunction" which is the same for all blocks. Thus, it follows from the model that the life span of an organism is determined to an absolute accuracy by its state (the greatest of the values of "disfunction") at the initial moment in time. This deterministic model, in which the length of life is fatally predestined, is presented by its author as a "probabilistic model of aging", and the quantity $m_j$, which grows in a strictly linear deterministic fashion with age, is declared to be a random variable (Koltover, 1983).

Since this "probabilistic" model leads to the simultaneous death of a homogeneous population, its author introduces the hypothesis that each population is heterogeneous with respect to its initial values of "disfunction". This is done in the following manner: "Since complex systems are characterised by an exponential distribution of 'failure-free' operating periods, it has been assumed that the set of values of $m_j$ represents a random sample from a truncated exponential distribution". As a result, the author obtains the desired distribution with an increasing force of mortality, i.e. a distribution describing the life span of aging organisms. The reader is asked to witness a logical "miracle": from the absence of aging (an exponential distribution of "failure-free" operating periods), what is deduced is the existence of aging! However, on closer examination it can be seen that this "miracle" is merely a feeble trick, since the exponential distribution of "failure-free" operating periods and the truncated exponential distribution of values of "disfunction" have nothing in common except the word "exponential". Moreover, it has become clear that this model, which contains mutually exclusive assertions, does not agree with the actual data (Gavrilov, 1984a; 1984c; Gavrilov et al., 1984).

An entertaining example of mathematical hocus pocus is contained in Witten's model (Witten, 1985), which was published in the above-mentioned international journal Mechanisms of Ageing and Development. The author starts from the assumption that the life span distribution of elements in a system is described by an exponential function with parameter $\alpha$:

$$Prob\ [T \leq t] = 1 - \exp[-\alpha t]. \tag{59}$$

Both the text of the article and common sense require that $\alpha > 0$ (otherwise the probability would be negative). After a series of transformations, the Gompertz law is obtained in the form:

$$\mu = h_0 \exp(\gamma t). \tag{60}$$

What is most surprising is the fact that the parameter $\gamma$ is introduced in the following manner:

$$\gamma = -\alpha. \tag{61}$$

It can easily be observed that the parameters $\gamma$ and $\alpha$ cannot be simultaneously positive, although the proposed model would be worthy of attention only if this were the case.

Of course, these examples should not be generalised to all mathematical models of life span, nor should the fruitfulness of mathematical modelling as a whole be put in doubt. The only purpose of this discussion has been to forewarn the reader to adopt a critical attitude towards mathematical models of life span, no matter where they might be published.

## 6.3. Limiting distributions of the life spans of biological systems

One of the promising approaches to modelling the survival of organisms is the use of techniques from the statistics of extremes (Barlow and Proschan, 1975; Galambos, 1978).

The idea behind this approach is simple and obvious. It is a fact that the death of an organism can occur as a result of damage to a wide variety of its subsystems (for example, heart, lungs, brain, kidneys, etc.). Therefore the life span of the whole organism is determined by the life span of the subsystem which is the first to fail. In other words, the length of an organism's life is equal to the shortest "life" of any of its subsystems. Consequently, the life span distribution law for organisms must be sought amongst the class of distributions of the *smallest* extreme values.

The second assumption made by this approach is that the number of vitally important structures in the organism is very large. In favour of this assumption, it is possible to point to the huge variety of diseases and causes of death reflected in multi-volumed medical texts. We are therefore justified in looking for simple *limiting* distributions of the least values of the lifetimes of a huge number of elements.

It turns out that for non-negative values, as is the case with life spans, there are only two types of limiting distributions of least values (Galambos, 1978):

$$L_{2,\gamma}(x) = 1 - \exp(-x^{\gamma}); \qquad x > 0,\ \gamma > 0 \qquad (62)$$

$$L_{3,0}(x) = 1 - \exp(-e^{x}); \qquad x > 0,$$

where $L_{2,\gamma}(x)$ and $L_{3,0}(x)$ are functions describing the life span distribution of organisms (the probability that the length of life will be less than $x$).

The distribution function $L_{2,\gamma}(x)$ is characterised by an increase in the force of mortality with age according to the Weibull law:

$$\mu(x) = \gamma x^{(\gamma-1)}. \tag{63}$$

In the particular case in which $\gamma = 1$, the force of mortality does not depend on age, corresponding to an absence of aging.

The distribution function $L_{3,0}(x)$ is characterised by an increase in the force of mortality with age according to the Gompertz law:

$$\mu(x) = e^{x}. \tag{64}$$

Aging of organisms according to the Weibull law will be observed if the wear and tear on the organism's subsystems satisfies the following condition (Galambos, 1978):

$$\lim_{t \to 0} \frac{F(x_0 + tx)}{F(x_0 + t)} = x^{\gamma}, \tag{65}$$

where $F(x)$ is a function describing the distribution of the lifetimes of an organism's subsystems, and $x_0$ is the age at which $F(x_0) = 0$ ($x_0$ is usually equal to zero). For example, if the subsystems of organisms fail according to the Weibull law, the organisms themselves will die out according to the Weibull law with the same parameter $\gamma$.

The Weibull distribution can also be derived from the following simple model. Suppose that each subsystem of the organism can suffer $\alpha$ occurrences of damage, all random and independent, with a constant intensity $\lambda$. Then the lifetime of an individual subsystem has a gamma-distribution (Barlow and Proschan, 1975):

$$F(x) = 1 - \sum_{i=0}^{\alpha-1} \frac{(\lambda x)^i}{i!} e^{-\lambda x} \qquad (66)$$

In this case, the force of mortality of the organisms grows with age according to the Weibull law, with $\gamma = \alpha$ (Barlow and Proschan, 1975). It can easily be noted that this textbook example was also described in one of the mathematical models of aging discussed above in section 6.2 (Skurnick and Kemeny, 1978a).

Let us examine another more interesting case, in which each subsystem of the organism consists of $n$ identical blocks which are connected in parallel in the context of reliability theory. In other words, the subsystem will function for as long as one block is in working order (an example is the backing-up of one kidney by the other). If the failure rate of the blocks is constant and equal to $\lambda$, the function describing the distribution of a subsystem's lifetime is:

$$F(x) = (1 - e^{-\lambda x})^n. \qquad (67)$$

It can be shown that in this case the force of mortality of the organisms grows with age according to the Weibull law, with $\gamma = n$. For example, if all the subsystems of the organism are backed-up ($n = 2$), the force of mortality grows linearly with age: $\gamma - 1 = 1$.

The Weibull law follows from many other models of the failure of the organism's subsystems. For example, if the lifetime of the subsystems has a standard Planck distribution with a probability density

$$f(x) = \frac{kx^3}{e^x - 1}, \qquad (68)$$

the force of mortality of the organisms grows with the square of the age (Galambos, 1978).

It is significantly more difficult to construct models in which the force of mortality of organisms grows with age according to the Gompertz law. In this case, the distribution function $F(x)$ for the lifetime of an organism's subsystems must satisfy the condition (Barlow and Proschan,

1975):

$$\lim_{x \to x_0} \frac{d}{dx} \frac{F(x)}{f(x)} = 0 \qquad (69)$$

where $f(x) = \left( \frac{d\ F(x)}{dx} \right)$.

In particular, if the organism's subsystems fail according to the Gompertz law, the force of mortality of the organisms themeselves will also grow with age according to the Gompertz law (Barlow and Proschan, 1975).

What is more, the Gompertz law is also obtained whenever the lifetimes of the organism's subsytems follow a normal or lognormal distribution (Galambos, 1978). In both these cases, however, we observe an exceptionally slow convergence to the Gompertz law, requiring a huge number of irreplaceable and vitally important structures in the organism (Galambos, 1978).

Unfortunately, it is impossible to find in the recent literature any descriptions of more meaningful extreme value models which lead to the Gompertz law.

It might seem that the Weibull law, which follows from a wide range of natural assumptions, ought to provide a significantly better description of the survival of organisms than the Gompertz law. However, it has been firmly established that the opposite is the case (Semyonova et al., 1985; Gavrilov and Gavrilova, 1986). Therefore, we have to reject the many plausible models for the destruction of organisms which follow the Weibull law, and pay particular attention to the search for models of destruction which lead to the Gompertz law.

One result of the research we have carried out is the discovery of two new scenarios for the destruction of organisms which lead to the Gompertz-Makeham law. One of these scenarios corresponds to the avalanche-like mechanism of the destruction of the organism, and is described in section 6.4 of this chapter. The second scenario for the destruction of the organism encompasses all the cases described above which lead to the Weibull distribution, but with one very significant correction: the initial state of the organism is not considered ideal (free of defects) as in the previously described cases, but on the contrary, is characterised by a huge number of defects. This small correction turns out to be sufficient for the Gompertz distribution, instead of the Weibull distribution, to be derived from the very simplest models of the destruction of the organism.

It is possible to propose the following meaningful interpretation to justify an assumption of this kind. In the case of technical systems, it is both possible and necessary to check the quality of each element in the system during the assembly process. For this reason, technical systems may indeed initially consist only of functioning elements, and consequently break down according to the Weibull law. However in the case of biological systems, which are formed in a process of self-assembly, the possibility of external quality control of the component parts of the system is practically excluded, and consequently the reliability of a biosystem has to be achieved not by the high quality of its component parts, but by the colossal redundancy in the system permitted by the miniaturisation of its elements. If it is assumed that the probability of an element functioning normally is very small and the number of redundant elements in the system is very large, the number of functioning elements in the system should follow the Poisson distribution. If this assumption is introduced into the models of the destruction of organisms which were described above, it is not difficult to derive the Gompertz law in all the cases which previously led to the Weibull law (see section 6.5 for details).

Turning now to a critical review of the conclusions which can be drawn from the statistics of extremes, we must examine a number of limitations which are inherent in the theory.

One of these limitations is connected with the hypothesis that the failures of an organism's subsystems are statistically independent. Indeed, this hypothesis may not be satisfied when the elements of the system are connected in parallel in the context of reliability theory. In this case, the failure of a certain number of elements can accelerate the wear and tear of the remaining elements as the load on them increases. However this model corresponds to a distribution of *maximum* rather than *minimum* values. A distribution of *minimum* values corresponds to the case in which the subsystems are connected in series in the context of reliability theory, when the failure of any of the subsystems entails the death of the organism. It is not difficult to see that in this case the failure dependence issue ceases to be relevant, since the first failure is equivalent to the collapse of the system. Moreover, further development of the theory of extreme values to include dependent values has shown that taking dependence into account fails in a number of cases to affect the final conclusions (Galambos, 1978).

A second limitation of the theory of extreme values is connected with the assumption that different subsystems have identical lifetime distributions. This hypothesis may in fact not be satisfied in extreme situations, when the primary cause of most cases of death is damage to the most sensitive systems of the organism (for example, in the case of the bone-marrow form of acute radiation sickness). However under normal conditions, because of the continual preferential elimination of individuals with low-reliability subsystems (natural selection), we should observe the evening out of the failure rates of different subsystems of the organism. Moreover, it is worth noting the conclusion of one study (Abernethy, 1979) that differences in the reliability of subsystems in a number of cases do not affect the results obtained using the statistics of the smallest extreme values.

The basic weakness of the statistics of smallest extreme values seems to be that it nevertheless fails to explain the phenomenon of aging, simply reducing the problem of the aging of an organism to the problem of the aging of its subsystems. Clearly, this way of using aging to "explain" aging must lead up a logical blind alley, since moving in succession from the aging of the organism to the aging of organs, tissues and cells, we eventually come to atoms, which are known not to age.

Thus, the key to the explanation of aging is the question: how do we explain the aging of a system constructed out of elements which do not age? From this point of view, systems whose elements are connected in parallel, rather than in series, are of special interest (Gavrilov, 1978; Gavrilov et al., 1978). Of fundamental importance is the assertion that only redundancy in the number of vital structures can lead to the phenomenon of aging (Kozlovsky and Gavrilov, 1983).

Another weakness of the asymptotic theory of extreme values is connected with the assumption that the system contains an infinitely large number of elements. In real systems, the number of elements may be huge, but it is always finite. Of course, the hypothesis that the number of elements is infinite greatly simplifies the analysis of models via obtaining limiting distributions. However in a number of cases this hypothesis leads to conclusions which are inapplicable to real systems. Thus, the asymptotic theory permits the existence of an unaging system constructed of irreplaceable aging elements! For example, in the case when the lifetimes of the elements follow a uniform distribution ($F(x) = x$), a system consisting of an infinite number of such aging elements connected in series will not age (Barlow and Proschan, 1975). The same statement is true in the case:

$$F(x) = 2\Phi(x) - 1, \tag{70}$$

where $\Phi(x)$ is the standard normal distribution function (Galambos, 1978). A system made up of such aging elements connected in series has an exponential life distribution, i.e. it does not age (Galambos, 1978). It is clear that wonders of this kind, contradicting common sense, are only possible for imaginary systems with an infinite number of elements. These conclusions do not apply to real systems consisting of a finite number of elements.

In conclusion, it must be noted that the potentiality of the statistics of extremes for modelling the survival of organisms is at present far from being exhausted. However, this approach has definite limitations which must be known before it can be correctly applied.

### 6.4. The model of the avalanche-like destruction of an organism in natural aging

For want of a nail
the shoe was lost,
For want of a shoe
the horse was lost,
For want of a horse
the rider was lost,
For want of a rider
the battle was lost,
For want of a battle
the kingdom was lost,
And all for the want
of a horseshoe nail.

(English nursery rhyme)

In 1978, we proposed the hypothesis that the aging of organisms is caused by a "cascade of dependent failures" which occurs when one of the organism's systems randomly fails (Gavrilov, 1978; Gavrilov et al., 1978). This idea that an avalanche-like mechanism is involved in the breakdown of an organism during natural aging is worth further development. In fact, it is well-known that defects in an organism have a tendency to multiply following an avalanche-like mechanism: for example, if there are $n$ cancer cells in the organism, each of which is capable of division, the rate at which the organism is transformed into a state with $n + 1$ cancer cells increases with the growth of the number of cancer cells ($n$) already accumulated. Infections of the organism follow similar regularities. The positive feedback between the degree and the rate of an organism's breakdown also follows from the fact that when parts of the structure fail, the load on the remaining structures increases, accelerating the wearing-out. A clear example of this cascade-like destruction of the organism is the development of uncompensated diabetes mellitus: prolonged hyperglycaemia resulting from a relative insulin insufficiency leads to the total exhaustion of the insulin-producing capability of the islet of Langerhans in the pancreas and the conversion of the relative insulin insufficiency into an absolute one. This in its turn leads to renal insufficiency (diabetic glomerulosclerosis) and lesions of the cardiovascular system (diabetic angiopathy). Induced renal failure in its turn induces hypertension, while hypertension may precipitate a fatal stroke. The list of examples of this kind could be continued. It seems that aging is caused by precisely such cascades of dependent failures developing over long periods in a hidden, preclinical form. Therefore mathematical models of the avalanche-like destruction of the organism are of particular interest.

We shall now examine the simplest model of the avalanche-like destruction of the organism. Let $s_0$, $s_1$, $s_2$, ..., $s_n$ ... denote the states of an organism with 0, 1, 2, ..., $n$ ... defects. Let $\lambda_0$ be the background rate at which defects accumulate (the background rate of destruction), independent of the stage of destruction which the organism has reached. Correspondingly, let $\mu_0$ be the background force of mortality. In the simplest case, both these quantities arise from random harmful effects of the external environment. In tandem, there is also an induced rate of destruction and an induced force of mortality which grow as the number of defects increases. At a first approximation, it can be assumed that both the induced rate of destruction and the induced force of mortality are directly proportional to the number of defects, so that for an organism with $n$

defects the induced rate of destruction is equal to $n\lambda$, and the induced force of mortality is $n\mu$.

With these assumptions, we can diagram the avalanche-like destruction of the organism as follows:

$$\begin{array}{ccccccccccc} & \lambda_o & & \lambda_o + \lambda & & \lambda_o + 2\lambda & & \lambda_o + 3\lambda & & & \lambda_o + n\lambda & \\ s_0 & \rightarrow & s_1 & \rightarrow & s_2 & \rightarrow & s_3 & \rightarrow & \ldots & s_n & \rightarrow & \ldots \\ \downarrow & & \downarrow & & \downarrow & & \downarrow & & & \downarrow & & \\ \mu_o & & \mu_o + \mu & & \mu_o + 2\mu & & \mu_o + 3\mu & & & \mu_o + n\mu & & \end{array}$$

This diagram corresponds to a system of differential equations:

$$\frac{ds_0}{dx} = -(\lambda_o + \mu_o)s_0 \qquad (71)$$

$$\frac{ds_1}{dx} = \lambda_o s_0 - (\lambda_o + \mu_o + \lambda + \mu)s_1$$

..................................

$$\frac{ds_n}{dx} = [\lambda_o + (n - 1)\lambda]s_{n-1} - [\lambda_o + \mu_o + n(\lambda + \mu)]s_n$$

..................................................

A similar system of equations (not taking into account the background force of mortality) was obtained and solved in a mathematical model linking the survival of organisms with chromosome damage (Le Bras, 1976).

If at the initial moment in time the number of defects in the organism is equal to zero, the proportion of organisms with 0, 1, 2, ..., $n$ ... defects changes with time according to the following formulae:

$$s_0 = e^{-(\lambda_0+\mu_0)x} \tag{72}$$

$$s_1 = s_0 \left[\frac{\lambda - \lambda e^{-(\lambda+\mu)x}}{\lambda + \mu}\right] \cdot \frac{\lambda_0}{\lambda}$$

$$s_2 = \frac{s_0}{2}\left[\frac{\lambda - \lambda e^{-(\lambda+\mu)x}}{\lambda + \mu}\right]^2 \cdot \frac{\lambda_0}{\lambda}\left(\frac{\lambda_0}{\lambda} + 1\right)$$

.......................................

$$s_n = \frac{s_0}{n!}\left[\frac{\lambda - \lambda e^{-(\lambda+\mu)x}}{\lambda + \mu}\right]^n \cdot \frac{\lambda_0}{\lambda}\left(\frac{\lambda_0}{\lambda} + 1\right)\left(\frac{\lambda_0}{\lambda} + 2\right)\cdots\left(\frac{\lambda_0}{\lambda} + (n-1)\right)$$

..................................................

If the number of defects is allowed to grow unrestrictedly, the relationship between the number of survivors and age is determined as follows:

$$l(x) = \sum_{n=0}^{\infty} s_n = s_0\left(1 + kz + \frac{k(k+1)}{2!} \cdot z^2 + \frac{k(k+1)(k+2)}{3!} \cdot z^3\right.$$

$$\left. + \ldots\right) = s_0(1 - z)^{-k} , \tag{73}$$

where $k = \dfrac{\lambda_0}{\lambda}$, and $z = \dfrac{\lambda - \lambda e^{-(\lambda+\mu)x}}{\lambda + \mu}$ .

Returning to the initial variables, we obtain:

$$l(x) = s_0(1 - z)^{-k} = e^{-(\lambda_0+\mu_0)x}\left(\frac{\lambda + \mu}{\mu + \lambda e^{-(\lambda+\mu)x}}\right)^{\frac{\lambda_0}{\lambda}} . \qquad (74)$$

The force of mortality is accordingly:

$$\mu(x) = -\frac{dl(x)}{l(x)dx} = \mu_0 + \frac{\mu\lambda_0(1-e^{-(\lambda+\mu)x})}{\mu + \lambda e^{-(\lambda+\mu)x}} . \qquad (75)$$

In the particular case when the rate at which defects multiply is significantly greater than the induced force of mortality ($\lambda >> \mu$), the growth in the force of mortality during the initial stage (with low values of $x$) is described by the Gompertz-Makeham law:

$$\mu(x) \approx \mu_0 + \frac{\mu\lambda_0(1 - e^{-(\lambda+\mu)x})}{\lambda e^{-(\lambda+\mu)x}} \approx A + Re^{\alpha x}, \qquad (76)$$

where $A = \mu_0 - \dfrac{\mu\lambda_0}{\lambda}$ ; $R = \dfrac{\mu\lambda_0}{\lambda}$ ; $\alpha = \lambda + \mu$ .

This model of the avalanche-like destruction of the organism not only provides a theoretical justification for the well-known Gompertz-Makeham law, but also explains why the values of parameter $A$ sometimes turn out to be negative. Within the framework of the model, parameter $A$ represents the background force of mortality ($\mu_0$) minus a quantity equal to $R$ (see above). Therefore, if the background force of mortality $\mu_0$ is small (as for populations in the developed countries and populations of laboratory animals), and the value of parameter $R$ is significant (when the background rate of destruction $\lambda_0$ is large), parameter $A$ can be negative.

Another important virtue of the avalanche-like destruction model is that it correctly predicts deviations from the Gompertz-Makeham law towards lesser values of mortality at very old ages. In this extreme age-range, the force of mortality grows according to the formula:

$$\mu(x) \approx \mu_0 + \lambda_0(1 - e^{-(\lambda+\mu)x}). \qquad (77)$$

Thus the model predicts an asymptotic growth in the force of mortality with an upper limit of $\mu_0+\lambda_0$. Putting this conclusion together with data on the survival of long-lived people, it can be determined that for human beings $\lambda_0$ is approximately 0.5 - 1.0 $\text{year}^{-1}$. The value of $\lambda$ for human beings is 0.05-0.12 $\text{year}^{-1}$ and $\mu = 10^{-6} - 10^{-7}$ $\text{year}^{-1}$ (these estimates are based on typical values of $R$ and $\alpha$ for human beings).

Alongside the virtues already listed, the avalanche-like destruction model has one significant weakness: it does not conform to the well-known phenomenon known as the compensation effect of mortality (see section 4.5). Although the model predicts an inverse relationship between the parameters $R$ and $\alpha$ as parameter $\lambda$ (the rate of multiplication of defects) is varied, this relationship takes a rather different form:

$$\ln R = \ln(\mu\lambda_0) - \ln(\alpha-\mu). \qquad (78)$$

Thus, this particular version of the avalanche-like destruction model unfortunately does not agree quantitatively with the compensation effect of mortality, and consequently needs to be significantly improved. Nevertheless, in our opinion the idea that organisms undergo cascade destruction is one of the most promising ideas in the mathematical modelling of life span.

## 6.5. The model of a multiply redundant system saturated with defects

It will be shown in this section that the exponential growth in the force of mortality of organisms, as well as other mortality phenomena, follow naturally from a very simple reliability model and a number of fundamental features of biosystems. The first fundamental feature of biosystems is the fact that, in distinction to technical (artificial) devices which are constructed out of previously manufactured and *tested* components, organisms form themselves in ontogenesis through a process of self-assembly out of de novo forming and *untested* elements (cells). The second property of organisms is the high degree of miniaturisation of their constituent elements (the microscopic dimensions of cells, as well as the molecular dimensions of information carriers like DNA and RNA), permitting the creation of a massive redundancy in the number of elements. Thus, it is to be expected that for living organisms, in distinction to many technical (artificial) devices, the reliability of the system is achieved not by the high initial quality of all the elements, but by their huge numbers. As will be shown below, it is this feature of organisms which provides an explanation why the force of mortality grows as an exponential rather than a power function of age, and which also enables us to understand the other mortality phenomena.

Let us examine in greater detail the reasons why significant differences exist between technical devices and biological systems. The fundamental difference in the manner the system is formed (assembly in the case of technical devices and self-assembly in the case of biosystems) has two important consequences. Firstly, it leads to the macroscopicity of technical devices in comparison with biosystems, since technical devices are assembled "top-down" with the participation of a macroscopic system (man), and have to be suitable for this macroscopic system to use (i.e. commensurate with man). Biosystems, on the other hand, are assembled "bottom-up" from molecules and cells, resulting in an exceptionally high degree of miniaturisation of the component parts. Secondly, since technical devices are assembled under the control of man, the possibilities for a preliminary check of the components are incomparably greater than in the self-assembly of biosystems, and this inevitably leads to biosystems being "littered" with a greater number of defective elements. Consequently, the reliability of technical devices can be assured by high quality components in strictly controlled numbers, while the reliability of biosystems can be more easily assured by an exceptionally high degree of redundancy. It must also be noted that the individual uniqueness of biosystems, which biologists take such a delight in, may be based on the saturation of the system with defects forming a unique pattern of individual damage. Attempts to make use of dilapidated computer equipment give rise to the same thought: the behaviour of this equipment can only be described by resorting to such "human" concepts as character, eccentricity, and change of mood. As will be shown below, ideas of this kind can be successfully used in constructing a mathematical model of life span.

In order to gain a better understanding of the proposed model, we shall first examine the special case in which all the elements of the system are initially functional and have a constant failure rate $k$. If these unaging elements are organised into blocks of $n$ mutually substitutable elements so that the failure of a block occurs only when all the elements of the block fail (parallel construction in the reliability theory context), the failure rate of block $\mu_b(n,k,x)$ depends on age $x$ in the following way:

$$\mu_b(n,k,x) = \frac{nke^{-kx}[1 - e^{-kx}]^{n-1}}{1 - [1 - e^{-kx}]^{n}} \tag{79}$$

$$\approx nk^n x^{n-1} \quad \text{when } x \ll 1/k$$

$$\approx k \quad \text{when } x \gg 1/k .$$

Thus, the failure rate of a block at first grows as a power function of age (the Weibull law), the tempo at which the failure rate grows then drops, and the failure rate tends asymptotically towards an upper limit equal to $k$. At this stage, it is worth paying attention to two significant points. Firstly, a block constructed out of unaging elements is now behaving like an aging object, i.e. *aging is a direct consequence of the redundancy* of the system (redundancy in the number of elements). Secondly, at very high ages the phenomenon of aging disappears at the same time as redundancy in the number of elements: the failure rate approaches an upper limit which is totally independent of the original number of elements, but coincides with their constant failure rate ($k$).

If in its turn a system is constructed out of $m$ irreplaceable blocks in such a way that the failure of any of the blocks leads to the failure of the whole system (series construction in the reliability theory context), the failure rate of the system is equal to the sum of the failure rates of all the blocks:

$$\mu_s(x) = \sum \mu_b = m\mu_b = \frac{mkne^{-kx}[1 - e^{-kx}]^{n-1}}{1 - [1 - e^{-kx}]^n} \tag{80}$$

$$\approx mnk^n x^{n-1} \quad \text{when } x \ll 1/k$$

$$\approx mk \quad \text{when } x \gg 1/k \ .$$

The reliability model discussed above leads to the failure rate growing as a power function rather than as an exponential function of age. Therefore this kind of model for a long time seemed incapable of describing the exponential growth of the force of mortality in biological systems, and attention was basically drawn towards the search for more complex destruction scenarios like the avalanche-like destruction model described in section 6.4.

In this section, we shall show that the reliability model presented above was undeservedly rejected merely because it started by analysing initially ideal structures in which all the elements are functional from the outset. As has already been noted, this assumptiom may be justified for technical devices, but it is not justified for living organisms saturated with defects. It is therefore necessary to examine a further particular case of the model in which initially functioning elements occur very rarely with a low probability $q$. In this case, the distribution of the blocks in the organism according to the number $i$ of initially functioning elements they contain is described by the Poisson law with parameter $\lambda = nq$, corresponding to the mean number of initially functioning elements in a block. Strictly speaking, this distribution ought to be truncated on the right, since the number of functioning elements ($i$) cannot exceed the total number ($n$) of elements in a block. In addition, for initially living organisms the distribution ought also to be truncated on the left, since according to the model an organism which contains a block without any functioning elements ($i = 0$) cannot be alive. Therefore the distribution of blocks according to the number $i$ of initially functioning elements within initially living organisms is determined by the following probabilities $P_i$:

$$P_i = 0, \qquad \text{for } i = 0,\ n+1,\ n+2,\ n+3 \ldots$$

$$P_i = c\, e^{-\lambda} \frac{(\lambda)^i}{i!}, \qquad \text{for } i = 1,\ 2,\ 3 \ldots n;$$

$$c = \frac{1}{1 - e^{-\lambda} - e^{-\lambda} \sum_{i=n+1}^{\infty} \frac{(\lambda)^i}{i!}} \tag{81}$$

where $c$ is a normalising factor which ensures that the sum of the probabilities of all the events is equal to unity. For sufficiently high values of $n$ and $\lambda$, the normalising factor turns out to be hardly greater than unity.

As has already been noted, the failure rate of a system constructed out of $m$ blocks connected in series is equal to the sum of the failure rates of these blocks $\mu_b(i)$:

$$\mu_s = \sum \mu_b(i) = \sum_{i=1}^{n} m\, P_i\, \mu_b(i) = mce^{-\lambda} \sum_{i=1}^{n} \frac{(\lambda)^i\, \mu_b(i)}{i!} \tag{82}$$

In its turn, the failure rate of blocks with $i$ initially functioning elements is given by:

$$\mu_b(i) = \frac{i\, k\, e^{-kx} [1 - e^{-kx}]^{i-1}}{1 - [1 - e^{-kx}]^i} \tag{83}$$

$$\approx \quad ik(kx)^{i-1} \qquad \text{when} \quad x \ll 1/k$$

$$\approx \quad k \qquad \text{when} \quad x \gg 1/k$$

Putting together these two formulae, we obtain:

$$\mu_s(x) \approx k\lambda mce^{-\lambda} \sum_{i=1}^{n} \frac{(\lambda kx)^{i-1}}{(i-1)!} = R(e^{\alpha x} - \varepsilon(x)) \approx Re^{\alpha x}$$

when $x \ll 1/k$, where $R = cm\lambda ke^{-\lambda}$

$$\alpha = \lambda k \tag{84}$$

$$\varepsilon(x) = \sum_{i=n+1}^{\infty} \frac{(\lambda kx)^{i-1}}{(i-1)!} \approx 0 \text{ for large } n$$

$\mu_s(x) \approx mk$ when $x \gg 1/k$

Thus, the force of mortality of an organism at first (for $x \ll 1/k$) grows exponentially with age (following the Gompertz law), then the rate of the growth in the force of mortality slackens and the force of mortality tends asymptotically to an upper limit equal to $mk$. If there exists a background component of mortality ($A$), we arrive at the well-known Gompertz-Makeham law.

The proposed model can explain not only the exponential increase in the force of mortality with age and the subsequent slowing in the rate of this increase, but also the compensation effect of mortality described in section 4.5.

Indeed, according to the definitions given:

$$\ln R = \ln(k\lambda cm) - \lambda \tag{85}$$

$$\alpha = \lambda k$$

i.e. the quantities $\ln R$ and $\alpha$ are parametrically linked via the parameter $\lambda$, which allows us to represent $\ln R$ as a function of $\alpha$:

$$\ln R = \ln(cm\alpha) - \frac{\alpha}{k} \qquad (86)$$

Thus, the compensation effect of mortality is observed whenever differences in mortality arise from interpopulation differences in parameter $\lambda$ (the mean number of initially functional elements in the block), and the rate of aging (the rate of irreversible breakdown of elements $k$) is practically identical for all the compared populations of a single species. In this case, the magnitude of the species-specific life span ($1/k$) turns out to be equal to the mean lifetime of the elements.

It also follows from the proposed model that even small progress in optimising the processes of ontogenesis and an increase in the probability $q$ can lead to a sharp fall in mortality and an increase in the life span of organisms. Experimental evidence in support of this prediction may be found in section 5.3 of this book.

Another consequence of the proposed model is the prediction that most of the elements in the system (for example, 99%) are initially non-functioning. It must however be noted that the testing and refutation of this prediction would not entail the incorrectness of the model, since most non-functioning elements may have died by the time the adult organism is formed. In fact, the model is based on the hypothesis that the number of functioning elements in the blocks is described by the Poisson distribution, and the fate of non-functioning elements and their death in no way affects the conclusions of the model. It is also important to note that the conclusions of the model are valid for any value of parameter $\lambda$, no matter how large. This is significant, since it is well-known that as parameter $\lambda$ increases, the Poisson distribution approximates to the normal distribution (Feller, 1968). Thus, the proposed model in fact encompasses a wide spectrum of distributions of blocks according to their degree of redundancy, starting with a marked positive (right-sided) asymmetry (for low values of $\lambda$) and ending with distributions which are close to the symmetrical normal distribution (for large values of $\lambda$). In other words, the proposed model might also be called the model of series-connected blocks with varying degrees of redundancy. This rather general model will be generalised even further in the next section.

## 6.6. The model of a redundant system with an arbitrary number of defects

In the preceding section, we examined a reliability model for a system consisting of $m$ series-connected blocks, each of which contains $n$ parallel-connected elements. It was shown that the behaviour of such a system depends decisively on the initial conditions. If the system is initially ideal, i.e. if the probability $q$ that the elements are initially functional is unity, the model leads to a power function for the growth in the force of mortality (the Weibull law). On the other hand, if the system is from the very outset saturated with defects, and the probability that the elements are initially functional is close to zero, the model leads to an exponential function for the growth in the force of mortality (the Gompertz law). In both cases, however, there exists an upper limit to the growth in the force of mortality which is determined by the product of the number of blocks ($m$) and the failure rate of the elements ($k$). In this section, we shall examine the more general case in which the probability of an element being initially functional can assume any positive value: $0 < q \leq 1$.

In the general case, the distribution of blocks in the organism according to the number of initially functioning elements is described by the binomial law rather than the Poisson law. For an initially living organism, this distribution has to be truncated on the left, since according to the model an organism which contains a block without any functioning elements ($i = 0$) cannot be alive.

Therefore, the distribution of blocks according to the number $i$ of initially functioning elements within initially living organisms is given by the following probabilities $P_i$:

$$P_i = 0 \qquad \text{for } i = 0 \tag{87}$$

$$P = c\binom{n}{i} q^i (1-q)^{n-i} \qquad \text{for } i = 1,2,3 \ldots n$$

$$\binom{n}{i} = \frac{n!}{i!(n-i)!} = \frac{n}{i}\binom{n-1}{i-1} \qquad \text{for } i = 1,2,3 \ldots n$$

$$c = \frac{1}{1 - (1-q)^n} \geq 1,$$

where $c$ is a normalising factor which ensures that the sum of the probabilities of all the events is unity.

The failure rate of a system constructed of $m$ series-connected blocks is equal to the sum of the failure rates of the blocks:

$$\mu_s = \sum \mu_b(i) = \sum_{i=1}^{n} m\, P_i\, \mu_b(i) = cm \sum \binom{n}{i} q^i (1-q)^{n-i} \mu_b(i) \quad (88)$$

As has been already noted, at the initial moment in time, when $x << 1/k$, the failure rate of a block with $i$ initially functioning elements is given by:

$$\mu_b(i) \approx ik(kx)^{i-1} \quad (89)$$

Putting together these two formulae, we obtain:

$$\mu_s \approx cm \sum_{i=1}^{n} \binom{n}{i} q^i (1-q)^{n-i} ik(kx)^{i-1} \quad (90)$$

Since $i\binom{n}{i} = n\binom{n-1}{i-1}$ for $i = 1,2,3 \ldots n$, we obtain:

$$\mu_s \approx cmknq \sum_{i=1}^{n} \binom{n-1}{i-1} (qkx)^{i-1} (1-q)^{(n-1)-(i-1)} \quad (91)$$

The sum represented in this expression is the Newtonian binomial for the expression $[(1-q) + qkx]^{n-1}$. It is therefore possible to write:

$$\mu_s \approx cmn(qk)^n[\frac{1-q}{qk} + x]^{n-1} = cmn(qk)^n(x_0 + x)^{n-1} \quad (92)$$

where $x_0 = \frac{1-q}{qk}$, a parameter which we propose to call the *initial virtual age of the system*. Indeed, this parameter has the dimension of time, and corresponds to the age by which an initially ideal system would have accumulated as many defects as a real system already possesses at the initial moment in time (at $x = 0$). In particular, when $q = 1$, i.e. when all the elements are functional at the outset, the initial virtual age of the system is zero and the force of mortality grows as a power function of age (the Weibull law), this being the case described in section 6.5. However when the system is not initially ideal ($q < 1$), we arrive at the so-called *binomial law of mortality*, first proposed in 1986 (Gavrilov and Gavrilova, 1986) and mentioned in section 2.4 of the present book. Thus, the generalised reliability model proposed here is a theoretical justification for the binomial law of mortality.

In the case when $x_0 > 0$, there is always an initial period of time such that $x \ll x_0$ and the following approximation to the binomial law is valid:

$$\mu_s \approx cmn(qk)^n \, x_0^{\,n-1}(1 + \frac{x}{x_0})^{n-1} \approx cmn(qk)^n \, x_0^{\,n-1} \exp[(\frac{n-1}{x_0})x] \quad (93)$$

Consequently, for any value of $q < 1$ there is always a period of time $x$ when the number of newly formed defects is much less than the original number, and the force of mortality grows exponentially with age:

$$\mu_s = R\, e^{\alpha x} \quad (94)$$

where $R = cmn(qk)^n \, x_0^{\,n-1} = cmnqk(1-q)^{n-1}$

$$\alpha = \frac{n-1}{x_0} = \frac{kq(n-1)}{1-q}$$

So, *if the system is not initially ideal ($q<1$), the force of mortality in the initial period of time grows exponentially with age (according to the Gompertz law).*

The model under discussion not only provides an explanation for the exponential increase in the force of mortality with age, but also for the compensation effect of mortality described in section 4.5. Indeed, according to the above formulae:

$$\ln R = \ln(cmnqk) - (n-1)\ln\left(\frac{1}{1-q}\right) \tag{95}$$

$$\alpha = \frac{kq}{1-q}(n-1),$$

i.e. the quantities $\ln R$ and $\alpha$ are parametrically linked via the quantity $(n-1)$, allowing $\ln R$ to be represented as a function of $\alpha$:

$$\ln R = \ln[cm\alpha(1-q) + cmkq] - \frac{\alpha(1-q)}{kq}\ln\left(\frac{1}{1-q}\right) \tag{96}$$

Thus, the compensation effect of mortality is observed whenever differences in mortality are brought about by interpopulation differences in the number of elements in a block ($n$), while the other parameters, including the rate of aging (the rate of irreversible element failure $k$) are practically identical for all compared populations of a single species. It is not difficult to see the similarity between this explanation for the compensation effect of mortality and the explanation which emerges from the preceding model described in section 6.5.

It must be noted that the length of the period when the force of mortality grows exponentially depends on the value of $q$. In the general case, the $q$-dependent behaviour of the system in the age interval $0 < x \ll 1/k$ can be reduced to the following three scenarios:

1. $0 < q \leq 1/2$. This case corresponds to the situation in which less than half the total number of elements are

initially functional. In this case $\frac{1-q}{q} \geq 1$, and consequently $x_0 \geq 1/k$. Therefore over the entire interval, when $x \ll 1/k$, the condition $x \ll x_0$ is also fulfilled. In this case, the force of mortality grows exponentially with age throughout the interval under consideration.

2. $1/2 < q < 1$. This case corresponds to the situation in which more than half of all elements are initially functional. In this case $\frac{1-q}{q} < 1$ and $x_0 < 1/k$. In these circumstances, the age-dependence of mortality in the age interval under consideration ($0 < x \ll 1/k$) consists of two stages:

(a) the first stage of the initial period, when $x \ll x_0$ and consequently the binomial law of mortality reduces to the Gompertz law.

(b) the second stage of the initial period, when $x \approx x_0$ and only the binomial law of mortality in its full form is valid, without any approximations.

3. $1/2 \ll q < 1$. This case corresponds to the situation only a small proportion of elements are initially defective. In this case $\frac{1-q}{q} \ll 1$ and $x_0 \ll 1/k$. The age-dependence of mortality then consists of three stages:

(a) the first stage of the initial period, when $x \ll x_0$ and the binomial law of mortality reduces to the Gompertz law.

(b) the second stage of the initial period, when $x \approx x_0$ and only the binomial law of mortality is applicable.

(c) the third stage of the initial period, when $x_0 \ll x \ll 1/k$ and the binomial law of mortality reduces to the power law for the growth in the force of mortality (the Weibull law).

As $q$ tends to unity, the length of the first stage of the initial period, with exponential growth in the force of mortality, is sharply reduced, and the length of the third stage is sharply increased. At the limit, in the case of an initially ideal system ($q = 1$), we arrive at the Weibull law, valid over the entire age interval $0 < x \ll 1/k$. This case has already been described in section 6.5.

As has already been noted, with increased age the failure rate of the blocks tends asymptotically to an upper limit independent of the number of initially functioning elements, and equal to $k$. Therefore the force of mortality of a system consisting of $m$ blocks in series tends asymptotically with increased age to an upper limit $mk$, independently of the values of $n$ and $q$.

Thus the reliability model under consideration provides an explanation for the basic regularities of the mortality of organisms: the exponential growth of the force of mortality in the initial period, with the subsequent slowing down of the rate of its increase, as well as the compensation effect of mortality. In addition, the model under consideration clarifies the conditions under which we observe not the exponential law, but the power law for the growth in the force of mortality (the Weibull law). Finally, the proposed model allows us to treat two at first sight mutually exclusive laws, the Gompertz law and the Weibull law, as special cases of a more general binomial law of mortality for which the model provides a rigorous theoretical justification.

According to the proposed model, the fate of non-functioning elements and their death has no effect whatsoever on the model's conclusions. Therefore the model will be valid even when all the non-functioning elements have already died by the time the adult organism has been formed, and the adult organism consists only of functioning elements (cells). What is important is that a *trace* nevertheless remains in the form of the binomial distribution of blocks according to the number of functioning elements within the organism. In fact, this is the essence of the model, and considerations of initially defective elements are only one of the possible explanations for the existence of variability in the degree of redundancy. For this reason, the proposed model might also be called *the model of series-connected blocks with varying degrees of redundancy*.

Bearing in mind these observations, the basic conclusion of the model might be reformulated as follows: *"If the vitally important blocks of a system differ in their degree of redundancy, the force of mortality in the initial period grows exponentially with age (according to the Gompertz law)"*. This assertion is valid regardless of the shape of the binomial distribution of blocks in the organism according to their degree of redundancy: whether there is negative (left-sided) asymmetry, total symmetry (a normal distribution), or positive (right-sided) asymmetry. The only effect of the shape of the distribution is that in the case of a negative (left-sided) asymmetry of the distribution the exponential growth in the force of mortality may last only a short time, while in the case of a symmetrical distribution or a distribution with a positive asymmetry the period of exponential growth in the force of mortality is significantly longer.

The proposed model also resolves the contradiction between exponential growth in the force of total mortality and power growth in the force of mortality from particular diseases. Indeed, the classification of diseases and causes of death is largely based on the anatomical principle and registration of the damage to individual organs (blocks) of the organism. One of the consequences of the model is that each block fails according to the power law, and this in turn leads to exponential growth in the force of total mortality. It must however be noted that this explanation, like the model itself, presupposes the initial homogeneity of organisms according to the number of blocks they contain with different degrees of redundancy. It is therefore worth analysing an initially heterogeneous population, and this will be the subject of the next section.

## 6.7. The heterogeneous population model

In the two preceding sections, we examined a situation in which blocks have varying degrees of redundancy within each organism, while the organisms themselves are considered to be initially identical to each other and to have an equal risk of death. This latter assumption can be justified in two cases.

Firstly, population homogeneity can be observed when there is a rigid genetic program which determines the initial degree of redundancy for each block (organ) in the organism. In other words, it is possible that the mechanism for generating variability of blocks is not at all random, and the models presented in the two preceding sections are applicable only because the genetically programmed distribution of blocks according to their degree of redundancy can be approximated by the Poisson distribution or the binomial distribution. We note that in this case the population of organisms is initially homogeneous with respect to the risk of failure of particular blocks (homogeneity with respect to mortality from individual causes).

Secondly, organisms may have practically identical distributions of the number of blocks with different degrees of redundancy even when the redundancy formation mechanism is random, provided that the number $m$ of irreplaceable (vital) blocks is very large. As a result, the population will be practically homogeneous with respect to the risk of each organism dying, despite the fact that the population may be exceptionally heterogeneous with respect to the risk of failure of particular blocks (corresponding to individual causes of death). If the number of irreplaceable (vital) blocks in the organism is comparatively small, or if we are interested in generalisations about mortality resulting from the failure of particular blocks (modelling mortality from individual causes), it becomes an urgent necessity to take into account the heterogeneity of the population.

In this section, we shall demonstrate that taking into account the heterogeneity of the population provides an explanation for all the basic laws of mortality, even in the simplest case when the organism consists of a single vital block with $n$ elements.

Let $q$ be the probability that an element is initially functional. Then the probability of encountering an organism with $i$ initially functional elements out of a total number $n$ of elements is given by the binomial distribution law:

$$P_i = \binom{n}{i} q^i (1-q)^{n-i} \tag{97}$$

The probability of an organism with $i$ initially functioning parallel elements dying before age $x$ is given by:

$$F_i(x) = (1 - e^{-kx})^i \tag{98}$$

where $k$ is constant failure rate of the elements.

The lifetime distribution function $F(x)$ for a mixture of organisms with $i$ = 0,1,2, ... $n$ initially functioning elements is determined as follows:

$$F(x) = \sum_{i=0}^{n} P_i\ F_i(x) = \sum_{i=0}^{n} \binom{n}{i}(1-q)^{n-i}\ q^i(1-e^{-kx})^i$$

$$= [(1-q) + q(1-e^{-kx})]^n = [1 - qe^{-kx}]^n \quad (99)$$

The force of mortality in this case is:

$$\mu(x) = \frac{nqke^{-kx}[1 - qe^{-kx}]^{n-1}}{1 - [1 - qe^{-kx}]^n} \approx cnqk[1-q+qkx]^{n-1} \text{ for } x \ll 1/k$$

$$\text{where } c = \frac{1}{1 - [1-q]^n} \geq 1$$

$$\approx k \qquad \text{for } x \gg 1/k \quad (100)$$

Thus the force of mortality in a heterogeneous population of organisms at first grows with age according to the binomial law of mortality, then asymptotically tends towards an upper limit $k$:

$$\mu(x) \approx cn(qk)^n[\frac{1-q}{qk} + x]^{n-1} = cn(qk)^n[x_0 + x]^{n-1} \text{ for } x \ll 1/k$$

$$\mu(x) \approx k \qquad \text{for } x \gg 1/k \quad (101)$$

where $x_0 = \frac{1-q}{qk}$, a parameter which we propose to call the *initial virtual age of the population*. This parameter has the dimension of time, and corresponds to the age by which an initially homogeneous population would have accumulated as many defective organisms as a real population actually possesses at the initial moment in time (at $x = 0$). In particular, when $q = 1$, i.e. when all the elements in each organism are functional at the outset, the initial virtual age of the population is zero and the force of mortality grows as a power function of age (the Weibull law), this being the case described in section 6.5. However when the population is not initially homogeneous ($q < 1$), we arrive at the already mentioned binomial law of mortality. Thus, the heterogeneous population model proposed here can also provide a theoretical justification for the binomial law of mortality.

In the case when $x_0 > 0$, there is always an initial period of time such that $x \ll x_0$ and the following approximation to the binomial law is valid:

$$\mu(x) \approx cn(qk)^n \, x_0^{\,n-1}\left(1 + \frac{x}{x_0}\right)^{n-1} \approx cn(qk)^n \, x_0^{\,n-1} \exp\left[\left(\frac{n-1}{x_0}\right)x\right] \tag{102}$$

Consequently, for any value of $q < 1$ there always exists an initial period of time $x$ when the force of mortality grows exponentially with age:

$$\mu(x) = R\, e^{\alpha x}, \tag{103}$$

$$\text{where } R = cn(qk)^n \, x_0^{\,n-1} = cnqk(1-q)^{n-1}$$

$$\alpha = \frac{n-1}{x_0} = \frac{kq(n-1)}{1-q}$$

So, *if a population is initially heterogeneous ($q < 1$), the force of mortality in the initial period of time grows exponentially with age (according to the Gompertz law).*

It is easy to see the striking similarity between the formulae and comments in this section and the preceding section of the book. This similarity is the result of two fundamentally different models nevertheless leading to virtually coincident final conclusions. It must however be noted that we are only dealing with a superficial similarity in behaviour between two fundamentally different and competing models, each of which deserves careful analysis despite the the striking similarity of its rival's behaviour. We should not therefore be surprised if the further analysis of the heterogeneous population model resembles the analysis of the model of series-connected blocks with varying degrees of redundancy undertaken in the preceding section of the book.

The heterogeneous population model not only provides an explanation for the exponential growth in the force of mortality with age, but also the compensation effect of mortality described in section 4.5. Indeed, according to the above formulae:

$$\ln R = \ln(cnqk) - (n-1)\ln\left(\frac{1}{1-q}\right) \qquad (104)$$

$$\alpha = \frac{kq}{1-q}(n-1),$$

i.e. the quantities $\ln R$ and $\alpha$ are parametrically linked via the quantity $(n-1)$, allowing $\ln R$ to be represented as a function of $\alpha$:

$$\ln R = \ln[c\alpha(1-q) + ckq] - \frac{\alpha(1-q)}{kq}\ln\left(\frac{1}{1-q}\right) \qquad (105)$$

Thus, the compensation effect of mortality is observed whenever differences in mortality are brought about by interpopulation differences in the number of elements in the organism ($n$), while the other parameters, including the rate of aging (the rate of irreversible element failure $k$) are practically identical for all compared populations of a single species. It is not difficult to see the similarity between this explanation for the compensation effect of mortality and the explanations which emerge from the preceding models described in sections 6.5 and 6.6.

As has already been noted in section 6.6, the length of the period when the force of mortality grows exponentially depends on the value of $q$. In the general case, the age dependence of the mortality of the population in the age interval $0 < x \ll 1/k$ can be reduced to the following three scenarios, depending on the value of $q$:

1. $0 < q \leq 1/2$. This case corresponds to the situation in which less than half the total number of elements are initially functional. In this case $\frac{1-q}{q} \geq 1$, and consequently $x_0 \geq 1/k$. Therefore over the entire interval, when $x \ll 1/k$, the condition $x \ll x_0$ is also fulfilled. In

this case, the force of mortality grows exponentially with age throughout the interval under consideration.

2. $1/2 < q < 1$. This case corresponds to the situation in which more than half of all elements are initially functional. In this case $\frac{1-q}{q} < 1$ and $x_0 < 1/k$. In these circumstances, the age-dependence of mortality in the age interval under consideration ($0 < x \ll 1/k$) consists of two stages:

(a) the first stage of the initial period, when $x \ll x_0$ and consequently the binomial law of mortality reduces to the Gompertz law.

(b) the second stage of the initial period, when $x \approx x_0$ and only the binomial law of mortality in its full form is valid, without any approximations.

3. $1/2 \ll q < 1$. This case corresponds to the situation when only a small proportion of elements are initially defective. In this case $\frac{1-q}{q} \ll 1$ and $x_0 \ll 1/k$. The age-dependence of mortality then consists of three stages:

(a) the first stage of the initial period, when $x \ll x_0$ and the binomial law of mortality reduces to the Gompertz law.

(b) the second stage of the initial period, when $x \approx x_0$ and only the binomial law of mortality is applicable.

(c) the third stage of the initial period, when $x_0 \ll x \ll 1/k$ and the binomial law of mortality reduces to the power law for the growth in the force of mortality (the Weibull law).

As $q$ tends to unity, the length of the first stage of the initial period, with exponential growth in the force of mortality, is sharply reduced, and the length of the third stage is sharply increased. At the limit, in the case of an initially ideal system ($q = 1$), we arrive at the Weibull law, valid over the entire age interval $0 < x \ll 1/k$.

Thus the heterogeneous population model provides an explanation for the basic regularities of the mortality of organisms: the exponential growth of the force of mortality in the initial period, with the subsequent slowing down of the rate of increase, as well as the compensation effect of mortality. In addition, the model under consideration clarifies the conditions under which we observe not the exponential law, but the power law for the growth in the force of mortality (the Weibull law). Finally, the proposed model allows us to treat two at first sight mutually exclusive laws, the Gompertz law and the Weibull law, as special cases of a more general binomial law of mortality for which the model provides a rigorous theoretical justification.

According to the heterogeneous population model, the fate of non-functioning elements and their death has no effect whatsoever on the model's conclusions. Therefore the model will be valid even when all the non-functioning elements have already died by the time the adult organism has been formed, and the adult organism consists only of functioning elements (cells). What is important is that a *trace* nevertheless remains in the form of the binomial distribution of organisms in the population according to the number of initially functioning elements. In fact, this is the essence of the model, and considerations of initially defective elements are only one of the possible explanations for the existence of variability between organisms in the degree of redundancy. For this reason, the proposed model might also be called the *model of populations of organisms with varying degrees of redundancy*.

Bearing in mind these observations, the basic conclusion of the model might be reformulated as follows: *"If the individuals in the population vary in their degree of redundancy, the force of mortality in the initial period grows exponentially with age (according to the Gompertz law)"*. This assertion is valid regardless of the shape of the binomial distribution of organisms: whether there is negative (left-sided) asymmetry, total symmetry (a normal distribution), or positive (right-sided) asymmetry. The only effect of the shape of the distribution is that in the case of a negative (left-sided) asymmetry of the distribution the exponential growth in the force of mortality may last only a short time, while in the case of a symmetrical distribution or a distribution with a positive asymmetry the period of exponential growth in the force of mortality is significantly longer and encompasses the whole initial period when $x \ll 1/k$.

So, the heterogeneous population model leads in principle to the same conclusions as the previously discussed model of series-connected blocks with varying degrees of redundancy. However, we are dealing with two fundamentally different models: whereas in the first model the individual risk of death is the same for all organisms and grows exponentially with age, in the second model there initially exist $n$ subpopulations of living organisms with different risks of death which grow as a power function rather than exponential function of age.

The coincidence of the final conclusions obtained from these two fundamentally different models means that it is impossible on the basis of the established mortality phenomena to select the true mechanism behind the age-related destruction of organisms, and further research is essential to test the competing models. One can of course derive no pleasure from this circumstance, but there are two reasons which give ground for optimism. Firstly, the different models seem to lead to virtually coincident interpretations of certain mortality phenomena. For example, the compensation effect of mortality is only possible, according to any of the models, when the rate of irreversible age changes is constant within a species. This treatment of the compensation effect of mortality is not only a feature of the three models examined sequentially in sections 6.5, 6.6 and 6.7 of the present book, but also of other models (Strehler and Mildvan, 1960; Gavrilov, 1978; Gavrilov et al., 1978). The existence of a multitude of competing theories is therefore compatible with the reliable and meaningful interpretation of a number of mortality phenomena, since pluralism of models does not preclude their agreement on a number of issues. Secondly, if the most varied models lead to the same formulae, for example the binomial law of mortality, this merely makes the problem of interpreting results more complicated for the theoretician, but significantly eases the work of the experimenter. Indeed, for the analysis of data it is preferable to use a formula which is supported not by a single model, but by a whole family of fundamentally different models which encompass a wide spectrum of possible situations.

The heterogeneous population model presented in this section provides an explanation for all the basic mortality phenomena even in the simplest case when the organism consists of a single vital block with $n$ parallel elements. Generalising the model to the case of $m$ blocks connected in series in each organism does not present any problems if the blocks are independent of each other with respect to their reliability. In this case, it is naturally expected that the force of mortality for a heterogeneous population is simply the sum of the failure rates of all blocks. In the particular case when the parameters defining the reliability of blocks ($n$, $q$, and $k$) are identical for all blocks, the force of mortality equals the product of the number of blocks ($m$) and the failure rate of one block. As a result, we arrive at the formulae obtained in section 6.6 for the homogeneous population model.

### 6.8. The model of accumulation of defects with constant intensity of the flow of damage.

In the preceding sections (6.4 - 6.7) of the book, models were examined in which the intensity of the accumulation of new damage depends on the degree of the organism's destruction. For instance, in the model of avalanche-like destruction (section 6.4), the intensity of the flow of damage grows in line with the increase in the number of defects in the organism. By contrast, in the other models (sections 6.5 - 6.7), the intensity of the flow of damage decreases in line with the reduction in the number of still undamaged elements (this follows from the fact that the intensity of the flow of damage is equal to the product of the number of still undamaged elements and the constant failure rate $k$). Thus, we have already examined models with increasing and decreasing intensity of the flow of damage. For the sake of completeness, we must naturally now also examine the third fundamental possibility, when the intensity of the flow of damage, numerically equal to the mean number of "hits" (effects) per unit of time, is practically independent of the state of the organism and is on average constant in time.

A wide variety of concepts about the destruction of the organism can lead to this kind of model of the accumulation of defects. In the simplest case, the model corresponds to a situation in which the organism is affected by a random flow of traumatic pressures with an on average constant intensity independent of the state of the organism (exogenous environmental damage like cosmic radiation, etc.) However, there is also the possibility of other mechanisms of destruction leading to this particular model of the accumulation of defects. In particular, this model can also be arrived at via a critical reinterpretation of the propositions underlying the previously described models (sections 6.5 - 6.7). In fact, these models contain the somewhat contentious assertion that the death of the organism occurs only when *all* the elements in a block fail. It is possible that this hypothesis may be justified in a number of cases for certain of the organism's blocks. However, in the majority of cases the hypothesis seems contentious. For example, it is hard to imagine that a single surviving liver cell (hepatocyte) can assume the functions of an entire destroyed liver. Significantly more realistic is the hypothesis that the system initially contains an enormous number of elements which greatly exceeds the critical number of lesions leading to the death of the organism. In this case also we arrive at a schema for the accumulation of damage in which the intensity of the flow of damage (equal to the product of the number of elements and their failure rate) turns out to be practically constant in view of the incommensurability of the number of elements and the permitted number of lesions.

Another virtue of the model under discussion is the fact that it allows account to be taken of the influence of living conditions on the value for the critical number of lesions incompatible with the survival of the organism. This possibility was excluded in the previous models (sections 6.5 - 6.7), since the hypothesis that the elements were connected in parallel presupposed that their number was initially fixed anatomically for each organism. Within these models the ***initial*** number of elements could vary depending on sex and population characteristics, but could not change in the course of an organism's life as its living conditions altered. In principle, this feature of the models agrees well with the fact that the parameters of the Gompertz function ($R$ and $\alpha$), which depend on the initial number of elements, indeed show historical stability. Moreover, the variation of these parameters depending on sex and population characteristics turns out to be coordinated in a way predicted by the mathematical models (the compensation effect of mortality, c.f. sections 6.5 - 6.7). However, inasmuch as the historical stability of the parameters of the Gompertz function is not absolute, the problem arises that the parallel connection models fail to correspond with the actually observed phenomena.

The key to the solution of this problem is the replacement of the parallel connection hypothesis with the more realistic assumption that there exists a critical number of lesions incompatible with the survival of the organism. In this case, it is natural to expect that under harsher conditions the critical number of lesions leading to death might be less than under more comfortable living conditions. In particular, in the wild, when an animal is deprived of care and forced to acquire its own food, as well as to defend itself against predators, the first serious damage to the organism can lead to death. It is therefore not surprising that the mortality of many animals (in particular, birds) is practically independent of age in the wild. This follows directly from the single-stage destruction of the organism model. On the basis of such facts concerning the absence of aging (age-related increase in mortality) in the wild, some investigators draw the conclusion that it is possible to conquer aging by a return to natural "healthy" living conditions without hypodynamia and overeating. It is clear, however, that fighting against the accumulation of damage by mercilessly and constantly annihilating all damaged organisms is an effective but not particularly attractive method of combatting aging. On the contrary, the greater the number of defects the organism can accumulate while remaining alive, the greater its life span will be, *ceteris paribus*. Considerations of this kind lead us to the necessity of analysing more deeply an accumulation of defects model which includes a constant intensity for the flow of damage and a critical number of lesions dependent on living conditions.

If the intensity of the flow of damage equals $k$, and an organism dies after the accumulation of $n$ lesions, the density of the life span distribution is identical to the density of the gamma function (see Barlow and Proschan, 1965; 1975):

$$f(x) = \frac{k \cdot (kx)^{n-1}}{(n-1)!} e^{-kx} \qquad (106)$$

At the initial moment in time, this distribution corresponds to a power function for the increase in the force of mortality (the Weibull law) with a mortality index equal to $(n - 1)$.

A fundamentally different result is obtained when the initially damaged nature of organisms is taken into account. If at the initial moment in time the average number of random lesions in the population equals $\lambda$, the probability of encountering a live organism with $i$ lesions is given by:

$$P_i = c \cdot e^{-\lambda} \frac{(\lambda)^i}{i!}, \quad \text{for } i = 0,1,2,3 \ldots n-1 \qquad (107)$$

$$c = \frac{1}{1 - e^{-\lambda} \sum_{i=n}^{\infty} \frac{(\lambda)^i}{i!}}$$

where $c$ is a normalising factor which ensures that the sum $P_i$ is equal to unity.

Since the death of an organism with $i$ lesions occurs after $n$ - 1 additional lesions, the density of the life span distribution for such organisms is given by:

$$f_i(x) = \frac{k\cdot(kx)^{n-i-1}}{(n-i-1)!} e^{-kx} \qquad \text{where } i < n \qquad (108)$$

The density of the life span distribution for the whole population, which is a mixture of organisms with $i$ = 0,1,2 ... $n$-1 initial lesions, equals:

$$f(x) = \sum_{i=0}^{n-1} P_i f_i(x) = ck\cdot e^{-(\lambda+kx)} \sum_{i=0}^{n-1} \frac{\lambda^i\cdot(kx)^{n-i-1}}{i!\cdot(n-i-1)!}$$

$$= \frac{ck\cdot(\lambda + kx)^{n-1}}{(n - 1)!} e^{-(\lambda+kx)} \qquad (109)$$

It is not difficult to see that at the initial moment in time this model leads to *the binomial law of mortality*, with an *initial virtual age of the population* equal to $\lambda/k$. A more detailed analysis of the model is superfluous, since it coincides formally with the analysis of the other models carried out in sections 6.6 - 6.7. We merely note that during the initial time period when $x \ll \lambda/k$, the model leads to an exponential function for the increase in the force of mortality (the Gompertz law) with an exponent of $\frac{k(n-1)}{\lambda}$ and a pre-exponential factor of $\frac{ck\cdot\lambda^{(n-1)}}{(n - 1)!}$. It is easy to see that an inverse relationship between these Gompertz parameters (the compensation effect of mortality) can arise both as a result of variation in parameter $\lambda$ (the degree to which the organisms are initially damaged), and of variation in parameter $n$ (the critical number of lesions, dependent on the harshness of living conditions).

Thus the basic mortality phenomena can equally be explained within the framework of the model of accumulation of defects with constant intensity of the flow of damage, as long as the organisms initially contain a significant number of defects.

## 6.9. The problem of the diversity of causes of death and their interaction.

The great diversity of diseases and causes of death creates a serious problem for the mathematical modelling of aging, mortality and life span.

There are two extreme positions on this issue: either to take into account all the observed diversity of causes of death, or to ignore it altogether. The first approach presupposes the creation of a multitude of individual models for the age dependence of each disease and cause of death. Some of these models have already been created (Sutherland and Bailar, 1978; Woodbury and Manton, 1983). The undoubted virtue of this approach is the possibility of taking into account the specific medico-biological facts concerning the actual mechanisms involved in the origin and development of each kind of pathology. It is clear that this approach is the one which creates the most favourable conditions for the development of worthwhile explanatory mathematical theories of the age dependence of individual diseases. However, this kind of approach fails to create a sufficiently simple model of total mortality, and hence of life span. In fact, inasmuch as total mortality is composed of a multitude of individual causes, its age dependence might be expected to have a very complex form devoid of all content from the point of view of mathematical modelling. However, the analysis of real data shows that this is not the case. For example, the age-dependence of total mortality can usually and with sufficient accuracy be described by the simple three-parameter Gompertz-Makeham model, no matter how diverse the ratio between individual causes of death (Gavrilov and Gavrilova, 1986).

The second approach involves the direct modelling of the age-dependence of total mortality, as if no individual causes of death existed. At present, this approach is the most wide-spread (Skurnick and Kemeny, 1978a; 1978b; Abernethy, 1979; Ďoubal, 1982; Economos, 1982; Witten, 1985;

Piantanelli, 1986), although it is totally unclear why the multitude of causes of death can be ignored for the sake of the simplicity of the mathematical description. The declaration by some authors that they are concerned with the development of a fundamental mathematical theory of life span, rather than particular medical issues, can hardly be considered a comprehensive theoretical justification for the approach. It is highly doubtful whether simply ignoring issues of age-related pathology can ensure the fundamental correctness of the proposed models of life span.

One of the possible explanations for the comparatively simple behaviour of total mortality is connected precisely with the diversity of causes of death. If there are indeed many causes of death, and the contribution of each of them to total mortality is relatively small, the preconditions are created for trying to find simple *limiting* distributions of the sum of the causes of death. As was noted previously, the Gompertz model can be deduced in the theory of extremes as a *limiting* distribution of minimal values (see section 6.3). An approach of this kind does not however provide an explanation for two further important phenomena in the biology of life span: the phenomenon of the historical stability of the age-dependent component of mortality, and the compensation effect of mortality (Gavrilov and Gavrilova, 1986). In the former case, the age-dependent component of mortality in the Gompertz-Makeham model remains stable throughout the 20th century, despite a sharp drop in total mortality and a radical change in the ratio between the causes of death. In the latter case, the graphs of the relationship between the biological component of mortality and age, constructed separately for men and women from different regions of the world, turn out to be organised in such a way that when extrapolated they meet at a single point, irrespective of sex and regional differences in the structure of causes of death. In both cases, the impression is created that there is a direct interaction (concordance) between individual causes of death which generates the observed regularities.

One of the hypotheses which can explain these observations has already been examined (see section 3.5), and is connected with the notion of a concordance between causes of death. Such a correlation might in principle be expected if the destruction of the organism is a multi-stage process, and the early stages of destruction, corresponding to the transition of the organism into a state of non-specific vulnerability, are the slowest (limiting) stages in the whole process (Gavrilov et al., 1982). Then, in conformity with the principles governing the kinetics of biological processes, the rate of this first stage determines the rate of the whole process of dying, i.e. in effect establishes a magnitude for the "death quota" which is independent of individual causes of death (Gavrilov et al., 1982). Clearly, the correctness of this hypothesis needs to be carefully tested. However, it is also clear that ignoring the diversity of causes of death in the mathematical modelling of life span will have to be justified in some way, whatever the model.

One of the paths to a solution of this problem might be a comparison of the age-specific mortality rates for particular causes in different countries of the world. If there really is a total "death quota" whose value varies from country to country, such a comparison, carried out using the *R*-technique of factor analysis (see Uberla, 1977), ought to reveal the existence of a total mortality factor, together with factors which do not affect total mortality despite their strong link with individual causes of death. Below we present the results which confirm the correctnesss of this assumption.

It must be noted that at present the accepted point of view is exactly the opposite one that individual causes of death are statistically independent. This hypothesis is behind all calculations of gains in life span when particular causes of death are eliminated (Greville, 1948). Techniques for estimating the contribution of each cause of death to total mortality are likewise based on the assumption that causes of death are independent (Greville, 1948). Therefore the question as to whether particular causes of death are dependent or independent of each other is not only of fundamental significance for the construction of mathematical models, but is also exceptionally important in determining a right strategy for health care.

We attempted to test the hypothesis that particular causes of death are statistically independent by undertaking a factor analysis of data from 25 countries published by the WHO (World Health Statistics Annual, 1972; 1977; 1980; 1982; 1983). In order to ensure the comparability of the data, the analysis includes only data compiled in accordance with the Eighth Revision of the International Statistical Classification of Diseases, Injuries and Causes of Death. Mortality values in the age group 55-64 years were used, since at these ages the rates of mortality from particular causes are already quite high, and on the other hand the data are not yet distorted by the effect of multiple pathology which is characteristic of older ages. From the WHO publications, we only selected information on the countries and regions where there was least doubt about the registration of causes of death: Australia, Austria, Belgium, Bulgaria, Canada, Czechoslovakia, Denmark, England and Wales, Finland, France, Greece, Hungary, Ireland, Israel, Italy, Japan, Netherlands, New Zealand, Norway, Portugal, Scotland, Spain, Sweden, Switzerland, USA, and West Germany. In the majority of cases, data for several points in time were used for each country (from 1972 to 1981), so that the total number of cases was 95. With this number of observations, the correlation coefficients between the variables whose modulus is greater than 0.2 are already significantly distinct from zero ($P < 0.05$).

The data were processed by factor analysis using the direct quartimin method of axis rotation. The calculations were performed with the aid of the BMDP package.

The calculations demonstrated the total failure of the hypothesis that causes of death are statistically independent. The falsity of this hypothesis is evident even from the matrix of correlations between individual causes of death. Thus, for females we observe a strong "competition" between mortality from stomach cancer and mortality from breast cancer: the correlation coefficient between them was -0.72. Between mortality from breast cancer and mortality from tuberculosis of the respiratory system, there is also an inverse correlation with a correlation coefficient of -0.71. Amongst the positive correlations, it is worth noting the relationship between mortality from stomach cancer and mortality from tuberculosis of the respiratory system ($r$ = +0.83); between mortality from pneumonia and mortality from bronchitis, emphysema and asthma ($r$ = +0.78); between mortality from stomach cancer and mortality from strokes ($r$ = +0.74); between mortality from malignant tumours of the trachea, bronchus and lung and mortality from ischaemic heart disease ($r$ = +0.74); between mortality from malignant tumours of the oesophagus and mortality from pneumonia ($r$ = +0.74). The list of these positive and negative correlations between particular causes of death might easily be extended.

Strong correlations between particular causes of death are also observed for males. Amongst the positive correlations, it is worth noting the correlation of mortality from stomach cancer with mortality from strokes ($r$ = +0.85) and tuberculosis of the respiratory system ($r$ = +0.80). Mortality from malignant tumours of the buccal cavity and pharynx correlates positively with mortality from malignant tumours of the larynx ($r$ = +0.73). Mortality from strokes correlates positively with mortality from "other nephritis and nephrosis" ($r$ = +0.75), as well as with tuberculosis of the respiratory system ($r$ = +0.79). Amongst the negative correlations, of special interest is the "competition" between mortality from "other neoplasm of the lymphatic and haematopoietic tissue" and mortality from tuberculosis of the respiratory system ($r$ = -0.74). Mortality from ischaemic heart disease "competes" with mortality from cirrhosis of the liver ($r$ = -0.60).

The interaction between causes of death leads to a paradoxical phenomenon: mortality from some causes of death has no effect on total mortality. For example, for males the correlation coefficient between total mortality and mortality from cirrhosis of the liver is only -0.22. However, if the effect of all the remaining causes of death is excluded by calculating the partial correlation coefficient, a strong correlation appears between the two variables ($r$ = +0.92). An analogous picture can be observed for women: the paired correlation coefficient between total mortality and mortality from cirrhosis of the liver amounts to only -0.08 (no correlation exists). When the effect of all the remaining causes of death is removed, the partial correlation coefficient between the two variables reaches +0.94. It seems that the contribution of mortality from cirrhosis of the liver to the total mortality is completely compensated by the effect of competing causes of death (for example, mortality from ischaemic heart disease).

These data demonstrate that in any consideration of the consequences of removing or reducing mortality from particular causes it is essential to take into account the inevitable change in the force of mortality from the remaining diseases. Consequently, what is needed is a radically different approach towards evaluating the relative importance of causes of death, one which is based not on the amount of mortality from the given cause of death, but on the fraction of total mortality which correlates with the given cause of death. In the simplest case the new approach amounts to determining the parameters of the linear regression:

$$\mu = a_i \mu_i + b_i \qquad (110)$$

where $\mu$ is the force of total mortality in the age group in question; $\mu_i$ is the force of mortality from the $i$-th disease in the same age group; the parameter $a_i$ is the slope coefficient of the regression, which we call the mortality amplification coefficient for the disease in question; $b_i$ is the free term (intercept) of the regression, which we call residual mortality. If the causes of death were genuinely independent, parameter $a_i$ would always be equal to unity, corresponding to a decrease in total mortality due to the studied cause alone. However the existence of correlations between causes of death means that in principle the mortality amplification coefficient can assume any values. For example, in the case of the mortality of women aged 55-64 years from ischaemic heart disease, parameter $a_i$ equals 1.7. This means that preventive measures which eliminate 10 deaths from this disease will save not 10, but 17 lives through the additional elimination of a number of accompanying diseases. In the case of female mortality from bronchitis, emphysema and asthma, the amplification coefficient equals 8.6, i.e. preventive measures which eliminate only 10 deaths from these diseases will save as many as 86 women! Thus, although the mortality of women aged 55-64 years from ischaemic heart disease exceeds mortality from bronchitis, emphysema and asthma by 7-8 times on average, the fight against diseases of the respiratory organs may turn out to be a more effective method of reducing total mortality. At the same time, it must be noted that the value of residual mortality for bronchitis, emphysema and asthma is $7.3 \times 10^{-3}$ year$^{-1}$, and for ischaemic heart disease $6.6 \times 10^{-3}$ year$^{-1}$. Consequently, the total elimination of ischaemic heart disease might in the long run lead to lower values of residual mortality than the elimination of bronchitis, emphysema and asthma.

Of course, this example is basically illustrative in nature and merely elucidates the idea behind the new approach. It would be more correct to carry out a factor analysis of the causes of death and isolate those factors which correlate strongly with total mortality. Since in a factor analysis the sample size (in our study 95 cases) ought to be no less than 3 times greater than the number of variables (Uberla, 1977), we carried out a series of calculations for different combinations of mortality rates from particular causes of death. We shall now concentrate in detail on one variant of this kind of data analysis.

A factor analysis of male mortality from 27 causes of death (listed below) revealed the existence of 6 factors which explain 77% of the observed variance. It was not necessary to obtain the seventh (and subsequent) factors, since it accounted for only 3.5% of the total variance. Below we give a short description of the 6 factors (their correlation with the mortality from the causes of death under analysis), arranged in order of the variance they explain.

**Factor 1.**

1. "Other diseases of the genitourinary system"; the correlation coefficient between mortality from this cause and the magnitude of this factor (the factor score) equals +0.86.

2. "Other diseases of the respiratory system", $r$ = +0.72.

3. Malignant neoplasm of skin, $r$ = -0.71.

4. "Other forms of heart disease", $r$ = +0.69.

5. "Other diseases of the digestive system", $r$ = +0.67.

6. Malignant neoplasm of bone, $r$ = +0.58.

7. "Other diseases of the circulatory system", $r$ = +0.46.

8. "Other neoplasm of lymphatic and haematopoietic tissue", $r$ = -0.47.

9. Malignant neoplasm of larynx, $r$ = +0.42.

Attention is immediately drawn to the fact that factor analysis has "homed in on" all the so-called "other" causes of death, which belong to the most diverse types of pathology, and has established their interrelationship. The greatest values of this "other" causes factor are observed in Belgium, and the lowest in New Zealand.

**Factor 2.**

1. Hypertensive disease, $r$ = +0.85.

2. Malignant neoplasm of stomach, $r$ = +0.83.

3. Cerebrovascular disease, $r$ = +0.75.

4. Tuberculosis of the respiratory system, $r$ = 0.72.

5. Peptic ulcer, $r$ = +0.63.

6. Cholelithiasis and cholecystitis, $r$ = +0.63.

At first sight, uniting these very different diseases seems strange. It is well-known, however, that hypertension and peptic ulcers are typical manifestations of stress, and the former disease predisposes to the occurrence of strokes while the latter predisposes to stomach cancer. The greatest values of this apparently stress-related factor are observed in Hungary, and the least in Canada.

**Factor 3.**

1. Malignant neoplasm of intestine, except the rectum, $r$ = +0.79.

2. Malignant neoplasm of the trachea, bronchus and lung, $r$ = +0.78.

3. Diseases of the arteries, arterioles and capillaries, $r$ = +0.66.

4. Bronchitis, emphysema and asthma, $r$ = +0.59.

5. Venous thrombosis and embolism, $r$ = +0.57.

6. Ischaemic heart disease, $r$ = +0.55.

7. "Other neoplasm of lymphatic and haematopoietic tissue", $r$ = +0.46.

The greatest values of this factor are recorded in Scotland, and the least in Japan.

**Factor 4.**

1. Malignant neoplasm of buccal cavity and pharynx, $r$ = +0.95.

2. Malignant neoplasm of oesophagus, $r$ = +0.83.

3. Malignant neoplasm of larynx, $r$ = +0.60.

4. Cirrhosis of liver, $r$ = +0.59.

5. "Other diseases of the circulatory system", $r$ = +0.60.

The highest values of this factor are characteristic of France, and the lowest are characteristic of the Netherlands.

**Factor 5.**

1. Pneumonia, $r$ = +0.84.

2. Bronchitis, emphysema and asthma, $r$ = +0.52.

3. Active rheumatic fever, $r$ = +0.73.

4. Cholelithiasis and cholecystitis, $r$ = -0.60.

The greatest values of this factor are observed in Bulgaria, and the least in Hungary.

**Factor 6.**

1. Chronic rheumatic heart disease, $r$ = +0.91.

2. Peptic ulcer, $r$ = +0.41.

3. Cholelithiasis and cholecystitis, $r$ = +0.40.

The highest values of this factor are observed in Hungary, and the lowest in Switzerland.

It is interesting to note that the above-listed factors turn out to be virtually independent of each other, although the postulate of their statistical independence was not introduced into the calculations. The highest value for the correlation coefficients between the factors (factor 1 and factor 4) turns out to be +0.3.

The practical significance of these kinds of calculation is obvious. In fact, having determined which of the calculated factors are closely connected (correlate) with total mortality, we thereby determine the strategic direction for reducing mortality. *It turns out that of the 6 factors only factor 3 correlates significantly with total mortality!* Thus, the vast multiplicity of diseases and causes of death can be reduced to a small number of groups, and the causes which are genuinely worth careful epidemiological investigation can be revealed. Unfortunately, a detailed exposition of this approach is beyond the bounds of this book. Returning to the problem of modelling mortality and life span mathematically, we must note the necessity of taking into account the interaction of diseases when constructing models of particular pathologies. Factor analysis can reveal the groups of diseases whose interaction must be taken into account in the first instance.

As far as the modelling of total mortality is concerned, it must be borne in mind that total mortality can be predicted knowing only a small proportion of the factors (causes of death) involved, and information concerning the remaining causes of death turns out to be superfluous. One particular manifestation of this phenomenon, called the paradox of superfluous causes of death (see Gavrilov, 1989), is the fact that information about mortality from a mere two causes (diseases of the arteries, arterioles and capillaries; bronchitis, emphysema and asthma) seems sufficient to account for 62% of all the interpopulation variation (dispersion) of total mortality. If causes of death were statistically independent, this proportion of explained dispersion would amount to only 2%. Information concerning 7 causes of death allows interpopulation differences in total mortality to be described with an accuracy commensurate with the accuracy of the data themselves: the proportion of unexplained dispersion is only 5% instead of the 28% to be expected if causes of death were independent (Gavrilov, 1989).

While modelling the age-related dynamics of total mortality, it is also worth bearing in mind that the aging rate (the magnitude of the absolute age-related increment in the force of mortality in a cohort of individuals of the same age) is also linked to a small number of diseases. The closest link with the aging rate is shown by mortality from hypertension, as well as from diseases of the arteries, arterioles and capillaries (Gavrilov, 1988).

Thus, taking into account the interconnections between individual causes of death and the aging rate is an important prerequisite for the successful construction of a fully valid quantitative theory of life span.

## 6.10. Concluding remarks

To summarise this survey of mathematical models of life span, we shall formulate the conditions which the proposed mathematical models must satisfy, and discuss two problems which arise in mathematical modelling: the problem of the interrelation between heterogeneity and stochasticity, and the problem of the origin of aging.

*The problem of the interrelation between heterogeneity and stochasticity.* Most existing models of life span can clearly be divided into two groups. The first group of mathematical models connects the observed life distribution with the heterogeneity of the population (Koltover, 1983). These models agree well with the fact that there are high risk groups, and with the clear lack of uniformity amongst most of the studied populations with respect to many of the studied characteristics. In essence, however, these models assert that the life span of each individual is fatally predetermined. What is more, they often assume the genetic nature of the observed heterogeneity, in contradiction with well-known facts concerning the low heritability of life span (see section 2.3).

The second group of models explains the observed life span distribution by the stochastic nature of the dying process, the analogy being the decay of identical atoms of a radioactive element. In the framework of these models, individuals are in principle indistinguishable, in contradiction with the fact that high risk groups exist.

It is clear that the two groups of models represent two extreme points of view, and it would appear that a fully adequate model of life span ought to combine the ideas of stochasticity and heterogeneity in a natural manner. This can be done if the dying process is conceived as a series of random sequential transitions from an initial (healthy) state of the organism to a final (lethal) state. Then, even in an initially homogeneous population, heterogeneity of death risk will inevitably arise with the passing of time, in conformity with the existence of high risk groups (Kozlovsky and Gavrilov, 1983; Gavrilov, 1987).

*The problem of the origin of aging.* As is well-known, the aging period (the period when the force of mortality grows with age) occupies the greater part of the life span of organisms, therefore any model of life span must explain the existence of this period. At the present time, it has already become a tradition to explain the aging of organisms through the aging of their subsystems. However, it is obvious that this traditional explanation of aging through aging inevitably leads up a logical blind alley, since after passing in turn from the aging of organisms to the aging of organs, tissues and cells, we eventually arrive at atoms, which are known not to age. Thus the key to explaining aging is the question: how can we explain the aging of a system made out of non-aging elements? An analogous problem arises in a comparative evolutionary approach to aging. Many of the simplest forms of life, for example spores and the non-dividing vegetative cells of bacteria, do not age (see Kozlovsky and Gavrilov, 1983). The question arises: why does the property of aging appear in organisms with more complex structures?

A schema in which the organism is sequentially destroyed also provides an answer to this question (Kozlovsky and Gavrilov, 1983). It turns out that if the death of an organism occurs not in one, but two or more sequential stages, this is sufficient for the phenomenon of aging to appear. The biological sense of this assertion is the following. Each stage of destruction corresponds to one of the organism's vitally important structures being damaged. In the simplest organisms with unique vitally important structures, this damage usually leads to their death. This is why defects in such organisms do not accumulate, and why the organisms themselves do not age. In more complexly structured organisms, when there is a significant redundancy in the number of vitally important structures, by no means every occurrence of damage leads to death. However in this case defects accumulate, giving rise to the phenomenon of ageing. Thus aging is a direct consequence of the increased reliability and life span of organisms which result from the redundancy of vitally important structures. This schema also provides an explanation for the seemingly paradoxical observation that the most prominent aging occurs in the most long-lived species.

In addition to the problems listed above, we can also formulate the conditions which the proposed mathematical models of life span must satisfy.

1. Above all, there must be a correspondence between the observed and theoretically predicted life span distributions. Although this criterion is rather obvious and is widely applied (Strehler, 1962; 1978), its potential is not always fully exploited. In particular, by no means every model agrees with the Gompertz-Makeham law. Of course, it is not required of a model that it coincide absolutely with this law, since the real data also sometimes deviate from it. However, every model ought to include the Gompertz-Makeham law as a special case of the theoretically expected distribution with a certain ratio of the parameters. In this connection, it is interesting to note that the well-known model of Strehler and Mildvan fails to correspond with the Gompertz-Makeham law for the simple reason that it does not observe the principle that the two types of mortality should be summed. This failing of the model arises from its basic postulate, according to which the internal and external fluctuations are ***amalgamated***, and what is subsequently examined is the distribution of this sum "challenge" (see Strehler and Mildvan, 1960, p.15). It is interesting to note that the previously described model of random

sequential transitions satisfies the Gompertz-Makeham law.

2. Models of life span ought to admit the possibility of a significant change in lifetimes while the biological constitution of the organisms remains unaltered, and in particular, they ought to be consistent with the historical fact of a reduction in human mortality. This condition is especially relevant in the case of pseudofundamental biological models, in which socially controllable life span factors are "forgotten" (Koltover, 1983). It is easy to test these models by calculating their parameters using life tables for a population before and after a demographic revolution (transition). Clearly, if in the course of this comparatively short period there is a significant change in the coefficient interpretable as the number of human releasing factors, such a model can only be of historical interest.

3. An important criterion for testing the validity of a model is its ability to explain the phenomenon of the historical stability of the age-dependent component of mortality. In this case, there is again no necessity that the model should presuppose the absolute historical invariance of the age-dependent component of mortality, since in practice it may show insignificant changes. It is important, however, for the model to explain the comparative stability of the age-dependent component of mortality against the background of a sharp drop in total mortality itself. As has already been noted, the hypothesis of random sequential transitions seems to be a framework in which this can be achieved.

4. A model of life span ought to be consistent with data concerning regional and sex differences in the age-dependent component of human mortality. In addition, the interpretation of these differences ought to be consistent with known epidemiological and physiological data obtained by independent methods.

5. The existence of species-specific life span invariants for organisms (the compensation effect of mortality) is a very important fact which is difficult to explain. It must be noted that an analogous phenomenon, with the same name of "compensation effect", has long been known in physical chemistry, as well as in chemical and biological kinetics (see Munblit and Trofimov, 1985). Nevertheless, despite the fact that these sciences are incomparably more developed, a satisfactory explanation is still lacking for the

compensation effect observed in them.

6. Finally, it is worth noting that models which predict the existence of an absolute limit to life span usually turn out to be unviable, since they are inconsistent with the observed dynamics of mortality at older ages (Gavrilov, 1984c; Gavrilov et al., 1984). Models of life span ought to describe the retardation which takes place in the rates of growth of the force of mortality with age, and its constancy at extremely high ages (Gavrilova and Gavrilov, 1982). It is interesting to observe that the hypothesis of random sequential transitions predicts the constancy of the force of mortality in all its concrete instantiations, no matter what the relationship between the constants. Qualitatively, this corresponds to the fact that as defects are accumulated, the redundancy of the organism in the number of vitally important elements finally disappears. As a result, the organism degenerates into a system with no redundancy, i.e. a system with elements connected in series in the reliability theory sense, when any new defect leads to death. Thus, the use of reliability theory turns out to be fruitful in this case also.

Summarising this survey of mathematical models of life span, it must again be emphasised that they are not a goal, but a method of investigation. Therefore attention was basically paid to a discussion of the interpretative aspects of the mathematical modelling of life span rather than a discussion of concrete models, formulae and mathematical transformations. The wide application of mathematical models and methods in a study of the mechanisms which determine life span is a new and inevitable stage in the development of the biology of life span. This stage has been accompanied by a significant number of errors and blunders in the published literature, which must therefore be carefully analysed. Mistakes of this kind are however an unavoidable illness of growing-up and should not discredit the concept of a quantitative approach. An analysis of the modern mathematical models and methods which have been used in the biology of life span shows that they are adequate to the complexity of the phenomena under investigation. The creation of a complete mathematical theory of life span will probably take some time, but there can be no doubt about the fruitfulness of mathematical modelling based on a distinctive synthesis of the methods of reliability theory and the kinetics of biological processes.

# CONCLUSION

In summary, it has to be admitted that many generally accepted and at first sight obvious notions about life span have been put into doubt and even refuted in this book. The following is a far from complete list of the myths which have been dispelled: the myth that there is a species-specific life span limit, the myth that there is a genetic program which determines life span, the myth that the greater life span of females is a general biological law, etc. The dispelling of these myths is however not at all the result of an aspiration for originality or a desire for self-assertion in seeking out others' errors. The reader has the right to draw his own conclusions on the basis of the arguments presented, and we will only be grateful for any critical comments. In the book, we adduced rather a large number of analogies, and to conclude we would like to make one more comparison. The biology of life span at present resembles a building for which a strong and powerful foundation has been constructed. However, the completion of the upper section of the building has been delayed because the majority of investigators turned in their time to other objects, mostly of a molecular-genetic nature, which seemed to them more promising. As a result, what we see on the foundation is a number of temporary barns, and quite a large amount of builders' rubble in the form of various scientific myths and prejudices. Naturally, further construction would have been impossible without clearing the foundation of its accumulated debris. Now, it seems to us, the original conception of the building has been restored, and we can start constructing on the already prepared foundation. Unfortunately, we still do not know the most important thing. The concrete mechanisms which determine life span remain as yet unknown. Nevertheless, there is every ground for optimism, since we can see the path to the goal. This path, in our view, lies at the crossroads of biology and reliability theory.

If we are to speak of the positive results of the investigations that have been carried out, the basic positive result must be that progress can be achieved in a number of fundamental problems of the biology of life span simply by carrying out a quantitative analysis of data which have been available for a long time. In this way, simple and quite general phenomena for the survivorship of organisms (the Gompertz-Makeham law, the phenomenon of the historical stability of the age-dependent component of mortality, the compensation effect of mortality, species-specific invariants of life span, etc.) have been either confirmed or rediscovered. On the basis of the regularities which were revealed, it was possible to distinguish the biological component of human mortality, and to outline epidemiological approaches to the search for ways of reducing it.

Of course, the quantitative analysis of already accumulated data is merely the first stage in the revival of systematic research in the biology of life span, one which is essential for the preparation and organisation of large scale epidemiological studies, and for the completion of key experiments in the given field. Naturally, all these further research projects will require considerable expenditure, and the question arises: where should the necessary funding come from?

An answer to this question was given as early as 1978 by the well-known American gerontologist, Johan Bjorkstan. Bjorkstan asked his senator, William Proxmire, to place before the US Congress a proposal that the military expenditure of the USA and USSR should be cut by 50% on both sides, with the aim that the funds which were freed could be used for the opposite purpose: the development of methods of increasing active life span. Bjorkstan also proposed that a joint Soviet-American scientific committee, based on equal representation, be set up to determine the best ways of distributing funds for the organisation of a human life extension project (see Pryakhin, 1985). Unfortunately, this proposal was not supported by the American administration in view of the complications in superpower relations which existed at that time. However, it generated a reciprocal peace movement amongst Soviet scientists (Gavrilov, 1985; 1989; 1990; Pryakhin, 1985), and prepared the ground for the project to be ready to start at the present time.

In 1986, a scientific conference on "Mechanisms of ageing and longevity" was held in the USSR with the participation of scientists from the USA, Japan, Hungary, Bulgaria, Czechoslovakia and East Germany. At this conference, in which the authors of this book also participated, a resolution was adopted which stated, in particular:

"The conference participants note that the resources which are at present used for military expenditure and the arms race are totally adequate to achieve the opposite goal: the extension of human life. In this case, the fruits of what is achieved by gerontology might be enjoyed even by the present generation.

"The conference participants wish to suggest the motto 'Replace the means of life annihilation with the means of life prolongation', and appeal to the general public of all countries of the world to support the proposed initiative by all possible and available methods. This will not only permit humanity to be saved from the threat of self-annihilation, but also open up before humanity the captivating vista of control over aging and the extension of life span" (see Gavrilov, 1990).

In 1988, the delegates of the 5th All-Union Congress of Gerontologists and Geriatricians (Tbilisi, USSR) unanimously supported the above appeal, and resolved to forward it to the leaderships of the USSR and the USA. In the resolution adopted at the congress, it was noted that "the development of methods of extending active life span is now a fully realisable project in which all humanity has an interest". The congress participants also came to the conclusion that for the project to get off the ground "... it is essential for the level of financing to correspond at last to the scale of the task" (see Gavrilov, 1989; 1990) .

All that remains is for the authors of this book to agree with these propositions, and to call on the readers to assist as best they can the successful development of the most humanitarian of scientific projects: the project to increase the active life span of human beings. It is our deep conviction that this is a project which is capable of uniting humanity in the 21st century, and becoming the basis of a new constructive humanitarian ideology (a new manner of thinking) which can successfully oppose destructive nationalistic, religious and political ambitions. The fruition of the life extension project would be a concrete manifestation of the idea that humanitarian values have priority - a fulfilment of the human right to a long and worthwhile life.

# APPENDIX 1

## List of published animal life tables.

(This list includes only the most reliable data obtained for sufficiently large samples.)

| Biological species | Number of tables | Source of the data |
|---|---|---|
| **1. ASCHELMINTHES** | | |
| **1.1. Nematoda** (round worms) | | |
| *Caenorhabditis elegans* | 1 | Findeis et al., 1983 |
| **2. Rotatoria** (wheel animalicules) | | |
| *Hydantyna senta* mictic and amictic females | 3 | Ferris, 1932 |
| *Proales sordida* | 5 | Jennings and Lynch, 1928a |
| " | 8 | Jennings and Lynch, 1928b |
| *Proales decipiens* | 2 | Noyes, 1922 |
| **2. MOLLUSCA** (molluscs) | | |
| **2.1. Gastropoda** (snails) | | |
| 2.1.1. *Pulmonata* | | |
| *Biomphalaria pfeifferi* different temperatures | 6 | De Kock and van Eeden, 1981 |

| Biological species | Number of tables | Source of the data |
|---|---|---|
| *Bulinus globosus* | | |
| different temperatures | 4 | Shiff, 1964 |
| *Bulinus truncatus* | 2 | Barlow and Muench, 1951 |
| *Bulinus tropicus* | | |
| effect of cohort size | 20 | De Kock and van Eeden, 1976 |
| different temperature regimes | 3 | De Kock, 1985 |
| different temperatures | 6 | De Kock and van Eeden, 1985 |
| successive generations | 10 | De Kock et al., 1986 |
| *Planorbis boissyi* | 2 | Barlow and Muench, 1951 |
| *Lymnaea columella* (pond snail) | 2 | Winsor and Winsor, 1935 |
| *Lymnaea natalensis* | | |
| different temperature regimes | 3 | De Kock, 1985 |
| different temperatures | 6 | De Kock and van Eeden, 1985 |
| *Agriolimax agrestis* | 4 | Szabo and Szabo, 1929 |
| 3.**ARTHROPODA** (arthropods) | | |
| 3.1.**Crustacea** (crustaceans) | | |
| *Daphnia pulex* different population densities | 7 | Frank et al., 1957 |
| *Daphnia magna* | | |
| different temperatures | 8 | MacArthur and Baillie, 1929 |

| Biological species | Number of tables | Source of the data |
|---|---|---|
| *Balanus balanoides* | 1 | Bodenheimer, 1958 |
| *Arthemia salina* irradiation | 2 | Suyama and Iwasaki, 1976 |
| **3.2. Arachnida** (spiders) | | |
| *Latrodectus mactans* (black widow spider) | 2 | Deevey and Deevey, 1945 |
| *Paruroctonus mesaensis* (scorpion) | 2 | Polis and Farley, 1980 |
| **3.3. Insecta** (insects) | | |
| 3.3.1. *Lepidoptera* (butterflies and moths) | | |
| *Fumea crassiorella* | 1 | Matthes, 1951 |
| *Samia cercopia* | 12 | Rau and Rau, 1912 |
| *Cullosamia promethea* | 2 | Rau and Rau, 1912 |
| *Tropaea luna* | 2 | Rau and Rau, 1912 |
| *Bombyx mori* (silk worm) | 4 | Alpatov and Gordeenko, 1932 |
| *Acrobasis caryae* (pecan nut case bearer) | 8 | Pearl and Miner, 1936 |
| *Carpocapsa pomonella* (codling moth) | 8 | McArthur and Baillie, 1932 |
| 3.3.2. *Coleoptera* (beetles) | | |
| *Rhizopertha dominica* (lesser grain borer) | 2 | Birch, 1953 |

| Biological species | Number of tables | Source of the data |
|---|---|---|
| *Calandra oryzae* (rice weevil) | | |
| different humidity, temperature and diet | 10 | Birch, 1953 |
| *Trogoderma versicolor* | 4 | Davis, 1945 |
| *Tribolium castaneum* (rust-red flour beetle) | 8 | Mertz et al., 1965 |
| *Tribolium madens* (black flour beetle) | 2 | Park, 1945 |
| *Tribolium confusum* (confused flour beetle) irradiation | 3 | Cork, 1957 |
| " | 8 | Mertz et al., 1965 |
| " | 2 | Pearl and Miner, 1941 |
| 3.3.2.*Diptera* (flies and mosquitoes) | | |
| *Phormia terrae novae* | 1 | Collatz and Wilps, 1983 |
| *Calliphora erythrocephala* (bluebottle fly) | 2 | Feldman-Muhsam and Muhsam, 1945 |
| *Musca vicina* | 2 | Feldman-Muhsam and Muhsam, |
| *Musca domestica* (house fly) | 2 | Rockstein and Lieberman, 1959 |
| different diets | 10 | Rockstein, 1957 |
| *Similium venustum* (black fly) | 6 | Davies, 1953 |
| *Drosophila pseudoobscura* | 15 | Vetukhiv, 1957 |

| Biological species | Number of tables | Source of the data |
|---|---|---|
| *Drosophila melanogaster* | | |
| different population density in larval stage | 12 | Miller and Thomas, 1958 |
| different temperatures | 9 | Loeb and Northrop, 1917 |
| Canton-S strain | 4 | Hall, 1969 |
| wild strain and cinnabar, curled, vestigial, white, yellow mutants (aseptic conditions) | 30 | Steinfeld, 1929 |
| Old Falmouth and Quintuple strains and their $F_1$, $F_2$ and $F_3$-hybrids | 35 | Pearl et al., 1923 |
| Old Falmouth strain, different population density | 26 | Pearl et al., 1927 |
| Old Falmouth, New Falmouth, Sepia, Eagle Point and Quintuple strains | 8 | Pearl and Parker, 1921 |
| Old Falmouth and New Falmouth strains | 14 | Pearl and Parker, 1922a |
| successive etherisations | 11 | Pearl and Parker, 1922b |
| different population density | 50 | Pearl and Parker, 1922c |

| Biological species | Number of tables | Source of the data |
|---|---|---|
| *Drosophila melanogaster* | | |
| different air ventilation | 9 | Pearl and Parker, 1922d |
| strains Old Falmouth and vestigial | 4 | Pearl and Parker, 1924a |
| complete absence of food, different population density | 12 | Pearl and Parker, 1924b |
| wild strain and vestigial, nicked, notched-b-pr mutants and their hybrids | 14 | Lüers, 1935 |
| wild strain and ebony mutant, complete absence of food | 4 | Greiff, 1940 |
| males, females, triploid females and intersexes | 6 | Gowen, 1931 |
| vestigial, black speck, arc, purple mutants and combinations of these mutations | 55 | Gonzales, 1923 |
| different temperatures during the larval and imaginal stages | 12 | Alpatov and Pearl, 1929 |
| different feeding during the larval and imaginal stages | 12 | Alpatov, 1930 |
| *Anopheles dirus* (mosquito) | 6 | Klein et al., 1987 |

| Biological species | Number of tables | Source of the data |
|---|---|---|
| *Aedes aegypti* (yellow-fever mosquito) | 2 | Kershaw et al., 1953 |
| **3.3.4.*Hymenoptera*** | | |
| *Habrobracon juglandis* (parasitic wasp) | 4 | Georgiana, 1949 |
| *Bracon kirkpatricki* | 1 | Sivaprakasam et al., 1985 |
| **3.3.5.*Blattoptera* (roaches)** | | |
| *Blatta orientalis* (oriental cockroach) | 2 | Rau, 1924 |
| **3.3.6.*Homoptera*** | | |
| *Therioaphis maculata* different temperatures | 2 | Messenger, 1964 |
| *Fiorinia theae* (tea scale) | 2 | Munir and Sailer, 1985 |
| **3.3.7.*Anoplura* (lice)** | | |
| *Pediculus humanus* (human louse) | 2 | Evans and Smith, 1952 |
| **4.CHORDATA** | | |
| **4.1.Urochordata** | | |
| **4.1.1.*Ascidia* (ascidias)** | | |
| *Ascidia nigra* | 1 | Goodbody, 1962 |

| Biological species | Number of tables | Source of the data |
|---|---|---|
| | **4.2. Vertebrata** (vertebrates) | |
| | 4.2.1. *Pisces* (fish) | |
| *Lebiates reticulatus* (guppy) | 14 | Comfort, 1963 |
| *Oryzias latipes* (Japanese medaka) | 1 | Egami and Etoh, 1969 |
| X-irradiation | 8 | Egami and Etoh, 1973 |
| *Poeciliopsis monacha* | 2 | Moore, 1975 |
| *Cynolebias bellottii* (Argentine pearlfish) | 2 | Liu and Walford, 1969 |
| | 4.2.2. *Tetrapoda* (quadrupeds) | |
| | 4.2.2.1. *Amphibia* (amphibians) | |
| *Desmognathus fuscus* (dusky salamander) | 12 | Danstedt, 1975 |
| | 4.2.2.2. *Reptilia* (reptiles) | |
| *Uta stansburiana* (side-blotched lizard) | 1 | Tinkle, 1967 |
| *Sceloporus virgatus* (striped plateau lizard) | 2 | Vinegar, 1975 |
| *Sceloporus undulatus* (eastern fence lizard) | 1 | Ferguson et al., 1980 |
| *Chrysemys picta* (painted turtle) | 2 | Wilbur, 1975 |
| *Opheodrys aestivus* (green snake) | 1 | Plummer, 1985 |

| Biological species | Number of tables | Source of the data |
|---|---|---|
| | | 4.2.2.3. *Aves* (birds) |
| *Gallus gallus* (white leghorn chicken) | 1 | Gardner and Hurst, 1933 |
| *Lonchura striata* (Bengalese finch) | 2 | Eisner, 1967 |
| *Pygoscelis adeliae* (Adelie penguin) | 1 | Ainley and DeMaster, 1980 |
| *Megadyptes antipodes* (yellow-eyed penguin) | 1 | Richdale, 1957 |
| *Fringilla coelebs* (chaffinch) | 2 | Paevsky, 1985 |
| *Sturnus vulgaris* (starling) | 1 | Coulson, 1960 |
| " | 1 | Paevsky, 1985 |
| *Erithacus rubecula* (robin) | 1 | Paevsky, 1985 |
| *Corvus monedula* (jackdaw) | 2 | Beklová and Pikula, 1985 |
| *Anser anser* (grey-lag goose) | 1 | Beklová and Pikula, 1985 |
| *Cygnus olor* (mute swan) | 1 | Beklová and Pikula, 1985 |
| *Anas querquedula* (garganey) | 1 | Beklová and Pikula, 1985 |
| *Anas strepera* (gadwall) | 1 | Beklová and Pikula, 1985 |

| Biological species | Number of tables | Source of the data |
|---|---|---|
| *Anas clypeata* (shoveller) | 1 | Beklová and Pikula, 1985 |
| *Aythya ferina* (common pochard) | 1 | Beklová and Pikula, 1985 |
| *Accipiter gentilis* (goshawk) | 1 | Beklová and Pikula, 1985 |
| *Buteo buteo* (common buzzard) | 1 | Beklová and Pikula, 1985 |
| *Gallinula chloropus* | 1 | Beklová and Pikula, 1985 |
| *Gallinago gallinago* | 1 | Beklová and Pikula, 1985 |
| *Columba palumbus* (wood-pigeon) | 2 | Beklová and Pikula, 1985 |
| *Athene noctua* (little owl) | 1 | Beklová and Pikula, 1985 |
| *Asio otus* (long-eared owl) | 1 | Beklová and Pikula, 1985 |
| *Corvus corone cornix* (hooded crow) | 1 | Beklová and Pikula, 1985 |
| *Pica pica* (magpie) | 1 | Beklová and Pikula, 1985 |
| " | 1 | Holyoak, 1971 |
| " | 1 | Tatner, 1986 |
| *Garrulus glandarius* (jay) | 1 | Beklová and Pikula, 1985 |
| *Actitis hypoleucos* (common sandpiper) | 1 | Beklová and Pikula, 1985 |

| Biological species | Number of tables | Source of the data |
|---|---|---|
| *Nycticorax nycticorax* (night heron) | 1 | Beklová and Pikula, 1985 |
| *Anatinae* (ducks) | 1 | Beklová and Pikula, 1985 |
| *Striginae* (owls) | 1 | Beklová and Pikula, 1985 |
| *Accipitridae* | 1 | Beklová and Pikula, 1985 |
| *Anas platyrhynchos* (mallard) | 3 | Beklová and Pikula, 1985 |
| *Accipiter nisus* (sparrow-hawk) | 1 | Beklová and Pikula, 1985 |
| *Fulica atra* (bald coot) | 1 | Beklová and Pikula, 1985 |
| *Strix aluco* | 1 | Beklová and Pikula, 1985 |
| *Corvus frugilegus* (rook) | 2 | Beklová and Pikula, 1985 |
| 4.2.2.4. *Mammalia* (mammals) | | |
| *Insectivora* (insect-eaters) | | |
| *Blarina brevicauda* (short-tailed shrew) | 1 | Blus, 1971 |
| *Chiroptera* (bats) | | |
| *Myotis daubentonii* (water bat) | 1 | Bezen et al., 1960 |
| *Myotis mystacinus* (whiskered bat) | 1 | Sluiter et al., 1956 |
| " | 1 | Bezen et al., 1960 |

| Biological species | Number of tables | Source of the data |
|---|---|---|
| *Myotis emarginatus* (notch-eared bat) | 1 | Bezen et al., 1960 |
| *Rhinolophus hipposideros* (little horseshoe-nosed bat) | 1 | Bezen et al., 1960 |
| | *Rodentia* (rodents) | |
| *Microtus arvalis* (common vole) | 1 | Fenyuk and Sheikina, 1940 |
| *Microtus agrestis* (short-tailed vole) | 1 | Leslie and Ranson, 1940 |
| *Microtus orcadensis* (Orkney vole) | 2 | Leslie et al., 1955 |
| *Ochotona princeps* (pika) | 1 | Millar and Zwickel, 1972 |
| *Zapus princeps* (jumping mouse, American pika) | 1 | Falk and Millar, 1987 |
| *Spermophilus beldingi* (Belding's ground squirrel) | 2 | Sherman and Morton, 1984 |
| " | 1 | Barkalov et al., 1970 |
| *Spermophilus columbianus* (Columbian ground squirrel) | 6 | Zammuto, 1987 |
| *Cricetus auratus* (Syrian hamster) | 2 | Poiley, 1957 |
| " | 4 | Fortner, 1961 |

| Biological species | Number of tables | Source of the data |
|---|---|---|
| *Rattus rattus* (roof rat) | | |
| Albany strain | 1 | Wright et al., 1940 |
| Wistar strain | 2 | Weisner and Sheard, 1935 |
| Wistar strain | 20 | Schlettwein-Gsell, 1970 |
| Wistar strain, exercises | 4 | Goodrick, 1980 |
| syngenic parabionts of different age | 2 | Ludwig and Elashoff, 1972 |
| Wistar strain | 2 | Gilbert and Gillman, 1958 |
| Sprague-Dawley strain | 1 | Durbin et al., 1966 |
| Copenhagen, Fisher strains, and their hybrids | 22 | Dunning and Curtis, 1946 |
| Wistar strain | 1 | Boorman and Hollander, 1973 |
| Copenhagen, Fisher, August strains | 10 | Curtis et al., 1931 |
| " | 1 | Kovtun, 1976 |
| different diets | 5 | Ross, 1961 |
| effect of oestrogens | 1 | Eisen, 1942 |
| Wistar strain, effect of procaine | 6 | Aslan et al., 1965 |
| *Rattus norvegicus* (Norway rat) | 1 | Leslie et al., 1952 |

| Biological species | Number of tables | Source of the data |
|---|---|---|
| *Rattus (Mastomys) natalensis* (multimammate mouse) | 1 | Oliff, 1953 |
| *Mus musculus* (house mouse) | 2 | Greenwood, 1928 |
| irradiation | 4 | Storer, 1965 |
| C3H strain | 1 | Bittner, 1935b |
| " | 2 | Andervont and McEleney, 1941 |
| C strain | 2 | Andervont, 1945 |
| C3Hf strain | 3 | Vellisto and Bank, 1968 |
| irradiation | 3 | Storer, 1962 |
| DBA and C57BL strains, their $F_1$ and $F_2$-hybrids | 12 | Murray and Little, 1935 |
| DBA strain | 1 | Murray, 1934 |
| DBA and A strains | 6 | Murray and Hoffman, 1941 |
| DBA and C57BL strains and their hybrids | 16 | Murray and Little, 1939 |
| C57BL strain | 3 | Little et al., 1939 |
| SAS/4 strain, irradiation | 8 | Lindop and Rotblat, 1961 |
| oral contraceptives | 5 | Garg et al., 1970 |
| A x DBA hybrids | 1 | Bittner, 1936 |

| Biological species | Number of tables | Source of the data |
| --- | --- | --- |
| *Mus musculus* | | |
| albino strain | 1 | Bittner, 1935a |
| " | 3 | Bittner, 1937 |
| A strain, different diets | 2 | Bittner, 1935c |
| CC57BR strain | 2 | Medvedev, 1958 |
| " | 4 | Gelshtein and Dyad'kova, 1955 |
| " | 2 | Medvedev and Olchovskaya, 1974 |
| CC57W strain | 2 | Medvedev and Olchovskaya, 1974 |
| CC57W, CC57BR strains | 8 | Medvedev, 1961 |
| $C_3$HA strain | 3 | Malyugina and Prokof'eva, 1957 |
| DBR strain | 2 | Dyad'kova and Lotosh, 1962 |
| irradiation | 5 | Thomson et al., 1985a |
| irradiation | 10 | Thomson et al., 1985b |
| fission neutrons and $\gamma$-irradiation | 24 | Thomson and Grahn, 1988 |
| $C_{57}$ black strain | 1 | Dyad'kova and Medvedev, 1956 |
| RF-strain | 1 | Yuhas and Clapp, 1972 |

| Biological species | Number of tables | Source of the data |
|---|---|---|
| | *Carnivora* (carnivores) | |
| *Canis familiaris* (dog) | | |
| Irish wolfhounds | 2 | Comfort, 1956 |
| Pekinese | 1 | Comfort, 1960 |
| cocker spaniel | 1 | " |
| English mastiff | 1 | " |
| beagles, X-irradiation | 3 | Andersen and Rosenblatt, 1969 |
| *Canis lupus* (grey wolf) | 1 | Comfort, 1957 |
| | *Pinnipedia* (pinnipeds) | |
| *Halichoerus grypus* (grey seal) | 1 | Hewer, 1964 |
| | *Artiodactyla* (artiodactyles) | |
| *Capreolus capreolus* (roe deer) | 2 | Andersen, 1953 |
| *Odocoileus hemionus* (mule deer) | 2 | Taber and Dasmann, 1958 |
| *Rangifor arcticus arcticus* (barren ground caribou) | 1 | Banfield, 1955 |
| *Hemitragus jemlahicus* (tahr) | 1 | Caughley, 1966 |
| *Alces alces* (European elk)) | 1 | Yazan, 1961 |

| Biological species | Number of tables | Source of the data |
| --- | --- | --- |
| *Ammotragus lervia* (barbary sheep) | 1 | Comfort, 1957 |
| *Ovis musimon* (mouphlon sheep) | 1 | Comfort, 1957 |
| *Ovis dalli* (mountain sheep) | 2 | Deevey, 1947 |
| *Ovis aries* (sheep) | 1 | Hickey, 1960 |
| *Syncerus caffer* (African buffalo) | 1 | Sinslair, 1974 |
| *Boselaphus tragocamelus* (nilgai) | 1 | Comfort, 1957 |
| *Perissodactyla* (odd-toed ungulates) | | |
| *Diceros bicornis* (African black rhinoceros) | 1 | Goddard, 1970 |
| *Equus caballus* (horse) | 2 | Comfort, 1958 |
| Arabian mares | 1 | Comfort, 1961 |

## APPENDIX 2

**Table *A*  Features which show a positive correlation with the life span of organisms (interspecies correlations)**

| Feature | Source of the data |
|---|---|
| | *Morphological and Anatomical Parameters* |
| Body weight | Sacher, 1959<br>Korchagin et al., 1973<br>Allison and Cicchetti, 1976<br>Calder, 1976<br>Economos, 1980a; 1980c<br>Hofman, 1983; 1984<br>Zammuto, 1986<br>Prothero and Jürgens, 1987<br>Read and Harvey, 1989 |
| Body size | Golikov, 1985 |
| Female body weight | Harvey and Clutton-Brock, 1985<br>Harvey et al., 1987 |
| Male body weight | Harvey and Clutton-Brock, 1985<br>Harvey et al., 1987 |
| Weight of individual neonates | Harvey and Clutton-Brock, 1985<br>Harvey et al., 1987<br>Read and Harvey, 1989 |
| Brain weight | Friedental, 1910<br>Zepelin and Rechtschaffen, 1974<br>Allison and Cicchetti, 1976<br>Economos, 1980b<br>Cutler, 1984a |

Table *A* (continued)

| Feature | Source of the data |
|---|---|
| Brain weight | Harvey and Clutton-Brock, 1985<br>Harvey et al., 1987<br>Prothero and Jürgens, 1987<br>Read and Harvey, 1989 |
| Neonatal brain weight | Harvey and Clutton-Brock, 1985<br>Harvey et al., 1987 |
| Index of encephalisation | Sacher, 1959; 1977; 1978<br>Korchagin et al., 1973<br>Mallouk, 1975<br>Calder, 1976 |
| Surface of cerebral cortex | Hofman, 1983 |
| Adrenal weight | Economos, 1980c |
| Liver weight | Economos, 1980c<br>Frolkis and Muradian, 1988 |
| Kidney weight | Prothero and Jürgens, 1987<br>Frolkis and Muradian, 1988 |
| Heart weight | Frolkis and Muradian, 1988 |
| Spleen weight | Frolkis and Muradian, 1988 |
| Skeleton weight | Frolkis and Muradian, 1988 |
| Alimentary canal weight | Frolkis and Muradian, 1988 |
| *Ecological and Demographic parameters* | |
| Generation length (the average age of females giving birth to all offspring) | Millar and Zammuto, 1983 |

Table *A* (continued)

| Feature | Source of the data |
|---|---|
| Age at sexual maturation for females (age at maturity) | Korchagin et al., 1973<br>Millar and Zammuto, 1983<br>Harvey and Zammuto, 1985<br>Harvey and Clutton-Brock, 1985<br>Harvey et al., 1987<br>Read and Harvey, 1989<br>Nakano et al., 1990 |
| Age at first breeding for females (age at first reproduction) | Harvey and Clutton-Brock, 1985<br>Harvey and Zammuto, 1985<br>Harvey et al., 1987 |
| Age at sexual maturity for males | Harvey and Clutton-Brock, 1985<br>Harvey et al., 1987 |
| Post-sexual maturation period of life span | Korchagin et al., 1973<br>Read and Harvey, 1989 |
| Gestation time (period) | Korchagin et al., 1973<br>Zepelin and Rechtschaffen, 1974<br>Allison and Cicchetti, 1976<br>Harvey and Clutton-Brock, 1985<br>Harvey et al., 1987<br>Read and Harvey, 1989 |
| Age at weaning (average time from birth to independence from maternal milk) | Harvey and Clutton-Brock, 1985<br>Harvey et al., 1987 |
| Period of maternal investment (gestation period plus age at weaning) | Read and Harvey, 1989 |
| Period as an independent juvenile (average time from the end of weaning until sexual maturation) | Read and Harvey, 1989 |

Table *A* (continued)

| Feature | Source of the data |
|---|---|
| Interlitter interval (average time between the births of successive litters) | Harvey and Clutton-Brock, 1985<br>Harvey et al., 1987<br>Read and Harvey, 1989 |
| Litter weight (product of neonatal weight and number of offspring per litter) | Read and Harvey, 1989 |
| Number of litters (childbirths) during the life | Korchagin et al., 1973 |
| Annual biomass production (product of neonatal weight and annual fecundity) | Read and Harvey, 1989 |
| Life-style collectivity rated according to a four-point scale on the extent of species collectivism | Korchagin et al., 1973 |
| Sleep exposure (unsafety of sleeping place) | Allison and Cicchetti, 1976 |
| *Physiological and Biochemical parameters* | |
| Total diurnal duration of systole | Mineev and Gershkovich, 1979 |
| Length of lens development stage | Tréton and Courtois, 1989 |
| Lifespan of circulating mammalian erythrocytes *in vivo* | Röhme, 1981 |

Table *A* (continued)

| Feature | Source of the data |
|---|---|
| Duration of the sleep cycle (period of slow wave sleep + the succeeding period of paradoxical sleep time) | Zepelin and Rechtschaffen, 1974 |
| Replicative life span of mammalian fibroblasts *in vitro* (number of population doublings attained by cell cultures) | Röhme, 1981 |
| Superoxide dismutase activity | |
| in brain | Ono and Okada, 1984 |
| in liver of primate species | Cutler, 1986 |
| Lactate dehydrogenase activity | |
| in brain | Ono and Okada, 1984 |
| Diamine oxidase activity | |
| in liver | Cutler, 1984a |
| Urate level | |
| in blood plasma of primate species | Cutler, 1984c |
| in blood serum of primate species | Cutler, 1984e |
| in brain of primate species | Cutler, 1984e |
| α-tocopherol (vitamin E) level | |
| in blood plasma of mammalian species | Cutler, 1984c; 1984d; 1986 |
| Ascorbate (vitamin C) level | |
| in *adult* brain of mammalian species | Cutler, 1986 |

Table *A* (continued)

| Feature | Source of the data |
|---|---|
| Ceruloplasmin level in blood plasma of primate species | Cutler, 1984c |
| Retinol (vitamin A) level in blood serum of mammalian species | Cutler, 1984b |
| in blood serum of primate species | Cutler, 1984b |
| in brain normalised by the weight of total lipids (for primates only) | Cutler, 1984b |
| Cholesterol level in blood serum of mammalian species | Cutler, 1984a |
| Choline level in blood plasma of mammalian species | Cutler, 1984a |
| Inhibition activity of blood serum of the rate of autoxidation of rat liver homogenate | Cutler, 1984a |
| Induced DNA repair capacity | Hart and Setlow, 1974<br>Tréton and Courtois, 1982<br>Maslansky and Williams, 1985 |
| Liver superoxide dismutase activity/specific metabolic rate ratio | Tolmassoff et al., 1980 |
| Brain superoxide dismutase activity/specific metabolic rate ratio | Tolmassoff et al., 1980 |

Table *A* (continued)

| Feature | Source of the data |
| --- | --- |
| Heart superoxide dismutase activity/specific metabolic rate ratio | Tolmassoff et al., 1980 |
| Liver catalase activity/ specific metabolic rate ratio | Cutler, 1984a |
| Blood glucose level/specific metabolic rate level | Cutler, 1984a |

**Table B Features which show a negative correlation with the span of organisms (interspecies correlation)**

| Feature | Source of the data |
|---|---|
| *Anatomical and Morphological parameters* | |
| Spleen weight/body weight ratio | Calder, 1976 |
| *Ecological and Demographic parameters* | |
| Reproductive value at maturity (expectation of future offspring) | Millar and Zammuto, 1983 |
| Annual fecundity (number of offspring produced per year) | Read and Harvey, 1989 |
| Litter size (number born) | Korchagin et al., 1973<br>Millar and Zammuto, 1983<br>Zammuto, 1986<br>Read and Harvey, 1989 |
| Mortality rate in wild-type populations | Cutler, 1984a |
| *Physiological parameters* | |
| Total sleep time per day | Zepelin and Rechtschaffen, 1974 |
| Duration of slow-wave sleep stage per day | Zepelin and Rechtschaffen, 1974<br>Allison and Cicchetti, 1976 |
| Duration of paradoxical sleep stage per day | Zepelin and Rechtschaffen, 1974<br>Allison and Cicchetti, 1976 |

Table *B* (continued)

| Feature | Source of the data |
|---|---|
| Specific metabolic rate (cal/g per day) | Rubner, 1908<br>Zepelin and Rechtschaffen, 1974<br>Hansche, 1975<br>Cutler, 1984a |
| *Genetic and Biochemical parameters* | |
| Rate of chromosome aberration frequency | see Cutler, 1984a |
| Cytochrome P-448 levels in fibroblast cells of mammalian species | Pashko and Schwartz, 1982 |
| Cytochrome P-450 levels in fibroblast cells of mammalian species | Pashko and Schwartz, 1982<br>Cutler, 1984c |
| Cytochrome C level | |
| in liver | Cutler, 1984a |
| in kidney | " |
| in brain | " |
| Cytochrome oxidase activity | |
| in liver | Cutler, 1984a |
| in kidney | " |
| in brain | " |
| Rate of conversion of polycyclic hydrocarbon carcinogens to: | |
| mutagenic forms (measured by mutagenicity) | Schwartz, 1975<br>Schwartz and Moore, 1979 |
| forms capable of covalent binding to DNA | Schwartz and Moore, 1977<br>Schwartz and Moore, 1979 |
| water-soluble metabolites | Moore and Schwartz, 1978 |

Table *B* (continued)

| Feature | Source of the data |
|---|---|
| Catalase activity | |
| in liver of primate species | Cutler, 1984d |
| in kidney of primate species | Cutler, 1984d |
| Glutathione peroxidase (GPX) activity | |
| in blood (red blood cells) | Cutler, 1984c |
| in brain of mammalian species | Cutler, 1984a; 1984c; 1984d; 1986 |
| in liver of mammalian species | |
| total GPX activity (Se + non-Se) | Cutler, 1984a; 1984c; 1984d; 1986 |
| selenium dependent GPX activity | Cutler, 1984a; 1984c |
| 3-hydroxybutyrate dehydrogenase activity | |
| in liver | Cutler, 1984a |
| Choline esterase activity | |
| in brain | Ono and Okada, 1984 |
| Glutamic oxalacetic transaminase activity | |
| in brain | Ono and Okada, 1984 |
| Rate of protein turnover | |
| in whole body of mammalian species | Cutler, 1984a |
| in blood plasma | Spector, 1974 |

Table *B* (continued)

| Feature | Source of the data |
|---|---|
| Glutathione level | |
| in whole blood of mammalian species | Cutler, 1984a; 1986 |
| in brain of mammalian species | Cutler, 1984c; 1984d; 1986 |
| Ascorbate (vitamin C) level | |
| in liver of mammalian species | Cutler, 1984a; 1986 |
| in adrenal gland of mammalian species | Cutler, 1984a |
| in eye lens of mammalian species | Cutler, 1984a |
| in cerebrospinal fluid of mammalian species | Cutler, 1984a |
| Retinol (vitamin A) level | |
| in brain normalised by the weight of total lipids for primates only | Cutler, 1984b |
| Lipid peroxides level | |
| in blood serum of mammalian species | Cutler, 1984a; 1984c |
| Lipid peroxidation rate | |
| of erythrocytes exposed to hydrogen peroxide | Cutler, 1984a |
| Rate of superoxide anion | |
| radical ($O_2^{\bullet}$) generation | Sohal et al., 1989 |
| Free radicals level in tissues | Maréchal et al., 1973 |

Table *B* (continued)

| Feature | Source of the data |
|---|---|
| Rate of spontaneous autoxidation of aerobically incubated homogenate of: | |
| brain | Cutler, 1984a; 1984c; 1985a |
| kidney | Cutler, 1985a |
| Rate of oxidative damage to DNA | |
| measured by urine excretion of oxidised DNA bases | Adelman et al., 1988 |

**Table *C* Features which show a weak correlation with the life span of organisms (interspecies correlation coefficient $|r| \leq 0.30$)**

| Feature | Source of the data |
|---|---|
| *Ecological and Demographic parameters* | |
| Predation index (severity of predation) | Allison and Cicchetti, 1976 |
| Overall danger (general estimate of predatory danger) | Allison and Cicchetti, 1976 |
| Type of consumed food | Korchagin et al., 1973 |
| Total number of offspring born during the whole life | Korchagin et al., 1973 |
| Degree of self-dependency of new-born cubs (child) | Korchagin et al., 1973 |
| Residual reproductive value at maturity (reproductive value in the next age interval with respect to the probability of surviving to that interval) | Millar and Zammuto, 1983 |
| *Physiological and Biochemical parameters* | |
| Rate of paradoxical sleep time to total sleep time | Zepelin and Rechtschaffen, 1974 |
| Length of oestrous cycle | Harvey and Clutton-Brock, 1985 |
| Catalase activity in brain of primate species | Cutler, 1984d |

Table *C* (continued)

| Feature | Source of the data |
|---|---|
| Glutathione S-transferase activity in liver | Cutler, 1984c |
| Glutathione S-transferase (non-selenium glutathione peroxidase) activity in liver | Cutler, 1984a |
| Creatin phosphokinase activity in brain | Ono and Okada, 1984 |
| Glucose-6-phosphate dehydrogenase activity in brain | Ono and Okada, 1984 |
| D-aminoacid oxidase activity | |
| in liver | Cutler, 1984a |
| in kidney | Cutler, 1984a |
| Alkaline phosphatase activity | |
| in liver | Cutler, 1984a |
| in kidney | Cutler, 1984a |
| Diamine oxidase activity in kidney | Cutler, 1984a |
| Ceruloplasmin oxidase activity in blood | Cutler, 1984a |
| Ascorbate (vitamin C) level | |
| in blood plasma | Cutler, 1984c |
| in blood serum | Cutler, 1984e |
| in *young* brain of mammalian species | Cutler, 1986 |

Table *C* (continued)

| Feature | Source of the data |
|---|---|
| Retinol (vitamin A) level in blood serum of primate species | Cutler, 1984b |
| Urate level in blood plasma of nonprimate species | Cutler, 1984a |
| Blood chemistries: Albumin | Cutler, 1984a |
| alpha and beta globulins | " |
| minerals (Na, K, Ca, Mg) | " |
| dehydroepiandrosterone (DHEA) | " |
| thiamine | " |
| riboflavin | " |
| cyanocobalamine | " |
| nicotinic acid | " |
| pantothenic acid | " |
| selenium | " |
| glucose | " |
| blood urea nitrogen | " |
| Copper level in blood plasma | Cutler, 1984a |
| in erythrocyte | Cutler, 1984a |

Table *C* (continued)

| Feature | Source of the data |
|---|---|
| Selenium level in blood | Cutler, 1984a |
| in red blood cells | Cutler, 1984c |
| UV-induced DNA repair synthesis rate* | Kato, 1979<br>Kato et al., 1980 |
| *Other parameters* | |
| Number of population doublings attained by animal cell cultures *in vitro* | Stanley, 1975 |

* This result casts doubt on the previously obtained result of Hart and Setlow (1974) that there exists a positive interspecies correlation between life span and the rate of repair of induced DNA defects.

# REFERENCES


Abernethy, J.D. (1979). The exponential increase in mortality rate with age attributed to wearing-out of biological components. J.Theor.Biol., 80, pp. 333-354.

Abrams, H.L. (1985). Extreme human longevity and nutrition: an anthropological evaluation. J.Applied Nutrition, 37, pp. 78-88.

Acsádi, G. and J.Nemescéri (1970). History of human life span and mortality. Acad. Kiado, Budapest.

Adelman, R., R.L.Saul, and B.N.Ames (1988). Oxidation damage to DNA: relation to species metabolic rate and life span. Proc.Natl.Acad.Sci.USA, 85, pp. 2706-2708.

Aivazyan, S.A., I.S.Yenyukov, and L.D.Meshalkin (1985). Prikladnaya statistika [Applied statistics]. Finansy i Statistika, Moscow Univ.Press, Moscow. (In Russian).

Alexeeva, T.I. (1986). Adaptive reactions in human populations. Moscow. (In Russian).

Allemand, R., Y.Cohet and J.David (1973). Increase in the longevity of adult *Drosophila melanogaster* kept in permanent darkness. Exp.Gerontol., 8, 279-285.

Allison, T. and D.V.Cicchetti (1976). Sleep in mammals: Ecological and constitutional correlates. Science, 194, 732-734.

Alpatov, W.W. (1930). Experimental studies on the duration of life. XIII. The influence of different feeding during the larval and imaginal on the duration of life of the imago of *Drosophila melanogaster*. Amer.Natur., 64, 37-55.

Alpatov, V.V. and N.A.Gordeenko (1932). On the influence of mating on the duration of life in moths with special reference to the silk worm *Bombyx mori*. Zool.Zhurn., 11, 60-65. (In Russian).

Alpatov, W.W. and R.Pearl (1929). Experimental studies on the duration of life. XII. Influence of temperature during the larval period and adult life on the duration of life of the imago of *Drosophila melanogaster*. Amer.Natur., 63, 37-67.

Altukhov, Yu.P. (1989). Genetic processes in populations. Nauka, Moscow. (In Russian).

Altukhov, Yu.P., O.L.Kurbatova, O.K.Botviniev et al. (1981). Gene markers and diseases: genetic, anthropometric and clinical characteristics of children with acute pneumonia. Genetika, 17, 920-931. (In Russian).
Andersen, A.C. and L.S.Rosenblatt (1969). The effect of whole-body X-irradiation on the median lifespan of female dogs (beagles). Radiat.Res., 39, 177-200.
Andersen, J. (1953). Analysis of a Danish roe-deer population. Dan.Rev.Game Biol., 2, 127-155.
Andervont, H.B. (1945). Fate of the C3H milk influence in mice of strains C and C57Black. J.Nat.Cancer Inst., 5, 383-390.
Andervont, H.B. and W.J.McEleney (1941). Spontaneous tumors in a subline of strain C3H mice. J.Nat.Cancer Inst., 1, 737-744.
Andreev, E.M. (1987). The analysis of survival using data on causes of death. In E.B.Burlakova and L.A.Gavrilov (eds.) Populyatsionnaya gerontologiya [Population gerontology], vol.6. VINITI, Moscow, 190-229. (In Russian).
Andryszek, Cz. and T.Naganski (1979). Zastosowanie wybranych modeli teoretycznych w ocenie zapadalnosci na nowotwory zlasliwe wedlug wieku. Zrow.publ., N 90, 749-759.
Anisimov, V.N., A.S.Loktionov, V.G.Morozov et al. (1988). Increased life span and reduced tumour incidence in mice following the administration of polypeptide thymus and pineal factors, starting at different ages. Dokl.Akad.Nauk SSSR, 302, 473-476. (In Russian).
Antomonov, Yu.G. (1977). Modelirovanie biologicheskikh sistem [Modelling biological systems]. Naukova dumka, Kiev. (In Russian).
Arking, R. (1987). Successful selection for increased longevity in *Drosophila*: analysis of the survival data and presentation of a hypothesis on the genetic regulation of longevity. Exp.gerontol., 22, 199-220.
Arshavsky, I.A. (1982). The species-specific life span of mammals]. In Biologiya stareniya [Biology of aging]. Nauka, Leningrad, 24-38. (In Russian).
Aslan, A., A.Vrabiescu, C.Domilescu et al. (1965). Long-term treatment with procaine (Gerovital $H_3$) in albino rats. J.Gerontol., 20, 1-8.
Banfield, A.W.F. (1955). A provisional life table for the barren ground caribou. Canad.J.Zool., 33, 143-147.
Barkalow, F.S., R.B.Hamilton and R.F.Soots (1970). The vital statistics of an unexploited gray squirrel population. J.Wildlife Manag., 34, 489-500.
Barlow, C.H. and H.Muench (1951). Life span and monthly mortality rate of *Bulinus truncatus* and *Planorbis*

*boissyi*, the intermediate hosts of schistosomiasis in Egypt. J.Parasitol., 37, 165-173.

Barlow, R.E. and F.Proschan (1965). Mathematical theory of reliability. Wiley and Sons, NY etc.

Barlow, R.E. and F.Proschan (1975). Statistical theory of reliability and life testing probability models. Holt, Rinehart and Winston Inc., NY etc.

Barnett, H.A.R. (1982). The span of life. Abstract of the discussion. J.Inst.Actuaries, 109, 345-346.

Bartlett, P.M. and A.W.A.Murray (1986). Modelling adult survival in the laboratory of diapause and non-diapause Colorado beetle *Leptinotarsa decemlineata* (Coleoptera: Chrysomelidae) from Normandy, France. Ann.Appl.Biol., 108, 487-501.

Beard, R.E. (1959). Note on some mathematical mortality models. In G.E.W.Wolstenholme and M.O'Connor (eds.) The lifespan of animals. Little, Brown, Boston, 302-311.

Bednyi, M.S. (1979). Mediko-demograficheskoe izuchenie narodonaseleniya [Medico-demographic population research]. Statistika, Moscow. (In Russian).

Beeton, M. and K.Pearson (1901). On the inheritance of the duration of life and on the intensity of natural selection in man. Biometrika, 1, 50-89.

Befolkningsförändringar. D.3. Hela ricket och länen (1970; 1971; 1972; 1973; 1974; 1975; 1976; 1977; 1978; 1979; 1980; 1981. Sveriges Officiella Statististik, Stockholm.

Beier, W. (1980). A mathematical relationship between growth rate and life span. Mech.Ageing and Devel., 13, 401-406.

Beklová, M. and J.Pikula. (1985). Age composition of some game bird species in Czechoslovakia. Prirodovedne prace ustavu Ceskoslovenske akademie ved v Brne, 19, 1-52.

Bell, E., L.F.Marek, D.S.Levinstone et al. (1978). Loss of division potential in vitro: Aging or differentiation? Science, 202, 1158-1422.

Benjamin, B. (1982). The span of life. J.Inst.Actuaries, 109, 319-340.

Berdyshev, G.D. and Sh.A.Asadov (1985). On the maximal chronological age of man. Z.Alternsforsch., 40, 289-294.

Bever, K. and H.Sprankel (1986). A contribution to the longevity of *Tupaia glis* Diard, 1820 in captivity. Z.Versuchstierkd., 28, 3-5.

Bezen, J.J., J.W.Sluiter and P.F.Van Heerdt (1960). Population statistics of five species of the bat genus Myotis and one of the genus Rhinolophus, hibernating in the caves of S.Limburg. Arch.Néerland.Zool., 13, 511-539.

Bileva, D.S. and A.A.Malinovsky (1981). The duration of life of *Drosophila melanogaster* hybrids depending on composition of nutrient medium. Genetika, 17, 469-475. (In Russian).

Bileva, D.S., L.N.Zimina and A.A.Malinovsky (1978). Effect of genotype and environment on the longevity of *Drosophila melanogaster*. Genetika, 14, 848-852. (In Russian).

Birch, L.C. (1953). Experimental background to the study of the distribution and abundance of insects. Ecology, 34, 698-711.

Bittner, J.J. (1935a). The breeding behavior and tumor incidence of an inbred albino strain of mice. Amer.J.Cancer, 25, 113-121.

Bittner, J.J. (1935b). The breeding behavior and tumor incidence of a black agouti stock of mice. Amer.J.Cancer, 25, 614-620.

Bittner, J.J. (1935c). Differences observed in the tumor incidence of an albino strain of mice following a change in diet. Amer.J.Cancer, 25, 791-796.

Bittner, J.J. (1936). Tumor incidence in reciprocal $F_1$ hybrid mice - A x D high tumor stocks. Proc.Soc.Exp.Biol. and Med., 34, 42-48.

Bittner, J.J. (1937). Mammary tumors in mice in relation to nursing. Amer.J.Cancer, 30, 530-538.

Blokhinov, Yu.B. (1982). Describing the age-dependent mortality of a population. In A.N.Voroshchuk (ed.) Modelirovanie protsessov ekologicheskogo razvitiya [Modelling ecological development processes]. Collected papers. Issue 2. The Institute for System Studies, Moscow, 86-91. (In Russian).

Blus, L.J. (1971). Reproduction and survival of short-tailed shrews (*Blarina brevicauda*) in captivity. Lab.Anim.Sci., 21, 884-891.

Bodenheimer, F.S. (1958). Animal ecology today. Junk, Den Haag.

Boorman, G.A. and C.F.Hollander (1973). Spontaneous lesions in the female WAG/$R_{ij}$ (Wistar) rat. J.Gerontol., 28, 152-159.

Borgan, J.-K. (1983). Kohort-Dødeligheten i Norge, 1846-1980. Statist.Sentralbyrå, Oslo.

Bourgeois-Pichat, J. (1952). Essai sur la mortalité biologique de l'homme. Population, 3, 381-394.

Bourgeois-Pichat, J. (1979). Future outlook for mortality decline in the world. In Prospects of population: Methodology and assumptions. United Nations, NY, 227-266.

Bozcuk, A.N. (1978). The effect of some genotypes on the longevity of adult Drosophila. Exp.Gerontol., 13, 279-

285.
Bremermann, H.J. (1982). Reliability of proliferation controls. The Hayflick limit and its breakdown in cancer. J.Theor.Biol., 97, 641-662.
Briegel, H. and C.Kaiser (1973). Life-span of mosquitoes (Culicidae, Diptera) under laboratory conditions. Gerontologia, 19, 240-249.
Brillinger, D.R. (1961). A justification of some common laws of mortality. Trans.Amer.Soc.Actuaries, 13, 116-119.
Brown, K.S. and W.F.Forbes (1974). A mathematical model of aging processes. J.Gerontol., 29, 46-51.
Bruk, S.I. (1981). Naselenie mira [World population]. Nauka, Moscow. (In Russian).
Buldakov, L.A., S.I.Dementiev and T.I.Levdik (1987). The lifetime of animals subjected to the combined effect of whole-body $\gamma$-radiation and incorporated plutonium 239. Radiobiologiya, 27, 123-126. (In Russian).
Byrne, J. and D.Warburton (1987). Male excess among anatomically normal fetuses in spontaneous abortion. Am.J.Med.Genet., 26, 605-611.
Calder, W.A. (1976). Aging in vertebrates: Allometric considerations of spleen size and life span. Fed.Proc., 35, 96-97.
Campbell, R.C. (1967). Statistics for biologists. Cambridge Univ.Press, Cambridge.
Carrel, A. (1912). On the permanent life of tissues outside of the organism. J.Exp.Med., 15, 516-528.
Caughley, G. (1966). Mortality patterns in mammals. Ecology, 47, 906-918.
Caughley, G. (1976). Wildlife management and the dynamics of ungulate populations. In T.H.Coaker (ed.). Applied biology, vol.1. Acad.Press, London, 183-246.
Caughley, G. (1977). Analysis of vertebrate populations. Wiley and Sons. London, NY, Sydney, Toronto.
Chebotarev, D.F. (1978). Results and prospects for the development of gerontology in the USSR. In Gerontologiya i geriatriya. 1978. Sovremennye problemy gerontologii [Gerontology and geriatrics. 1978. Contemporary issues in gerontology]. Inst. of Gerontology, Kiev, 8-25. (In Russian).
Chebotarev, D.F. (1986). The "life extension" programme - an important stage in the development of Soviet gerontology. Vestn.AMN SSSR, 10, 3-8. (In Russian).
Chebotarev, D.F. (1987). Prospects for the development of research in gerontology. Vestn.AMN SSSR, 11, 47-52. (In Russian).
Cheney, K.E., R.K.Liu, G.S.Smith et al. (1980). Survival and disease patterns in C57BL/6J mice subjected to

undernutrion. Exp.Gerontol., 15, 237-258.
Chiang, Ch.L. (1978). Life table and mortality analysis. World Health Organisation, Geneva.
Clapp, N.K., L.C.Satterfield and N.D.Bowles (1979). Effects of the antioxidant butylated hydroxytoluene (BHT) on mortality in BALB/c mice. J.Gerontol., 34, 497-501.
Clark, A.M. and M.Rockstein (1964). Aging in insects. In Physiology of Insecta. Acad.Press, NY, vol.1, 227-281.
Clarke, R.D. (1982). The span of life. Abstract of the discussion. J.Inst.Actuaries, 109, 349-350.
Collatz, K.-G. and H.Wilps (1983). Alten bei Insekten. Naturwiss.Rdsch., 36, 57-63.
Comfort, A. (1956). Longevity and mortality of Irish wolfhounds. Proc.Zool.Soc.London, 127, 27-34.
Comfort, A. (1957a). The duration of life in molluscs. Proc.Malacol.Soc.London, 32, 6, 219-241.
Comfort, A. (1957b). Survival curves of mammals in captivity. Proc.Zool.Soc.London, 128, 349-364.
Comfort, A. (1958). The longevity and mortality of thoroughbred mares. J.Gerontol., 13, 342-350.
Comfort, A. (1960). Longevity and mortality in dogs of four breeds. J.Gerontol., 15, 126-129.
Comfort, A. (1961). A life table for Arabian mares. J.Gerontol., 17, 14.
Comfort, A. (1963). Effect of delayed and resumed growth on the longevity of a fish (*Lebistes reticulatus* Paters) in captivity. Gerontologia, 8, 150-155.
Comfort, A. (1964). Ageing. The Biology of Senescence. Holt, Rinehart and Winston, Inc., NY, Chicago, San Francisco.
Comfort, A. (1979). The Biology of Senescence. 3rd edition. Elsevier, London.
Comfort, A. (1984). The Biology of Senescence. Elsevier, NY.
Conti, C.J., N.Clapp, A.J.P.Klein-Szanto et al. (1985). Survival curves and incidence of neoplastic disease in SENCAR mice. Carcinogenesis, 6, 1649-1652.
Cork, J.M. (1957). Gamma-radiation and longevity of the flour beetle. Radiat.Res., 7, 551-557.
Coulson, J.C. (1960). A study of the mortality of the starling based on ringing recoveries. J.Anim.Ecology, 29, 257-271.
Cramer, F. (1981). Is ageing a biochemical problem? On the accuracy of protein biosynthesis. Science and Scientists Essays Biochem.Biol.and Chemists, 147-152.
Curtis, M.R., F.D.Bullock and W.F.Dunning (1931). A statistical study of the occurence of spontaneous tumors in a large colony of rats. Amer.J.Cancer, 15, 67-121.
Cutler, R.G. (1981). Life span extension. In J.L.McGaugh and S.B.Kiesler (eds.) Aging: Biology and behaviour.

Acad.Press, NY, 31-76.

Cutler, R.G. (1983). Superoxide dismutase, longevity and specific metabolic rate. Gerontology, 29, 113.

Cutler, R.G. (1984a). Evolutionary biology of aging and longevity in mammalian species. In J.E.Johnson (ed.) Aging and cell function. Plenum Press, NY, London, 1-147.

Cutler, R.G. (1984b). Carotenoids and retinol: Their possible importance in determining longevity of mammalian species. Proc.Natl.Acad.Sci.USA, 81, 7627-7631.

Cutler, R.G. (1984c). Antioxidants, aging and longevity. In W.Pryor (ed.) Free Radicals in Biology, vol.6. Acad.Press, NY, 371-428.

Cutler, R.G. (1984d). Antioxidants and longevity. In D.Armstrong, R.S.Sohal, R.G.Cutler, T.F.Slater (eds.). Free Radicals in Molecular Biology, Aging and Disease. Raven Press, NY, 235-266.

Cutler, R.G. (1984e). Urate and ascorbate: their possible roles as antioxidants in determining longevity of mammalian species. Arch.Gerontol.Geriatr., 3, 321-348.

Cutler, R.G. (1985a). Peroxide-producing potential of tissues: inverse correlation with longevity of mammalian species. Proc.Natl.Acad.Sci.USA, 82, 4798-4802.

Cutler, R.G. (1985b). Antioxidants and longevity of mammalian species. In A.D.Woodhead, A.D.Blackett and A.Hollander (eds.) Molecular Biology of Aging. Proc. Symp., Upton, NY, Sept.30-Oct.3, 1984, NY, 15-73.

Cutler, R.G. (1986). Aging and oxygen radicals. In A.E.Taylor, S.Matalon and P.Ward (eds.) Physiology of oxygen radicals. Clin.Monogr.Ser., Am.Physiol.Soc., Bethesda, 251-286.

Cutler, S.J. and F.Ederer (1958). Maximum utilization of the life table methodology in analyzing survival. J.Chron.Dis., 8, 699-712.

Danstedt, R.T. (1975). Local geographical variation in demographic parameters and body size of *Desmognathus fucsus* (Amphibia: Plethodontidae). Ecology, 56, 1054-1067.

Darwin, C. (1847). The descent of man. Murray, London.

Davidovsky, I.V. (1962). The problem of causality in medicine. Meditsina, Moscow. (In Russian).

Davies, D.M. (1953). Longevity of black flies in captivity. Can.J.Zool., 31, 304-312.

Davis, M.B. (1945). The effect of population density on longevity in *Trogoderma versicolor* Creutz. (-T.Inclusa Leo.). Ecology, 26, 353-362.

Deevey, E.S. (1947). Life tables for natural populations of

animals. Quart.Rev.Biol., 22, 283-314.
Deevey, G.B. and E.S.Deevey (1945). A life table for the black widow. Trans.Connecticut Acad.Arts and Sci., 36, 115-134.
De Kock, K.N. (1985). Effect of programmed circadian temperature fluctuations on population dynamics of *Bulinus tropicus* (Krauss) and *Lymnaea natalensis* Krauss (Gastropoda: Mollusca). J.Limnol.Soc.Sth.Afr., 11, 71-74.
De Kock, K.N. and J.A.van Eeden (1976). The way in which different cohort sizes affect the value of the population statistics $r_m$ of a fresh water snail species. Wetenskaplike Bydraes van die PU vir CHO, Reeks B: Natuurwetenskappe, 80, 1-20.
De Kock, K.N. and J.A.van Eeden (1981). Life table studies on freshwater snails. The effect of constant temperature on the population dynamics of *Biomphalaria pfeifferi* (Krauss). Wetenskaplike Bydraes van die PU vir CHO, Reeks B: Natuurwetenskappe, 105, 1-17.
De Kock, K.N. and J.A.van Eeden (1985). Effect of constant temperature on population dynamics of *Bulinus tropicus* (Krauss) and *Lymnaea natalensis* Krauss. J.Limnol.Soc.Sth.Afr., 11, 27-31.
De Kock, K.N., J.A.van Eeden and S.J.Pretorius (1986). Effect of laboratory breeding on the population dynamics of successive generations of the freshwater snail *Bulinus tropicus* (Krauss). South Afr.J.of Science, 82, 369-372.
Del Maestro, R.F. (1980). An approach to free radicals in medicine and biology. Acta physiol.scand., 492, suppl., 153-168.
Demographic yearbook. 1961, vol.13; 1967, vol.18, 1974, vol.26. United Nations, NY.
Depoid, F. (1973). La mortalité des grand viellards. Population, 28, 755-792.
Descartes, R. Izbrannye proizvedeniya [Selected works]. Gospolitizdat, Moscow, 1950, p.711. (Russian translation).
Diamond, J.M. (1987). Causes of death before birth. Nature, 329, 487-488.
Dilman, V.M. (1981). The law of deviation of homeostasis and diseases of ageing. John Wright PSG Inc., Boston.
Dilman, V.M. (1986). Ontogenetic model of ageing and disease formation and mechanisms of natural selection. J.Theor. Biol., 118, 73-81.
Dilman, V.M. (1987). Four models of medicine. Meditsina, Leningrad. (In Russian).
Doll, R. and R.Peto (1981). The causes of cancer. Oxford

Univ.Press, Oxford.

Données de démographie régionale, 1954. Étud.Démogr., 1964, N 8.

Ďoubal, S. (1982). Theory of reliability, biological systems and aging. Mech.Ageing and Devel., 18, 339-353.

Dubina, T.L. and A.N.Razumovich (1975). Vvedenie v eksperimental'nuyu gerontologiyu [Introduction to experimental gerontology]. Nauka i Tekhnika, Minsk. (In Russian).

Dublin,L.I. and A.J.Lotka (1936). Length of life. A study of the life table. The Ronald Press Co., NY.

Dunning, W.F. and M.R.Curtis (1946). The respective roles of longevity and genetic specificity in the occurence of spontaneous tumors in the hybrids between two inbred strains of rats. Cancer Res., 6, 61-81.

Durbin, P.W., M.H.Williams, N.Jeung et al. (1966). Development of spontaneous mammary-tumors over the life-span of the female Charles River (Sprague-Dawley) rat: The influence of ovariectomy, thyroidectomy and adrenalectomy-ovariectomy. Cancer Res., 26, 1, 400-411.

Dyad'kova, A.M. and E.A.Lotosh (1962). The oncological characteristics of a new line of laboratory mice DBR. Voprosy onkologii, 8, 46-47. (In Russian).

Dyad'kova, A.M. and N.N.Medvedev (1956). The oncological characteristics of mice of a low cancer line C57 Black. Voprosy onkologii, 2, 201-203. (In Russian).

Ebeling, A.H. (1913). The permanent life of connective tissue outside of the organism. J.Exp.Med., 17, 273-285.

Economos, A.C. (1980a). Taxonomic differences in the mammalian life span-body weight relationship and the problem of brain weight. Gerontology, 26, 90-98.

Economos, A.C. (1980b). Kinetics of metazoan mortality. J.Social Biol.Struct., 3, 317-329.

Economos, A.C. (1980c). Brain-life span conjecture: A reevaluation of the evidence. Gerontology, 26, 82-89.

Economos, A.C. (1982). Mammalian design and rate of living. Exp.Gerontol., 17, 145-152.

Economos, A.C. (1983). Rate of aging, rate of dying and the mechanism of mortality. Arch.Gerontol. and Geriatr., 1, 3-27.

Economos, A.C. (1985). Rate of aging, rate of dying and non-Gompertzian mortality - encore... Gerontology, 31, 106-111.

Egami, N. and H.Etoh (1969). Life span data for the small fish *Oryzias latipes*. Exp.Gerontol., 4, 127-129.

Egami, N. and H.Etoh (1973). Effect of x-irradiation during embryonic stage on life span in the fish, *Oryzias latipes*. Exp.Gerontol., 8, 219-222.

Eisen, M.J. (1942). The occurence of benign and malignant mammary lesions in rats treated with crystalline estrogen. Cancer Res., 2, 632-644.
Eisner, E. (1967). Actuarial data for the Bengalese finch (*Lonchura striata*: Fam.Estrildidae) in captivity. Exp.Gerontol., 2, 187-189.
Eklund, J. and G.E.Bradford (1977). Longevity and lifetime body weight in mice selected for rapid growth. Nature, 265, 48-49.
Ekonomov, A.L. and V.N.Yaryagin (1989). The age dynamics of mortality and the Gompertz-Makeham law. Zhurn.Obshch.Biol., 50, 236-243. (In Russian).
Ekonomov, A.L., C.L.Rudd and A.J.Lomakin (1989). Actuarial aging rate is not constant within the human life span. Gerontology, 35, 113-120.
El Shaarawi A., R.L.Prentice and W.F.Forbes (1974). the goodness of fit of certain aging models. J.Chron.Diseases, 27, 377-385.
Emanuel, N.M. (1982). Antioxidants in life prologation. In Biologiya stareniya [The biology of aging]. Nauka, Leningrad, 569-585. (In Russian).
Ermakov, S.P. and N.S.Gavrilova (1987). Primary statistical analysis of data on the survival of organisms. In E.B.Burlakova and L.A.Gavrilov (eds.) Populyatsionnaya gerontologiya [Population gerontology]. VINITI, Moscow, vol.6, 230-276. (In Russian).
Evans, F.C. and F.E.Smith (1952). The intrinsic rate of natural increase for the human louse, *Pediculus humanus* L. Amer.Natur., 86, 299-310.
Falk, J.W. and J.S.Millar (1987). Reproduction by female *Zapus princeps* in relation to age, size, and body fat. Can.J.Zool., 65, 568-571.
Feldman-Muhsam, B. and H.V.Muhsam (1945). Life tables for *Musca vicina* and *Callifora erythrocephala*. Proc.Zool.Soc.London, 115, 296-305.
Feller, W. (1968). An introduction to probability theory and its applications. Vol.1, Wiley and Sons, NY.
Fenyuk B. and M.Sheikina (1940). Length of life of *Microtus arvalis* Pall. Vestn.Microbiol.,Epidemiol.i Parasitol., 19, 571-589. (In Russian).
Fernandes, G., E.J.Yunis and R.A.Good (1976). Influence of diet on survival of mice. Proc.Nat.Acad.Sci.USA, 73, 1279-1283.
Ferris, J.C. (1932). A comparison of the life-histories of mictic and amictic females in the rotifer, *Hydantina senta*. Biol.Bull., 63, 442-455.
Finch, C.E. (1976). The regulation of physiological changes during mammalian aging. Quart.Rev.Biol., 51, 49-83.

Findeis, P.M., C.J.Barinaga, J.D.Willett et al. (1983). Age-synchronous culture of *Caenorhabditis elegans*: technique and applications. Exp.Gerontol., 18, 263-275.

Flourens, M.J.P. (1856). De la longevité humain et de quantité de la vie sur le globe. Paris. Quoted in Urlanis (1978), p.243.

Fortner, J.G. (1961). The influence of castration on spontaneous tumorigenesis in the Syrian (golden) hamster. Cancer Res., 21, 1491-1498.

Frank, P.W., C.D.Boll and R.W.Kelly (1957). Vital statistics of laboratory cultures of *Daphnia pulex* De Geer as related to density. Physiol.Zool., 30, 287-305.

Frick, M.H., O.Elo, K.Haapa et al. (1987). Helsinki Heart Study: primary-prevention trial with gemfibrozil in middle-aged men with dyslipidemia. N.Engl.J.Med., 317, 1237-1245.

Friedenthal, H. (1910). Uber die Gültigkeit der Massenwirkung für den Energieumsatz der lebendigen Substanz. Zbl.Physiol., 24, 321-327.

Friedman, D.B. and T.E.Johnson (1988). Three mutants that extend both mean and maximum life span of the nematode, *Caenorhabditis elegans*, define the age-1 gene. J.Gerontol., 43, B102-109.

Fries, J.F. (1980). Aging, natural death and the compression of morbidity. N.Engl.J.Med., 303, 130-135.

Fries, J.F. (1983). The compression of morbidity. Health and Society, 61, 397-419.

Frolov, I.T. (1983). Perspektivy cheloveka [Human prospects]. Politizdat, Moscow. (In Russian).

Frolkis, V.V. and Kh.K.Muradian (1988). Eksperimental'nye podkhody k prodleniyu zhizni [Experimental approaches to life extension]. Nauka, Leningrad. (In Russian).

Galambos, J. (1978). The asymptotic theory of extreme order statistics. Wiley and Sons, NY.

Gardner, G. and H.Hurst (1933). Life tables for white leghorn chickens in the State of Utah. Proc.of the Utah Acad.Sci., 10, 149-150.

Gardner, M.B., B.E.Henderson, R.W.Rongey et al. (1973). Spontaneous tumors of aging wild house mice. Incidence, pathology, and c-type virus expression. J.Natl.Cancer Inst., 56, 719-734.

Garg, M.L., B.R.Rao and C.K.Redmond (1970). Maximum-likelihood estimation of the parameters of the Gompertz survival function. Appl.Statist., 19, 152-159.

Gavrilov, L.A. (1978). A mathematical model of the aging of animals. Dokl.Akad.Nauk SSSR, 238, 490-492. (In Russian).

Gavrilov, L.A. (1980). Issledovanie genetiki

prodolzhitel'nosti zhizni s pomoshchu kineticheskogo analiza [Investigation of life span genetics using kinetic analysis]. Dissertation. Moscow State University, Moscow. (In Russian).

Gavrilov, L.A. (1984a). Population-statistical approaches to the study of life span. In Biologicheskie problemy stareniya [Biological problems of aging]. VINITI, Moscow, 135-171. (In Russian).

Gavrilov, L.A. (1984b). Kolichestvennye zakonomernosti prodolzhitel'nosti zhizni cheloveka i laboratornykh zhivotnykh [Quantitative regularities of the human and laboratory animals' life span]. Nauchnyi Tsentr Biologicheskikh Issledovanii AN SSSR, Pushchino. (In Russian).

Gavrilov, L.A. (1984c). Is there a limit to the life span of organisms? Biofizika, 29, 908-909. (In Russian).

Gavrilov, L.A. (1985). Mozhet li chelovek zhit' dol'she [Can humans live longer?]. Mysl', Moscow. (In Russian).

Gavrilov, L.A. (1987). Critical analysis of mathematical models of aging, mortality and life span. In E.B.Burlakova and L.A.Gavrilov (eds.) Populyatsionnaya gerontologiya [Population gerontology]. VINITI, Moscow, vol.6, 155-189. (In Russian).

Gavrilov, L.A. (1988a). Reliability and aging of biological systems. Zhurn.Obshch.Biol., 49, 571-572. (In Russian).

Gavrilov, L.A. (1988b). The biology of life span: the subject, methods and prospects for research. In Biologicheskie problemy stareniya i uvelicheniya prodolzhitel'nosti zhizni [Biological problems of aging and life span extension]. Nauka, Moscow, 106-110. (In Russian).

Gavrilov, L.A. (1988c). Biologico-demographic aspects of life span research. In Demograficheskie issledovaniya [Demographic Studies]. Moscow State University Press, Moscow, 105-121. (In Russian).

Gavrilov, L.A. (1988d). What determines the rate of human aging? In V Vsesoyuznyi S'ezd Gerontologov [Proc.5th All-Union Congress of Gerontologists and Geriatricians]. Institute of Gerontology, Kiev, 136. (In Russian).

Gavrilov, L.A. (1989a). 5th All-Union Congress of Gerontologists and Geriatricians. Voprosy Meditsinskoi Khimii, 35, 137-138. (In Russian).

Gavrilov, L.A. (1989b). "The paradox of superfluous causes of death" and its significance for the biology of life span. In Doklady MOIP 1987. Obshchaya Biologiya [MOIP Papers 1987. General Biology]. Nauka, Moscow, 116-119. (In Russian).

Gavrilov, L.A. (1990). ¿Podrá prolongarse la vida del

hombre? Mir, Moscow. (In Spanish).
Gavrilov, L.A. and N.S.Gavrilova (1978). Culture of diploid cells as an object for studying the molecular mechanisms of aging. Uspekhi sovrem.biol., 85, 267-283. (In Russian).
Gavrilov, L.A. and N.S.Gavrilova (1979a). The study of human mortality kinetics in historical aspect. Dokl.Akad.Nauk SSSR, 245, 1017-1020. (In Russian).
Gavrilov, L.A. and N.S.Gavrilova (1979b). Determination of species specific life span. Dokl.Akad.Nauk SSSR, 246, 465-469. (In Russian).
Gavrilov, L.A. and N.S.Gavrilova (1982). Aging in cell cultures. In Biologiya stareniya [The biology of aging]. Nauka, Leningrad, 236-247. (In Russian).
Gavrilov, L.A. and N.S.Gavrilova (1984). Historically stable indices of mortality and their use for medical geography. Geographia Medica, 14, 305-306.
Gavrilov, L.A. and N.S.Gavrilova (1985). Human species-specific life-span: Its estimation and interpretation. In Proc.1st Intern.Congr.Biomed.Gerontol., Amer.Aging Assoc., NY, 20. See also Age, 8, 94.
Gavrilov, L.A. and N.S.Gavrilova (1986). Biologiya prodolzhitel'nosti zhizni [The biology of life span]. Nauka, Moscow. (In Russian).
Gavrilov, L.A. and N.S.Gavrilova (1989). Life span is determined by resting metabolic rate of parents only! Age, 12, 113.
Gavrilov, L.A. and V.N.Nosov (1985). A new trend in human mortality decline: Derectangularization of the survival curve. In Proc.1st Intern.Congr.Biomed.Gerontol., Amer. Aging Assoc., NY, 19. See also Age, 8, 93.
Gavrilov, L.A. and L.S.Yaguzhinsky (1978). The kinetics of the growth of a human fibroblast culture. Biofizika, 23, 1041-1045. (In Russian).
Gavrilov, L.A., N.S.Gavrilova and V.N.Nosov (1983). Human life span stopped increasing: why? Gerontology, 29, 176-180.
Gavrilov, L.A., N.S.Gavrilova and V.G.Semyonova (1985). Epidemiological approach to the biology of human life span. Geographia medica, 15, 40-64.
Gavrilov, L.A., N.S.Gavrilova and V.G.Semyonova (1986). The life span. System analysis. In Demograficheskie protsessy i ikh zakonomernosti [Demographic processes and their regularities]. Mysl', Moscow, 89-109. (In Russian).
Gavrilov, L.A., N.S.Gavrilova and L.S.Yaguzhinsky (1978). The main regularities of animal aging and death viewed in terms of reliability theory. Zhurn.Obshch.Biol., 39,

734-742. (In Russian).

Gavrilov, L.A., Z.B.Mazurkevich and N.S.Gavrilova (1984). Testing aging hypotheses on the basis of long-livers' survival data. In Dolgozhiteli [Long-livers]. Metsniereba, Tbilisi, 67-68. (In Russian).

Gavrilov, L.A., V.G.Semyonova and N.S.Gavrilova (1982). On the possibility of forecasting mortality by methods of biological kinetics. In Voprosy vosproizvodstva naseleniya i demograficheskoi politiki [Issues of population reproduction and demographic policies]. Institute of Sociological Studies of the Soviet Academy of Sciences, Moscow, 108-123. (In Russian).

Gavrilova, N.S. (1982). Study of biological characteristics for the human life span. Dissertation. Moscow State University, Moscow. (In Russian).

Gavrilova, N.S. and L.A.Gavrilov (1982a). A new method of evaluating the prospects of influence which extend life-duration. Rejuvenation, 10, 14-15.

Gavrilova, N.S. and L.A.Gavrilov (1982b). Kinetic studies of survival in the elderly. In Aktual'nye voprosy gerontologii i geriatrii [Current issues in gerontology and geriatrics]. Metsniereba, Tbilisi. (In Russian).

Gavrilova, N.S., V.G.Semyonova and L.A.Gavrilov (1979). A new principle for testing mathematical theories of aging, mortality and life span. In Gerontologiya i geriatriya. Prodlenie zhizni: prognozy, mekhanizmy, kontrol' [Gerontology and geriatrics. Life extension: forecasts, mechanisms, verification]. Institute of Gerontology, Kiev, 30-34. (In Russian).

Gavrilova, N.S., V.G.Semyonova and L.A.Gavrilov (1983). The card-index of data on the human life span. In Problemy biologii stareniya [Problems in the biology of aging]. Nauka, Moscow, 71-76. (In Russian).

Gehan, E.A. (1975). Statistical methods for survival time studies. In M.J.Staquet (ed.): Cancer therapy, Prognostic factors and criteria of response. Raven Press, NY, 7-35.

Gehan, E.A. and M.M.Siddiqui (1973). Simple regression methods for survival time studies. J.Am.Stat.Assoc., 68, 848-856.

Geller, S.A. (1983). Autopsy. Sci.American, 248, 110-122.

Gelman, R., A.Watson, R.Bronson et al. (1988). Murine chromosomal regions correlated with longevity. Genetics, 118, 693-704.

Gelshtein, V.I. and A.M.Dyad'kova (1955). Further data on the oncological characteristics of laboratory mice from strain $CC_{57}$ brown. Voprosy onkologii, 1, 32-39. (In Russian).

Geodakian, V.A. (1982). Sexual dimorphism and the evolution of duration of ontogenesis and its stages. In V.J.A.Novák and J.Mlikovsky (eds.) Evolution and Environment. Proc.Int.Symp.Brno, 1981. CSAV, Praha, 229-237.

Geodakian, V.A. (1983). Sexual dimorphism in the pattern of aging and human mortality. In Problemy biologii stareniya [Problems in the biology of aging]. Nauka, Moscow, 103-110. (In Russian).

Georgiana, S.M. (1949). Longevity of the parasitic wasp *Habrobracon Juglandis* Ashmead. Amer.Natur., 83, 39-48.

Gibbons, J.W. (1987). Why do turtles live so long? BioScience, 37, 262-269.

Giess, M.C. (1980). Differences between natural ageing and radio-induced shortening of the life expectancy in *Drosophila melanogaster*. Gerontology, 26, 301-310.

Gilbert, Ch. and J.Gillman (1958). Spontaneous neoplasms in the albino rat. S.Afr.J.Med.Sci., 23, 257-272.

Gläser, H.-J. (1986). Parthenogenese bei *Carassius*, sexuelle Fortpflanzung und Theorie des Alterns. Biol.Zentralbl., 105, 475-489.

Glasunov, I.S, V.Grabauskas, W.W.Holland et al. (1983). An integrated programme for the prevention and control of non-communicable diseases: a Kaunas report. J.Chron. Diseases, 36, 419-426.

Goddard, J. (1970). Age criteria and vital statistics of a black rhinoceros population. E.Afr.Wildlife J., 8, 105-121.

Goldstein, S., S.Niewarowski and D.P.Singal (1975). Pathological implications of cell aging in vitro. Fed.Proc., 34, 56-63.

Goldstein, S. and D.P.Singal (1974). Senescence of cultured human fibroblasts: Mitotic versus metabolic time. Exp.Cell Res., 88, 359-364.

Golikov, A.N. (1985). On the variability of size and life span in the course of evolution. Zh.Obshch.Biol., 46, 771-777. (In Russian).

Gompertz, B. (1825). On the nature of the function expressive of the law of human mortality and on a new mode of determining life contingencies. Philos.Trans.Roy.Soc.London A, 115, 513-585.

Gonzales, B.M. (1923). Experimental studies on the duration of life. VIII. The influence upon duration of life of certain mutant genes of *Drosophila melanogaster*. Amer.Natur., 57, 289-328.

Goodbody, I. (1962). The biology of *Ascidia nigra* (Savigny). 1. Survival and mortality in an adult population. Biol.Bull., 122, 40-51.

Goodrick, C.L. (1974). The effect of exercise on longevity and behavior of hybrid mice which differ in coat color. J.Gerontol., 29, 129-133.

Goodrick, C.L. (1980). Effects of long-term voluntary wheel exercise on male and female Wistar rats. 1.Longevity, body weight, and metabolic rate. Gerontology, 26, 22-33.

Gordon, H.A., E.Bruckner-Kardoss and B.S.Wostmann (1966). Aging in germ-free mice: Life tables and lesions observed at natural death. J.Gerontol., 21, 380-387.

Goskomstat SSSR (1989). Naselenie SSSR 1988: Statisticheskii ezhegodnik [USSR population 1988: Statistical yearbook]. Finansy i Statistika, Moscow. (In Russian).

Gould, A.B. and A.M.Clark (1977). X-ray induced mutations causing adult life-shortening in *Drosophila melanogaster*. Exp.Gerontol., 12, 107-112.

Gowen, J.W. (1931). Metabolism as related to chromosome structure and the duration of life. J.Gen.Physiol., 14, 463-472.

Grabauskas, V., R.Prochorskas and I.Miseviciene (1985). Risk factors as indicators of ill health. In Proc.Intern. Conf.Preventive Cardiology. Inst.Prev.Cardiol., Moscow, 24.

Greenwood, M. (1928). "Laws" of mortality from the biological point of view. J.Hyg., 28, 267-294.

Greenwood, M. and J.O.Irwin (1939). The biostatistics of senility. Hum.Biol., 11, 1-23.

Greiff, D. (1940). Longevity in *Drosophila melanogaster* and its ebony mutant in the absence of food. Amer.Natur., 74, 363-376.

Grenander, U. (1956). On the theory of mortality measurement. Skandinavisk Aktuarietidskrift, Haft 1-2, 70-153.

Greville, T.N.E. (1948). Mortality tables analyzed by cause of death. Am.Inst.of Actuaries, The Record, 37, 283-294.

Gubler, E.V. (1978). Vychislitel'nye metody analiza i raspoznavaniya patologicheskikh protsessov [Computational methods of analysis and recognition of pathological processes]. Meditsina, Moscow. (In Russian).

Guess, F. and M.Witten (1988). On the incompatibility of Gompertz or Weibull survival dynamics with exponentially distributed individual lifespans. Bull.Math.Biol., 50, 187-192.

Gumbel, E.J. (1958). Statistics of extremes. Columbia Univ.Press, NY.

Haff, R.F. and H.E.Swim (1956). Serial propagation of 3 strains of rabbit fibroblasts; their susceptibility to infection with Vaccinia virus. Proc.Soc.Exp.Biol.Med.,

93, 200-204.

Hall, J.C. (1969). Age-dependent enzyme changes in *Drosophila melanogaster*. Exp.Gerontol., 4, 207-222.

Hansche, W.J. (1975). Role of the nervous system in aging: Correlations among life span, brain-body weight and metabolism. Adv.Behav.Biol., 16, 23-25.

Harman, D. (1981). The aging process. Proc.Nat.Acad.Sci.USA, 78, 7124-7128.

Harman, D. (1985). Role of free radicals in aging and disease. In H.A.Johnson (ed.) Relations between normal aging and disease. Raven Press, NY, 45-84.

Harman, D. and D.E.Eddy (1979). Free radical theory of aging: beneficial effect of adding antioxidants to the maternal mouse diet on life span of offspring; possible explanation of the sex difference in longevity. Age, 2, 109-122.

Harrison, D.E. (1984). Do hemopoietic stem cells age? In H.W.Sauer (ed.) Monographs in developmental biology, vol.17, Cellular ageing. Karger, Basel, 21-41.

Hart, R.W. and R.B.Setlow (1974). Correlation between deoxyribonucleic acid excision repair and life-span in a number of mammalian species. Proc.Nat.Acad.Sci.USA, 71, 2169-2173.

Hart, R.W. and A.Turturro (1981). Evolution and longevity-assurance processes. Naturwissenschaften, 68, 552-557.

Harvey, P.H. and T.H.Clutton-Brock, 1985. Life history variation in primates. Evolution, 39, 559-581.

Harvey, P.H., R.D.Martin and T.H.Clutton-Brock (1987). Life histories in comparative perspective. In B.B.Smuts and T.T.Struhsaker (eds.) Primate societies. Chicago University Press, Chicago, 181-196.

Harvey, P.H. and R.M.Zammuto (1985). Patterns of mortality and age at first reproduction in natural populations of mammals. Nature, 319-320.

Hayflick, L. (1965). The limited in vitro lifetime of human diploid cell strains. Exp.Cell Res., 37, 614-636.

Hayflick, L. (1966). Senescence and cultured cells. In N.Shock (ed.) Perspectives in experimental gerontology. Charles C.Thomas (Publisher), Springfield, Illinois, 195-211.

Hayflick, L. (1968). Human cells and aging. Sci.American, 218, 32-37.

Hayflick, L. (1974). The longevity of cultured human cells. J.Am.Geriatr.Soc., 22, 1-12.

Hayflick, L. (1975). Current theories of biological aging. Federation Proc., 34, 9-13.

Hayflick, L. (1977). The cellular basis for biological aging. In C.Finch and L.Hayflick (eds.) Handbook of the

biology of aging. Van Nostrand Reinhold, NY, 159-186.
Hayflick, L. (1980). The cell biology of human aging. Sci.American, 242, 42-49.
Hayflick, L. (1982). Ageing and death of vertebrate cells. In A.Vidik (ed.) Lectures on gerontology, Vol.1: On biology of aging, Part A. Acad.Press, London, 59-97.
Hayflick, L. (1988). Why do we live so long? Geriatrics, 43, 77-87.
Hayflick L. and P.S.Moorhead (1961). The serial cultivation of human diploid cell strains. Exp.Cell Res., 25, 585-621.
Hazzard, W.R. (1986). Biological basis of the sex differential in longevity. J.Am.Geriatr.Soc., 34, 455-471.
Hazzard W.R. (1987). Response to Waldron. J.Am.Geriatr.Soc., 35, 366.
Heicklen, J. and E.Brown (1987). Increase in life expectancy for mice fed Diethylhydroxylamine (DEHA). J.Gerontol., 42, 674-680
Heligman, L. and J.H.Pollard (1980). The age pattern of mortality. J.Inst.Actuaries, 107, 49-80.
Henderson, R. (1915). Mortality laws and statistics. Wiley and Sons, Inc., NY.
Hewer, H.R. (1964). the determination of age, sexual maturity, longevity and a life-table in the grey seal (*Halichoerus Grypus*). Proc.Zool.Soc.London, 142, 593-624.
Hibbs, A.R. and R.L.Walford (1989). A mathematical model of physiological processes and its application to the study of aging. Mech.Ageing Devel., 50, 193-214.
Hickey, F. (1960). Death and reproductive rates of sheep in relation to flock culling and selection. N.Z.J.Agr.Res., 3, 332-344.
Hoel, D.G. and H.E.Walburg (1972). Statistical analysis of survival experiments. J.Natl.Cancer Inst., 49, 361-372.
Hofman, M.A. (1983). Energy metabolism, brain size and longevity in mammals. Quart.Rev.Biol., 58, 495-512.
Hofman, M.A. (1984). On the presumed coevolution of brain size and longevity in hominids. J.Human Evolution, 13, 371-376.
Holehan, A.M. and B.J.Merry (1986). The experimental manipulation of ageing by diet. Biol.Rev., 61, 329-368.
Hollicroft, J., E.Lorenz, E.Miller et al. (1967). Delayed effects in mice following acute total-body X-irradiation: Modification by experimental treatment. J.Nat.cancer Inst., 18, 615-640.
Holliday, R. and T.B.L.Kirkwood (1981). Predictions of the somatic mutation and mortalization theories of cellular

ageing are contrary to experimental observations. J.Theor.Biol., 93, 627-642.
Holliday, R. and A.Stevens (1978). The effect of an amino acid analogue, p-Fluorophenylalanine, on longevity of mice. Gerontology, 24, 417-425.
Holyoak, D. (1971). Movements and mortality of Corvidae. Bird Study, 18, 97-106.
Hopkins, G.A. and K.K.Carrol (1985). Role of diet in cancer prevention. J.Environ.Pathol.Toxicol.and Oncol., 5, 279-298.
Hsieh, J. (1985). Construction of expanded infant life tables: a method based on a new mortality law. Math.Biosci., 76, 221-242.
Hufeland, C.W. (1796). Die Kunst das menscheliche Leben zu verlangern. Berlin. Quoted in Zhirmunsky and Kuz'min (1980), p.251.
Hug, O. and A.M.Kellerer (1966). Stochastik der Stahlenwirkung. Springer, Berlin.
Huret, J.L., J.M.Delabar, F.Marlhens et al. (1987). Down syndrome with duplication of a region of chromosome 21 containing the CuZn superoxide dismutase gene without detectable karyoptic abnormality. Hum.Genet., 75, 251-257.
Hursch, J.B., T.R.Noonam, G.Casarett et al. (1955). Reduction of life span of rats by Roentgen irradiation. Amer.J.Roentgenol., 74, 130-134.
Hutchinson, G.E. (1978). An introduction to population ecology. Yale Univ.Press, New Haven, London.
Igbinosa, I.B. (1985). Life-table studies for the nettle caterpillar, *Latolia viridissima* Holland, on the oil palm, *Elaeis guineesis* Jacq., and the cocnut palm, *Cocos nucifer* L. Agriculture, Ecosystems and Environment, 14, 77-93.
Ingram, D.K., R.Weindruch, E.L.Spangler et al. (1987). Dietary restriction benefits learning and motor performance of aged mice. J.Gerontol., 42, 78-81.
Jacquard, A. (1982). Heritability of human longevity. In H.S.Preston (ed.) Biological and social aspects of mortality and the length of life. Liege.
Jacquard, A. (1983). Heritability: One word, three concepts. Biometrics, 39, 465-477.
Jennings, H.S. and R.S.Lynch (1928a). Age, mortality, fertility and individual diversities in the rotifer *Proales sordida* Gosse. I.Effect of age of the parent on characteristics of the offspring. J.Exp.Zool., 50, 345-407.
Jennings, H.S. and R.S.Lynch (1928b). Age, mortality, fertility and individual diversities in the rotifer

*Proales sordida* Gosse. II.Life history in relation to mortality and fecundity. J.Exp.Zool., 51, 339-381.
Johnson, H.D., L.D.Kintner and H.H.Kibler (1963). Effects of 48°F (8.9°C) and 83°F (28.4°C) on longevity and pathology of male rats. J.Gerontol., 18, 29-36.
Johnson, T.E. (1987). Aging can be genetically dissected into component processes using long-lived lines of *Caenorhabditis elegans*. Proc.Natl.Acad.Sci.USA, 84, 3777-3781.
Johnson, T.E., W.L.Conley and M.L.Keller (1988). Long-lived lines of *Caenorhabditis elegans* can be used to establish predictive biomarkers of aging. Exp.Gerontol., 23, 281-295.
Joubert, P.H., S.J.Pretorius, K.N.de Kock et al. (1986). Survival of *Bulinus africanus* (Krauss), *Bulinus globosus* (Morelet) and *Biomphalaria pfeifferi* (Krauss) at constant high temperatures. S.Afr.J.Zool., 21, 85-88.
Juckett, D.A. (1987). Cellular aging (the Hayflick limit) and species longevity: A unification model based on clonal succession. Mech.Ageing and Devel., 38, 49-71.
Juckett, D.A. and B.Rosenberg (1988). Integral differences among human survival distributions as a function of disease. Mech.Ageing and Devel., 43, 239-257.
Kalbfleisch, J.D. and R.L.Prentice (1980). The statistical analysis of failure time data. Wiley, NY.
Kase, S. (1953). A theoretical analysis of the distribution of tensile strength of vulcanized rubber. J.Polym.Sci., 11, 425-431.
Kato, H., M.Harada, K.Tsuchiya et al. (1980). Absence of correlation between DNA repair in ultraviolet irradiated mammalian cells and life span of the donor species. Jap.J.Genet., 55, 99-108.
Kaveggia, F.F. (1985). Survival analysis of the severely and profoundly mentally retarded. Am.J.Med.Genetics, 21, 213-223.
Kellog, E.W. and M.G.Yost (1986). The effects of long-term air ion and D.C. electric field exposures on survival characteristics in female NAMRU mice. J.Gerontol., 41, 147-153.
Kendall, M.G. (1962). Rank correlation methods, 3rd edition. Griffin, London.
Kershaw, W.E., M.M.J.Lavoipierre and T.A.Chalmers (1953). Studies on the intake of microfilariae by their insect vectors, their survival, and their effect on the survival of their vectors. I.*Dirofilaria immitis* and *Aedes Aegypti*. Ann.Tropical Med.and Parasitol., 47, 207-224.
Keyfitz, N. (1982), Choice of function for mortality

analysis: Effective forecasting depends on a minimum parameter representation. Theor.Pop.Biol., 21, 329-352.

Keyfitz, N. and W.Flieger (1968). World population: An analysis of vital data. Univ.of Chicago Press, Chicago, London.

Khalyavkin, A.V. (1983). One approach to the verification of life span. In Problemy biologii stareniya [Problems in the biology of aging]. Nauka, Moscow, 49-55. (In Russian).

Khalyavkin, A.V. (1985). The extreme human life span and its evaluation. In Isledovanie pervichnykh mekhanizmov biologicheskikh sistem [Investigation of the primary mechanisms of biological systems]. Nauka, Moscow, 43-46. (In Russian).

Kirkman, H. and P.K.S.Yau (1972). Longevity of male and female, intact and gonadectomized, untriated and hormone-triated, neoplastic and non-neoplastic Syrian hamsters. Amer.J.Anat., 135, 205-220.

Klein, T.A., B.A.Harrison, J.S.Grove et al. (1987). The correlation of survival characteristics of *Anopeles dirus* A mosquitoes (Diptera: Culicidae) with various indices of *Plasmodium cynomogli* density. Byull.VOZ., 64, 66-72.

Klevey, L.M. (1987). Cholesterol reduction: safety, other concerns. Ann.Intern.Med., 107, 421.

Knox, E.G. (1977). Foods and diseases. Brit.J.Prev. and Soc. Med., 31, 71-80.

Knuiman, J.T., C.E.West and J.Burema (1982). Serum total and high density lipoprotein cholesterol concentrations and body index in adult men from 13 countries. Amer.J.Epidemiol., 116, 631-643.

Kogan, A.B., N.P.Naumov, B.G.Rezhabek et al. (1977). Biologicheskaya kibernetika [Biological cybernetics]. Vysshaya shkola, Moscow. (In Russian).

Kogan, I.G. (1986). Heritability of life-span in *Drosophila melanogaster*. Genetika, 22, 2800-2804. (In Russian).

Kohn, H.I. and P.H.Guttman (1959). Latent period of X-ray induced ageing: A study based on mortality rate and tumor incidence. Nature, 184, 735-736.

Kohn, H.I. and P.H.Guttman (1963). Age at exposure and the late effects of X-rays: Survival and tumor incidence in $CAF_1$ mice irradiated at 1 to 2 years of age. Radiat.Res., 18, 348-373.

Koivisto, K. and P.Portin (1987). Induction of mutations which shorten the adult life span in *Drosophila melanogaster*. Hereditas, 106, 83-87.

Kok, F.J., A.M.De Bruijn, A.Hofman et al. (1987). Is serum selenium a risk factor for cancer in men only?

*Am.J.Epidemiol.*, 125, 12-16.

Koltover, V.K. (1983). Theory of reliability, superoxide radicals and aging. *Uspekhi Sovrem.Biol.*, 96, 85-100. (In Russian).

Koltover, V.K. (1986). Comments on the paper by L.A. Gavrilov. *Rejuvenation*, 14, 24-25.

Korchagin, N.V., A.A.Malinovsky, T.V.Anfalova et al. (1973). The interspecific correlations in mammals. In N.N.Iordansky and A.A.Malinovsky (eds.) *Nekotorye problemy teorii evolyutsii* [Some problems of the theory of evolution], vol.2. 2 MOLGMI, Moscow, 101-122. (In Russian).

Kovtun, A.I. (1976). Life span and some indices of biological age in laboratory rats. In *Gerontologiya i geriatriya 1975. Biologicheskie vozmozhnosti uvelicheniya prodolzhitel'nosti zhizni* [Gerontology and geriatrics 1975. Biological possibilities of increasing life span]. Inst.of Gerontology, Kiev, 215-216. (In Russian).

Kozlovsky, V.S. and L.A.Gavrilov (1983). The multistage character of death as the basis of aging (the hypothesis of random sequential transitions). In *Problemy biologii stareniya* [Problems in the biology of aging]. Nauka, Moscow, 19-23. (In Russian).

Kroes, R., J.M.Garbis-Berkvens, T.de Vries et al. (1981). Histopathological profile of a Wistar rat stock including a survey of the literature. *J.Gerontol.*, 36, 259-279.

Kron, A.A. (1980). *Insomnia: A novel*. Sovetsk.Pisatel', Moscow. (In Russian).

Kudryashov, Yu.B. and B.S.Berenfel'd (1982). *Osnovy radiatsionnoi biofiziki* [Foundations of radiation biophysics]. Moscow Univ.Press, Moscow. (In Russian).

Kunstýř I. and H.-G.W.Leuenberger (1975). Gerontological data of C57BL/6J mice. I.Sex differences in survival curves. *J.Gerontol.*, 30, 157-162.

Kurtzman, J. and P.Gordon (1976). *No more dying*. J.P.Tarcher, Inc., Los Angeles.

Labat, J.C. (1970). Données de démographie régionale, 1962. *Coll.INSEE*, N R3.

Lamb, M.J. (1977). *Biology of ageing*. Blackie, Glasgow and London.

Le Bras, H. (1976). Lois de mortalité et age limité. *Population*, 31, 655-692.

Lery, A. (1976). Mortalité des personnes tres agées. *Coll.INSEE*, N D49, 111-124.

Leslie, P.H. and R.M.Ranson (1940). The mortality, fertility and rate of natural increase in the vole (*Microtus agrestis*) as observed in the laboratory. J.Anim.Ecol., 9, 27-52.

Leslie, P.H., J.S.Tener, M.Vizoso et al. (1955). The longevity and fertility of the Orkney vole, *Microtus orcadensis*, as observed in the laboratory. Proc.Zool. Soc.London, 125, 115-125.

Leslie, P.H., U.M.Venables and N.S.V.Venables (1952). The fertility and population structure of the brown rat (*Rattus norvegicus*) in corn-ricks and some other habitats. Proc.Zool.Soc.London, 122, 187-238.

Lestienne, R. (1988). On the thermodynamical and biological interpretation of the Gompertzian mortality rate distribution. Mech.Ageing and Devel., 42, 197-214.

Levander, O.A. (1982). Selenium: Biochemical actions, interactions and some human health implications. In Clinical, biochemical and nutritional aspects of trace elements. Liss, NY, 345-368.

Lindop, P.J. and J.Rotblat (1961). Long-term effects of a single whole-body exposure of mice to ionizing radiations. I.Life shortening. Proc.Roy.Soc.London B, 154, 332-349.

Lints, F.A. (ed.) (1985). Non-mammalian models for research on aging. Karger, Basel.

Lints, F.A., M.Bourgois, A.Delalieux et al. (1983). Does the female life span exceed that of the male? A study in *Drosophila melanogaster*. Gerontology, 29, 336-352.

Lints, F.A., C.V.Lints, P.Bullens, M.Bourgois et al. (1989). Unexplained variations in life span of the Oregon-R strain of *Drosophila melanogaster* over a four-year period. Exp.Gerontol., 24, 265-271.

Lints, F.A., J.Stoll, G.Gruwez et al. (1979). An attempt to select for increased longevity in *Drosophila melanogaster*. Gerontology, 25, 192-204.

Lipid Research Clinics Program (1984). The Lipid Research Clinics coronary primary prevention trials results. I.Reduction incidence of coronary heart disease. J.Am.Med.Assoc., 251, 351-364.

Little, C.C., W.S.Murray and A.M.Cloudman (1939). The genetics of non-epithelial tumor formation in mice. Amer.J.Cancer, 36, 431-450.

Litton, J.R. (1987). Specificity of the α-tocopherol (Vitamin E) effect on lifespan and fecundity of bdelloid rotifers. Hydrobiologia, 147, 135-139.

Liu, R.K. and R.L.Walford (1969). Laboratory studies on life-span, growth, aging and pathology of the annual fish *Cynolebias bellottii* Steindachner. Zoologica, 54, 1-16.

Loeb, J. and J.H.Northrop (1917). On the influence of food

and temperature upon the duration of life. J.Biol.Chem., 32, 103-126.

Lohmann, W. (1977). Altersindex und Interpretation der Absterbekurven. Ztschr.Alternsforsch., 32, 461-466.

Loo, D.T., J.I.Fiquay, C.L.Rawson et al. (1987). Extended culture of mouse embryo cells without senescence: inhibition by serum. Science, 236, 200-202.

Lopez, A.D. (1983). The sex mortality differential in developed countries. In A.D.Lopez and L.T.Ruzicka (eds.) Sex differentials in mortality: Trends, determinants and consequences. Austral.National Univ., Canberra, 53-120.

Lorenz, E., J.W.Hollcroft, E.Miller et al. (1955). Long-term effects of acute and chronic irradiation in mice. I.Survival and tumor incidence following chronic irradiation of 0.11 R per day. J.Nat.Cancer Inst., 15, 1049-1058.

Luchnik, N.V. and Ya.B.Livchak (1963). On the interpretation of "time-effect" curves. In Primenenie matematicheskikh metodov v biologii [Application of mathematical methods in biology], Vol.2. Leningrad Univ.Press, Leningrad, 212-224. (In Russian).

Luckinbill, L.S. and M.J.Clare (1985). Selection for life span in *Drosophila melanogaster*. Heredity, 55, 9-18.

Luckinbill, L.S. and M.J.Clare (1986). A density threshold for the expression of longevity in *Drosophila melanogaster*. Heredity, 56, 329-335.

Luckinbill, L.S. and M.J.Clare (1987). Letter to the editor. Exp.Gerontol., 22, 221-222.

Ludwig, F.C. and R.M.Elashoff (1972). Mortality in syngenic rat parabionts of different chronological age. Trans.N.Y.Acad.Sci., Ser.2, 34, 582-587.

Lüers, H. (1935). Die Beinflussung der Vitalität durch multiple Allele untersucht an vestigial-Allelen von *Drosophila melanogaster*. W.Roux' Arch.Entwicklungsmech. Organism., 133, 88-117.

MacArthur, J.W. and W.H.T.Baillie (1929). Metabolic activity and duration of life. I.Influence of temperature on longevity in *Daphnia Magna*. J.Exp.Zool., 53, 221-242.

MacArthur, J.W. and W.H.T.Baillie (1932). Sex differences in mortality in abraxastype species. Quart.Rev.Biol., 7, 313-325.

Makeham, W.M. (1860). On the law of mortality and the construction of annuity tables. J.Inst.Actuaries, 8, 301-310.

Malinovsky, A.A. (1983). Evolutionary-genetic factors of life span. In Problemy biologii stareniya [Problems in the biology of aging]. Nauka, Moscow, 5-11. (In Russian).

Mallouk, R.S. (1975). Longevity in vertebrates is proportional to relative brain weight. Fed.Proc., 34, 2102-2103.

Malyugina, L.L. and O.G.Prokof'eva (1957). On the oncological characteristics of $C_3HA$ mice. Voprosy onkologii, 3, 197-203. (In Russian).

Mamaev, V.B. and T.L.Nadzharyan (1987). Kinetic analysis of mortality. In E.B.Burlakova and L.A.Gavrilov (eds.) Populyatsionnaya gerontologiya [Population gerontology]. VINITI, Moscow, 6, 106-154. (In Russian).

Man, E.H., M.E.Sandhouse, J.Burg et al. (1983). Accumulation of D-aspartic acid with age in the human brain. Science, 220, 1407-1408.

Mankovsky, N.B. and N.S.Verchratsky (1980). The "life extension" programme. Vestnik AMN SSSR, N 3, 85-87. (In Russian).

Manson, J.E., M.J.Stampfer, C.H.Hennekens et al. (1987). Body weight and longevity. A reassessment. J.Am.Med.Assoc., 257, 353-358.

Manton, K.G. and E.Stallard (1984). Recents trends in mortality analysis. Acad.Press, Orlando.

Manton, K.G., E.Stallard and J.W.Vaupel (1986). Alternative models for the heterogeneity of mortality risks among the aged. J.Am.Stat.Assoc., 81, 635-644.

Maréchal, R., Y.Lion and J.Duchesne (1973). Radicaux libres organiques et longevité maximale chez les mammifères et les oiseaux. C.R.Acad.Sci.Paris, Ser.D., 277, 1085-.

Maslansky, C.J. and G.M.Williams (1985). Ultraviolet light-induced DNA repair synthesis in hepatocytes from species of different longevities. Mech.Ageing Devel., 29, 191-203.

Masoro, E.J. (1988). Life span extension and food restriction. Comprehensive Therapy, 14, 9-13.

Masoro, E.J., M.S.Katz and C.A.McMahan (1989). Evidence for the glycation hypothesis of aging from the food-restricted rodent model. J.Gerontol., 44, 820-822.

Matsumura, T., Z.Zerrudo and L.Hayflick (1979). Senescent human diploid cells in culture: Survival, DNA synthesis and morphology. J.Gerontol., 34, 328-334.

Matthes, E. (1951). Der Einfluss der Fortpflanzung auf die Lebensdauer eines Schmetterlings (*Fumea Crassiorella*). Ztschr.Vergl.Physiol., 33, 1-13.

McCarter, R., E.J.Masoro and B.P.Yu (1985). Does food restriction retard aging by reducing the metabolic rate? Am.J.Physiol., 248, 488-490.

McMillen, M.M. (1979). Differential mortality by sex in fetal and neonatal deaths. Science, 204, 89-91.

Medawar, P.B. (1957). The uniqueness of the individual.

Methuen, London.
Medvedev, N.N. (1958). Oncological characteristics of the CC-57-Brown strain of mice. Byull.Eksperim.Biol.i Med., 45, 104-106. (In Russian).
Medvedev, N.N. (1961). The origin and comparative oncological characteristics of mice of CC57W and CC57BR strains. Voprosy onkologii, 7, 23-37.
Medvedev, N.N. and I.G.Ol'khovskaya (1974). On the oncological characteristics of mice of CC57BR and CC57W strains. Voprosy Onkologii, 20, 84-87. (In Russian).
Meites, J. (1988). Neuroendocrine biomarkers of aging in the rat. Exp.Gerontol., 23, 349-358.
Mertz, D.B., T.Park and W.J.Youden (1965). Mortality patterns in eight strains of flour beetles. Biometrics, 21, 99-114.
Messenger, P.S. (1964). Use of life tables in a bioclimatic study of an experimental aphid-braconid wasp host-parasite system. Ecology, 45, 119-121.
Millar, J.S. and R.M.Zammuto (1983). Life histories of mammals: an analysis of life tables. Ecology, 64, 631-635.
Millar, J.S. and F.C.Zwickel (1972). Determination of age, age structure, and mortality of the pike, *Ochotona princeps* (Richardson). Can.J.Zool., 50, 229-232.
Miller, R.S. and J.L.Thomas (1958). The effect of larval crowding and body size on the longevity of adult *Drosophila melanogaster*. Ecology, 39, 118-125.
Milne, C.P. (1985). An estimate of the heritability of worker longevity or length of life in the honeybee. J.Apicultural Res., 24, 140-143.
Mineev, I.F. and E.I.Gershkovich (1979). Total diurnal duration of systole and natural life length in birds and mammals. Izv.Akad.Nauk. GSSR, 5, 395-399. (In Russian).
Miquel, J., A.G.Economos and J.E.Johnson (1984). A systems analysis - thermodynamics view of cellular and organismic aging. In J.E.Johnson (ed.) Aging and Cell Function. Plenum Press, NY, London, 247-280.
Mode, C.J., R.J.Ashleigh, A.Zavodniack et al. (1984). On statistical tests of significance in studies of survivorship in laboratory animals. J.Gerontol., 39, 36-42.
Moore, C.J. and A.G.Schwartz (1978). Inverse correlation between species lifespan and capacity of cultured fibroblasts to convert benzo(a)pyrene to water-soluble metabolites. Exp.Cell Res., 116, 359-364.
Moore, W.S. (1975). Stability of small unisexual-bisexual populations of *Poeciliopsis*. (Pisces: Poecilidae). Ecology, 56, 791-808.

Mos, J. and C.F.Hollander (1987). Analysis of survival data on aging rat cohorts: pitfalls and some practical considerations. Mech.Ageing and Devel., 38, 89-105.

Muenchow, G. (1986). Ecological use of failure time analysis. Ecology, 67, 246-250.

Multiple risk factor intervention trial research group (1982). Multiple risk factor intervention trial: Risk factor changes and mortality results. J.Amer.Med.Assoc., 248, 1465-1477.

Munblit, V.Ya. and V.I.Trofimov (1985). Compensation phenomena in the thermoinactivation of microorganisms. Biofizika, 30, 98-102. (In Russian).

Munir, B. and R.I.Sailer (1985). Population dynamics of the tea scale *Fiorinia theae* (Homoptera: Diaspididae), with biology and life tables. Environ.Entomol., 14, 742-748.

Murphy, E.A. (1978). Genetics of longevity in man. In E.L.Schneider (ed.) The genetics of aging. Plenum Press, NY, 261-301.

Murray, W.S. (1934). The breeding behavior of the dilute brown stock of mice (Little dba). Amer.J.Cancer, 20, 573-593.

Murray, W.S. and J.G.Hoffman (1941). Physiological age as a basis for the comparison of strains of mice subject to spontaneous mammary carcinoma. Cancer Res., 1, 298-309.

Murray, W.S. and C.C.Little (1935). The genetics of mammary incidence in mice. Genetics (US), 20, 466-496.

Murray, W.S. and C.C.Little (1939). Chromosomal and extrachromosomal influence in relation to the incidence of mammary tumors in mice. Amer.J.Cancer, 37, 536-552.

Mutanen, M. and P.V.Koivistoinen (1983). The role of imported grain on the selenium intake of Finnish population in 1941-1981. Intern.J.Vitam.and Nutr.Res., 53, 102-108.

Myers, G.G. and K.G.Manton (1983). Compression of mortality: Myth or reality? Duke Univ., Cent.for Demogr.Stud., Durham.

Nakano, M., T.Mizuno and S.Gotoh (1990). Accumulation of cardiac lipofuscin in mammals: correlation between sexual maturation and the first appearance of lipofuscin. Mech.Ageing and Develop., 52, 93-106.

Neary, G.J., R.J.Munson and R.H.Mole (1957). Chronic radiation hazards. Pergamon Press, London etc.

Nekrasova, A.V. and V.G.Shakhbazov (1981). Length of ontogenesis and age-related changes in the fecundity and heat-resistance of *Drosophila melanogaster* in connection with the effect of heterosis. Tsitologiya i Genetika, 15, 49-53. (In Russian).

Nikitin, V.N. (1984). Experimental approaches to life

extension. In Biologicheskie problemy stareniya [Biological problems of aging], Vol.4. VINITI, Moscow, 6-43. (In Russian).

Nowell, P.C. and L.J.Cole (1959). Late effects of fast neutrons versus X-rays in mice: Nephrosclerosis, tumors, longevity. Radiat.Res., 11, 545-556.

Noyes, B. (1922). Experimental studies on the life-history of a rotifer reproducing parthenogenetically (*Proales decipiens*). J.Exp.Zool., 35, 225-255.

Obukhova, L.K. and N.M.Emanuel (1984). Molecular mechanisms of slowing down aging with antioxidants. In Biologicheskie problemy stareniya [Biological problems of aging], Vol.4. VINITI, Moscow, 44-80. (In Russian).

Oliff, W.D. (1953). The mortality, fecundity and intrinsic rate of natural increase of the multimammate mouse *Rattus (Mastomys) natalensis* (Smith) in the laboratory. J.Amer.Ecol., 22, 217-226.

Oliver, M.F. (1988). Reducing cholesterol does not reduce mortality. J.Am.College Cardiol., 12, 814-817.

Ono, T. and S.Okada (1984). Unique increase of superoxide dismutase level in brains of long living mammals. Exp.Gerontol., 19, 349-354.

Osanai, M. (1978). Longevity and body weight loss of silkworm moth, *Bombyx mori*, varied by different temperature treatments. Exp.Gerontol., 13, 375-388.

Owsley, D.W. and W.M.Bass (1979). A demographic analysis of skeletons from the Lirson site (39WW2) Walworth County, South Dakota: vital statistics. Am.J.Physical Anthropol., 51, 145-154.

Paevsky, V.A. (1985). Demografiya ptits [Demography of birds]. Nauka, Moscow. (In Russian).

Pakin, Yu.K. (1988). Tendencies of the age-dependent mortality of the adult population of economically developed capitalist countries. In Demograficheskie issledovaniya [Demographic studies]. Naukova Dumka, Kiev, 125-133. (In Russian).

Pakin, Yu.V. and S.M.Hrisanov (1984). Critical analysis of the applicability of the Gompertz-Makeham law in human populations. Gerontology, 30, 8-12.

Park, T. (1945). Life tables for the black flour-beetle *Tribolium madens* Charp. Amer.Natur., 79, 436-444.

Pashko, L.L. and A.G.Schwartz (1982). Inverse correlation between species life span and specific cytochrome P-448 content of cultured fibroblasts. J.Gerontol., 37, 38-41.

Pashko, L.L. and A.G.Schwartz (1983). Effects of food restriction, dehydroepiandrosterone, or obesity on the binding of $^{3}$H-7,12-dimethylbenz(a)anthracene to mouse skin DNA. J.Gerontol., 38, 8-12.

Pearl, R. (1922). The biology of death. Lippincott, Philadelphia.
Pearl, R. (1931). Studies on human longevity. IV.The inheritance of longevity. Hum.Biol., 3, 245-269.
Pearl, R. and J.R.Miner (1935). Experimental studies on the duration of life. XIV.The comparative mortality of certain lower organisms. Quart.Rev.Biol., 10, 60-79.
Pearl, R. and J.R.Miner (1936). Life tables for the pecan nut case bearer, *Acrobasis caryae* Grote. Mem.Mus.Roy. Hist.Natur.Belg., Ser.2, N 3, 169-184.
Pearl, R. and J.R.Miner (1941). Experimental studies on the duration of life. XVI.Life tables for the flour beetle *Tribolium confusum* Duval. Amer.Natur., 75, 5-19.
Pearl R. and S.L.Parker (1921). Experimental studies on the duration of life. I.Introductory discussion of the duration of life in *Drosophila*. Amer.Natur., 55, 481-500.
Pearl R. and S.L.Parker (1922a). Experimental studies on the duration of life. II.Hereditary differences in duration of life in line-bred strains of *Drosophila*. Amer.Natur., 56, 174-187.
Pearl R. and S.L.Parker (1922b). Experimental studies on the duration of life. III.The effect of successive etherizations on the duration of life of *Drosophila*. Amer.Natur., 56. 273-280.
Pearl R. and S.L.Parker (1922c). Experimental studies on the duration of life. IV.Data on the influence of density of population on duration of life in *Drosophila*. Amer.Natur., 56, 312-321.
Pearl R. and S.L.Parker (1922d). Experimental studies on the duration of life. V.On the influence of certain environmental factors on the duration of life in *Drosophila*. Amer.Natur., 56, 335-398.
Pearl R. and S.L.Parker (1924a). Experimental studies on the duration of life. IX.New life tables for *Drosophila*. Amer.Natur., 58, 71-82.
Pearl R. and S.L.Parker (1924b). Experimental studies on the duration of life. X.Studies on the duration of life of *Drosophila melanogaster* in the complete absence of food. Amer.Natur., 58, 193-218.
Pearl, R. and R.W.Pearl (1934). The ancestry of the long-lived. Milford, London.
Pearl, R., J.R.Miner and S.L.Parker (1927). Experimental studies on the duration of life. XI.Density of population and life duration in *Drosophila*. Amer.Natur., 61, 289-318.
Pearl, R., S.L.Parker and B.M.Gonzales (1923). Experimental studies on the duration of life. VII.The Mendelian

inheritance of duration of life in crosses of wild type and Quintuple stocks of *Drosophila melanogaster*. Amer.Natur., 57, 153-192.

Perks, W. (1932). On some experiments in the graduation of mortality statistics. J.Inst.Actuaries, 63, 12-40.

Petranka, J.W. (1985). Does age-specific mortality decrease with age in amphibian larvae? Copeia, No.4, 1080-1083.

Philippe, P. (1978). Familial correlations of longevity: An isolate-based study. Amer.J.Med.Genet., 2, 121-129.

Piantanelli, L. (1986). A mathematical model of survival kinetics. I.Theoretical basis. Arch.Gerontol.Geriatr., 5, 107-118.

Pioli, G.., E.Palummeri, G.Girasole et al. (1986). Senescenza: Aspetti biologici. Parte II. Maximum life span. Giornale di Gerontologia, 34, 442-451.

Plummer, M.V. (1985). Demography of green snakes (*Opheodrys aestivus*). Herpetologica, 41, 373-381.

Poiley, S.M. (1957). Breeding and care of the Syrian hamster *Cricetus auratus*. In E.J.Farris (ed.) The care and breeding of laboratory animals, NY.

Polis, G.A. and R.D.Farley (1980). Population biology of a desert scorpion: survivorship, microhabitat, and the evolution of life history strategy. Ecology, 61, 620-629).

Polskie tablice wymieralnósci, 1963-1965, wedlug województw (1967). Glówny Urzand Statyst., Warszawa.

Portier, C.J., J.C.Hedges and D.G.Hoel (1986). Age-specific models of mortality and tumor onset for historical control animals in the national toxicology program's carcinogenicity experiments. J.Natl.Cancer Inst., 46, 4372-4378.

Potapenko, A.I., A.P.Akifiev and V.I.Ivanov (1982). Radiation induced shortening of the life span of *D.melanogaster*. Report 1. Analysis of survival curves after γ-irradiation of adult insects. Radiobiologiya, 22, 203-208. (In Russian).

Preston, S.H. and V.E.Nelson (1974). Structure and change in causes of death: an international summary. Population Studies, 28, 19-51.

Prokof'ev, E.A., R.S.Zotina and A.I.Zotin (1982). Phenomenological equations of growth and their application to the determination of maximal life span. In A.I.Zotin and E.V.Presnov (eds.) Matematicheskaya biologiya razvitiya [Mathematical developmental biology]. Nauka, Moscow, 56-66. (In Russian).

Promislow, D.E.L. and P.H.Harvey (1990). Living fast and dying young: a comparative analysis of life-history variation among mammals. J.Zool.Lond., 220, 417-437.

Prothero, J. and K.D.Jürgens (1987). Scaling of maximal lifespan in mammals: a review. In A.D.Woodhead and K.H.Thompson (eds.) Evolution of longevity in animals: A comparative approach. Plenum Press, NY, 49-74.

Pryakhin, V.F. (1985). The international gerontological community and the fight to prevent nuclear war. Mir Nauki, 3, 14-17. (In Russian).

Rau, P. (1924). The biology of the roach *Blatta orientalis* Linn. Trans.Acad.Sci.St Louis, 25, 57-79.

Rau, P and N.Rau (1912). Longevity in saturnid moths: An experimental study. J.Exp.Zool., 12, 179-204.

Read, A.F. and P.H.Harvey (1989). Life history differences among the eutherian radiations radiations. J.Zool.Lond., 219, 329-353.

Redman, H.C., C.H.Hobbs and A.H.Rebar (1979). Survival distribution of Syrian hamsters (*Mesocricetus auratus*, Sch: SYR) used during 1972-1977. In F.Homburger (ed.) The Syrian hamster in toxicology and carcinogenesis research. Karger, Basel, 108-117.

Relman, A.S. (1988). Aspirin for the primary prevention of myocardial infarction. N.Engl.J.Med., 318, 245-246.

Report from the Committee of principal investigators (1978). A co-operative trial in the primary prevention of ischaemic heart disease using clofibrate. Brit.Heart.J., 40, 1069-1118.

Rheinwald, J.G. and H.Green (1977). Epidermal growth factor and the multiplication of cultured human epidermal keratinocytes. Nature, 265, 421-424.

Richdale, L.E. (1957). A population study of penguins. Clarendon Press, Oxford.

Richie, J.P., B.J.Mills and C.A.Lang (1987). Correction of a glutathione deficiency in the aging mosquito increases its longevity (42454). Proc.Soc.Exp.Biol.Med., 184, 113-117.

Richter, C., J.-W.Park and B.N.Ames (1988). Normal oxidative damage to mitochondrial and nuclear DNA is extensive. Proc.Natl.Acad.Sci.USA, 85, 6465-6467.

Risto, K. (1980). Cohort mortality in Finland from 1851. Tutkimuksia, N 57, 1-97.

Roberts, E.H. (ed.) (1972). Viability of seeds. Chapman and Hall Ltd., London.

Rockstein, M. (1957). Longevity of male and female house flies. J.Gerontol., 12, 253-256.

Rockstein, M. and H.M.Lieberman (1959). A life table for the common house fly, *Musca domestica*. Gerontologia, 3, 23-26.

Röhme, D. (1981). Evidence for a relationship between longevity of mammalian species and life spans of normal fibroblasts *in vitro* and erythrocytes *in vivo*. Proc.Natl.Acad.Sci.USA, 78, 5009-5013.

Rosenberg, B. and D.A.Juckett (1987). Preliminary studies of a new stochastic human death function involving small integers. Mech.Ageing and Devel., 40, 223-241.

Rosenberg, B., G.Kemeny, L.G.Smith et al. (1973). The kinetics and thermodynamics of death in multicellular organisms. Mech.Ageing and Devel., 2, 275-293.

Ross, M.H. (1961). Length of life and nutrition in the rat. J.Nutr., 75, 197-210.

Rosset, E. (1979). Trwanie zycia ludzkiego. Wyd-wo Polskiej Akad.Nauk, Warszawa (In Polish).

Rotstein, M. (1986). Biochemical studies of aging. Chemical and Engineering News, 64, 26-39.

Rubner, M. (1908). Probleme des Wachstums und der Lebensdauer. In Gesellschaft für Innere Medizin und Kinderheilkunde, vol.7, Mitteilungen, Beiblatt. Wien, 58-72.

Saboulin, M. (1981). Données de démographie régionale, 1975. Coll.INSEE, N D82.

Sacher, G.A. (1956). On the statistical nature of mortality with special reference to chronic radiation mortality. Radiology, 67, 250-258 (with discussion).

Sacher, G.A. (1959). Relation of lifespan to brain weight and body weight. In G.E.W.Wolstenholme and M.O'Connor (eds.) The lifespan of animals. Little, Brown, Boston, 115-141.

Sacher, G.A. (1966). The Gompertz transformation in the study of the injury-mortality relationship: Application to late radiation effects and ageing. In P.J.Lindop and G.A.Sacher (eds.) Radiation and Ageing. Taylor and Francis, London, 411-441.

Sacher, G.A. (1976a). Evaluation of the entropy and information terms governing mammalian longevity. Interdiscipl.Topics Gerontol., 9, 69-82.

Sacher, G.A. (1976b). Longevity in vertebrates: a further comment. Fed.Proc., 35, 1112.

Sacher, G.A. (1977). Life table modification and life prolongation. In C.E.Finch and L.Hayflick Handbook of the biology of aging. Reinhold, NY, London, 151-168.

Sacher G.A. (1978a). Evolution of longevity and survival characteristics in mammals. In E.L.Schneider (ed.) The genetics of aging. Plenum Press, NY, London, 151-168.

Sacher, G.A. (1978b). Longevity and aging in vertebrate evolution. BioScience, 28, 497-501.

Sacher, G.A. and P.H.Duffy (1979). Genetic relation of life span to metabolic rate for inbred mouse strains and their hybrids. Fed.Proc., 38, 184-188.

Sacher, G.A. and R.W.Hart (1978). Longevity, aging and comparative cellular and molecular biology of the house mouse, *Mus musculus*, and the white-footed mouse, *Peromyscus leucopus*. Birth Defects: Original Article Series, 14, 71-96.

Sacher, G.A. and E.Trucco (1962). The stochastic theory of mortality. Ann.N.Y.Acad.Sci., 96, 985-1007.

Salonen, J.T. (1987). Selenium in ischaemic heart disease. Int.J.Epidemiol., 16, 323-328.

Salonen, J.T., G.Alfthan, J.K.Huttunen et al. (1982). Association between cardiovascular death and myocardial infarction and serum selenium in a matched-pair longitudinal study. Lancet, 2, 175-179.

Sathe, T.V. (1986). Life table and intrinsic rate of increase of *Cotesia diurnii* Chalikwar and Rao (Hymenoptera: Braconidae), a larval parasitoid of *Exelastis atomosa* Wals. Entomon, 11, 281-283.

Sauvy, A. (1954). Le vieillissement des populations et l'allongement de la vie. Population, 4.

Sauvy, A. (1961). Les limites de la vie humaine. Hachette, Paris.

Sawada, M. and J.C.Carlson (1987). Association between lipid peroxidation and life-modifying factors in rotifers. J.Gerontol., 42, 451-456.

Schlettwein-Gsell, D. (1970). Survival curves of an old age rat colony. Gerontologia, 16, 111-115.

Schmidt-Nielsen, K. (1984). Scaling. Why is animal size so important? Cambridge Univ.Press, Cambridge etc.

Schneider, E.L. (1978). Cytogenetics of aging. In E.L.Schneider (ed.) The genetics of aging. Plenum Press, NY, London, 27-52.

Schneider, E.L. (1985). Aging research: Challenge of the twenty-first century. In Molecular Biology of Aging, Proc.Symp., Upton, NY, Sept.30-Oct.3, 1984. NY, London, 1-11.

Schneider, E.L. and J.A.Brody (1983). Aging, natural death and the compression of morbidity: Another view. N.Engl.J.Med., 309, 854-856.

Schwartz, A.G. (1975). Correlation between species life span and capacity to activate 7,12-dimethylbenz(a)anthracene to a form mutagenic to a mammalian cell. Exp.Cell Res., 94, 445-447.

Schwartz, A.G. and C.J.Moore (1977). Inverse correlation between species life span and capacity of cultured fibroblasts to bind 7,12-dimethylbenz(a)anthracene to DNA. Exp.Cell.Res., 109, 448-450.

Schwartz, A.G. and C.J.Moore (1979). Inverse correlation between species life span and capacity of cultured

fibroblasts to metabolize polycyclic hydrocarbon carcinogens. Fed.Proc., 38, 1989-1992.
Semyonova, V.G. (1983). Analysis of the life span distribution of laboratory animals. In Problemy biologii stareniya [Problems in the biology of aging]. Nauka, Moscow, 56-62. (In Russian).
Semyonova, V.G. (1989). Izuchenie kolichestvennykh zakonomernostei vliyaniya vozrasta, pola i geograficheskoi sredy na smertnost' v populyatsiyakh cheloveka [A study of qualitative laws for the influence of age, sex and geographical environment on mortality in human populations]. Ph.D. dissertation, Moscow State University, Moscow. (In Russian).
Semyonova, V.G., L.A.Gavrilov and A.M.Gofman (1985). Investigation of the age dynamics of human mortality: Analysis of deviations from the Gompertz-Makeham equation. In Issledovanie pervichnykh mekhanizmov biologicheskikh sistem [Investigation of the primary mechanisms of biological systems]. Nauka, Moscow, 57-60. (In Russian).
Service, P.M., E.W.Hutchinson and M.R.Rose (1988). Multiple genetic mechanisms for the evolution of senescence in *Drosophila melanogaster*. Evolution, 42, 708-716.
Shakhbazov, V.G., A.V.Nekrasova, G.E.Lapta et al. (1980). The influence of the age of the parental generation of *Drosophila melanogaster* on the viability of the descendants and the interaction of homologous chromosomes. Vestnik Khar'kovskogo Universiteta, 195, 42-44. (In Russian).
Shamberger, R.J., M.S.Gunsch, C.E.Willis and L.J.McCormack (1978). Selenium and heart disease. II.Selenium and other trace metal intakes and heart disease in 25 countries. In D.D.Hemphill (ed.) Trace substances in environmental health-XII. Univ.of Missoury., Columbia, 48-52.
Shamberger, R.J., C.E.Willis and L.J.McCormack (1979). Selenium and heart disease. III.Blood selenium and heart mortality in 19 states. In D.D.Hemphill (ed.) Trace substances in environmental health-XIII. Univ.of Missoury, Columbia, 59-63.
Sherman, P.W. and M.L.Morton (1984). Demography of Belding's ground squirrels. Ecology, 65, 1617-1628.
Shiff, C.J. (1964). Studies on *Bulinus (Physopsis) globosus* in Rhodesia. I.The influence of temperature on the intrinsic rate of natural increase. Ann.Trop.Med. Parasitol., 58, 94-105.
Shukailo, V.F. (1979). The principles of the mathematical representation of the essence of mortality processes. In

Prodolzhitel'nost' zhizni: analiz i modelirovanie [Life span: analysis and modelling]. Statistika, Moscow, 104-123. (In Russian).

Sinslair, A.R.E. (1974). The natural regulation of buffalo populations in East Africa. III.Population trends and mortality. E.Afr.Wildlife J., 12, 185-200.

Sivaprakasam, N., G.Balasubramanian, S.Venkatesan et al. (1985). Life-table and intrinsic rate of increase of *Bracon kirkpatricki* Wilkinson on *Corcyra cephalonica* Stainton. Indian J.of Agricult.Sciences, 55, 762-764.

Skurnick, I.D. and G.Kemeny (1978a). Stochastic studies of aging and mortality in multicellular organisms. I.The asymptotic theory. Mech.Ageing and Devel., 7, 65-80.

Skurnick, I.D. and G.Kemeny (1978b). Stochastic studies of aging and mortality in multicellular organisms. II.The finite theory. Mech.Ageing and Devel., 10, 157-172.

Slob, W. and C.Janse (1988). A quantitative method to evaluate the quality of interrupted animal cultures in aging studies. Mech.Ageing and Devel., 42, 275-290.

Sluiter, J.W., P.F.Van Heerdt and J.J.Bezem (1956). Population statistics of the bat *Myotis Mystacinus*, based on the marking-recapture method. Arch.Néerland.de Zool., 12, 64-88.

Smith, G.S., R.L.Walford and M.R.Mickey (1973). Lifespan and incidence of cancer and other diseases in selected long-lived inbred mice and their $F_1$ hybrids. J.Natl.Cancer Inst., 60, 1195-1213.

Social Security Administration. Office of the Actuary. Life tables for the United States, 1900-2050: Actuarial study N 87. Wash: Government Printing Office (1982). SSA Publ. N 11-11534.

Sohal, R.S., I.Svensson, B.H.Sohal and U.T.Brunk (1989). Superoxide anion radical production in different animal species. Mech.Ageing and Develop., 49, 129-135.

Solbrig O.T. (ed.) (1980). Demography and evolution in plant population. Harvard Univ.Press, Harvard.

Solomon, M. (1983). Future life (L'avenir de la vie), transl. by Guy Daniels. Macmillan, NY.

Spector, I.M. (1974). Animal longevity and protein turnover rate. Nature, 249, 66.

Stanley, J.F., D.Pye and A.MacGregor (1975). Comparison of doubling numbers attained by cultured animal cells with life span of species. Nature, 255, 158-159.

Starmer, W.T., W.B.Heed and E.S.Rocwood-Sluss (1977). Extension of longevity in *Drosophila mojavensis* by environmental ethanol: Differences between subraces. Proc.Nat.Acad.Sci.USA., 74, 387-391.

Statist.tidskr., 1959, N 5, s.310; N 10, s.592-593; 1960, N

3, s.202-203; 1961, N 3, s.200-201; 1962, N 2, s.124-125.

Statist.medd.B., 1963, N 18, s.4-5; 1964, N 17, s.2-3; 1965, N 21, s.3-6; 1966, N 20, s.5-6; 1968, N 4, s.3-4; 1969, N 5, s.3-4; 1971, N 13, s.4-9.

Steinfeld, H. (1929). Length of life of *Drosophila melanogaster* under aseptic conditions. Univ.Cal.Publ. Zool., 31, 131-178.

Storer, J.B. (1962). Evaluation of radiation response as an index of aging in mice. Radiat.Res., 17, 878-902.

Storer, J.B. (1965). Radiation resistance with age in normal and irradiated populations of mice. Radiat.Res., 25, 435-459.

Storer, J.B. (1979). Life shortening in RFM and BALB/c mice as a function of radiation quality, dose and dose rate. Radiat.Res., 78. 122-161.

Strehler, B.L. (1961). Studies on the comparative physiology of aging. II.On the mechanism of temperature life-shortening in *Drosophila melanogaster*. J.Gerontol., 16, 2-12.

Strehler, B.L. (1962). Time, cells, and ageing. Acad. Press, NY and London.

Strehler, B.L. (1978). Time, cells, and ageing, 2nd edition. Acad.Press, NY and London.

Strehler, B.L. and A.S.Mildvan (1960). General theory of mortality and aging. Science, 132, 14-21.

Sullivan, J.L. (1982). Superoxide dismutase, longevity and specific metabolic rate. Gerontology, 28, 242-244.

Sunlin, M. (1985). The evolution of the inherent lifespan and its implications in experimental gerontology. Rejuvenation, 13, 52-54.

Sutherland, V. and J.C.Bailar (1984). The multihit model of carcinogenesis: etiologic implications for colon cancer. J.Chron.Dis., 37, 465-480.

Suyama, I. and T.Iwasaki (1976). Radiation-induced life span shortening of Artemia under different temperature conditions. Exp.Gerontol., 11, 133-140.

Swim, H.E. (1959). Microbiological aspects of tissue culture. Annual Review of Microbiology, 13, 141-176.

Swim, H.E. and R.F.Parker (1957). Culture characteristics of human fibroblasts propagated serially. Am.J.Hyg., 66, 235-243.

Szabo, I. and M.Szabo (1929). Lebensdauer, Wachstum und Altern: Studdiert bei der Nachtschneckenart *Agriolimax agrestis* L. Biol.Gen., 7, 95-118.

Taber, R.D. and R.F.Dasmann (1958). The dynamics of three natural populations of the deer *Odocoileus hemionus columbianus*. Ecology, 38, 233-246.

Tablas de mortalidad provinciales (1969-1972 ano 1970) (1978). Inst.Nac.de Estadistica, Madrid.

Tabuas abreviadas de mortalidade distritais e regionais 1959-1962 e 1969-1972 (1976). Centro de Estudos Demographicos, Lisboa. Caderno N 4.

Tantawy, A.O. and M.R.El-Helw (1970). Studies of natural populations of *Drosophila*. IX.Some fitness components and their heritabilities in natural and mutant populations of *Drosophila melanogaster*. Genetics (US), 64, 79-91.

Tarkhanov, I.R. (1891). Longevity of animals, plants and people. Vestnik Evropy, 3, No.6, 538-552; No.9, 87-125. (In Russian).

Tatner, P. (1986). Survival rates of urban magpies. Ringing and Migration, 7, 112-118.

Thatcher, A.R. (1982). The span of life. Abstract of the discussion. J.Inst.Actuaries, 109, 346-349.

The 15th life tables (1984). Statist.and Inform.Dep., Minister's Secretariat, Ministry of Health and Welfare, Tokyo.

The Guinness Book of Records 1990 (1989). D.McFarlan (ed.). Guinness Publishing Ltd., Barcelona.

The New Encyclopaedia Britannica (1989), 15th edition. Encyclopaedia Britannica Inc., Chicago. In 32 volumes.

Thomson, J.F. and D.Grahn (1988). Life shortening in mice exposed to fission neutrons and $\gamma$ rays. VII.Effects of 60 once-weekly exposures. Radiat.Res., 115, 347-360.

Thomson, J.F., F.S.Williamson and D.Grahn (1981). Life shortening in mice exposed to fission neutrons and X-rays. I.Single and short-term fractionated exposures. Radiat.Res., 86, 559-572.

Thomson, J.F., F.S.Williamson and D.Grahn (1985a). Life shortening in mice exposed to fission neutrons and $\gamma$ rays. IV.Further studies with fractionated neutron exposures. Radiat.Res., 104, 420-428.

Thomson, J.F., F.S.Williamson and D.Grahn (1985b). Life shortening in mice exposed to fission neutrons and $\gamma$ rays. IV.Further studies with single low doses. Radiat. Res., 103, 77-88.

Timofeeff-Ressovsky, N.W. and K.G.Zimmer (1947). Biophysik. I.Das Trefferprinzip in der Biologie. Leipzig.

Timofeev-Ressovsky, N.W., V.I.Ivanov and V.I.Korogdin (1968). Primenenie printsipa popadaniya v radiobiologii [Application of the hit principle in radiobiology]. Atomizdat, Moscow. (In Russian).

Tinkl, D.W. (1967). The life and demography of the sideblotched lizard, *Uta stansburiana*. Misc.Publ.Mus. Zool.Univ.Mich., 132, 1-182.

Tolmasoff, J.M., T.Ono and R.G.Cutler (1980). Superoxide dismutase: Correlation with life-span and specific metabolic rate in primate species. Proc.Natl.Acad.Sci. USA, 77, 2777-2781.
Tréton, J.A. and Courtois (1982). Correlation between DNA excision repair and mammalian lifespan in lens epithelial cells. Cell Biol.Intern.Reports, 6, 253-260.
Tréton, J.A. and Courtois (1989). Evidence for a relationship between longevity of mammalian species and a lens growth parameter. Gerontology, 35, 88-94.
Uberla, K. (1977). Faktorenanalyse. Springer-Verlag, Berlin, Heidelberg, NY.
Unlü, H. and A.N.Bozcuk (1979). Genetics of longevity in *Drosophila*. I.The effects of w, m and f mutant genes in various genotype combinations. Exp.Gerontol., 14, 117-124.
Urlanis, B.Ts. (1978). Evolyutsiya prodolzhitel'nosti zhizni [Evolution of life span]. Statistika, Moscow. (In Russian).
Van Herrewege, J. and J.David (1970). Reduction of longevity in *Drosophila* by chronic injection of a bacterial toxin. Exp.Gerontol., 14, 117-124.
Van Leeuwen, C.J., W.J.Luttmer and P.S.Griffioen (1985). The use of cohorts and populations in chronic toxicity studies with *Daphnia magna*: a cadmium example. Ecotoxicology and Environ.Safety, 9, 26-39.
Vanyushin, B.F. and G.D.Berdyshev (1977). Molekulyarno-geneticheskie mekhanizmy stareniya [Molecular-genetic mechanisms of aging]. Meditsina, Moscow. (In Russian).
Vaupel, J.W., K.G.Manton and E.Stallard (1979). The impact of heterogeneity in individual frailty on the dynamics of mortality. Demography, 16, 439-454.
Vellisto, I.E. and F.B.Bang (1968). Effect of differential parity through five generations on the incidence of tumors in C3Hf mice. J.Natl.Cancer Inst., 40, 1213-1221.
Vetukhiv, M. (1957). Longevity of hybrids between geographic populations of *Drosophila pseudoobscura*. Evolution, 11, 348-360.
Vinegar, M.B. (1975). Demography of the striped plateau lizard *Sceleporus virgatus*. Ecology, 56, 172-182.
Vishnevsky, A.G. (1982). Vosproizvodstvo naseleniya i obshchestvo [Population reproduction and society]. Finansy i Statistika, Moscow. (In Russian).
Voitenko, V.P. (1987). Sex differences in aging and mortality. In E.B.Burlakova and L.A.Gavrilov (eds.) Populyatsionnaya gerontologiya [Population gerontology]. VINITI, Moscow, 6, 64-105. (In Russian).
Von Ardenne (1985). Stimulation of the defence mechanism

against diseases and the reduction of their perilousness in old age as actual ways to approach the biological limit of human life. Rejuvenation, 13, 66-78.

Von Hahn, H.P. (1983). The biological aging process. Experientia, 39, 47-49.

Wainscoat, J.S., T.E.A.Peto and A.Waswo (1987). HLA genes and longevity. Lancet, 2, 1399.

Waldron, I. (1986). What do we know about causes of sex differences in mortality? A review of the literature. In Population Bulletin of the United Nations, 18, 1985. United Nations, NY.

Waldron, I. (1987). Causes of the sex differential in longevity. J.Am.Geriatrics.Soc., 35, 365-366.

Walford, R.L. (1983). Maximum life span. Norton, NY, London.

Walford, R.L. (1986). The 120-year diet. Simon and Schuster, NY.

Walford, R.L., G.C.Smith, P.L.Meredith et al. (1978). Immunogenetics of aging. In E.L.Schneider (ed.) The genetics of aging. Plenum Press, NY, London, 383-401.

Walton, J. (1982). The role of limited cell replicative capacity in pathological age change. A review. Mech.Ageing Devel., 19, 217-244.

Waters, W.E. (1969). The life table approach to analysis of insect impact. J.Forestry, 67, 300-304.

Weibull, W.A. (1951). A statistical distribution function of wide applicability. J.Appl.Mech., 18, 293-297.

Weismann, A. (1882). Uber die Dauer des Lebens. Jena.

Weismann, A. (1884). Uber Leben und Tod. Jena.

Weismann, A. (1892). Uber Leben und Tod. Verlag von Gustav Fisher, Jena.

Weismann, A. (1914). On life and death (1884). In V.A.Vagner and E.A.Shul'ts (eds.) Smert' i bessmertie. (Novye idei v biologii; sb.3) [Death and immortality. (New ideas in biology; vol.3)], No.1, 1-66. (Russian translation).

Weisner, B.P. and N.M.Sheard (1935). The duration of life in an albino rat population. Proc.Roy.Soc.Edinburgh, 55, 1-22.

Werkö, L. (1987). The enigma of coronary heart disease and its prevention. Acta Med.Scand., 221, 323-333.

Williams, G.C. (1957). Pleiotropy, natural selection, and the evolution of senescence. Evolution, 11, 398-411.

Williams, W.P. (1967). The growth and mortality of four species of fish in the river Thames at Reading. J.Anim.Ecology, 36, 695-720.

Winsor, C.P. and A.A.Winsor (1935). Longevity and fertility in the pond snail, *Lymnaea columella*. J.Wash.Acad.Sci., 25, 302-307.

Witkowski, J.A. (1987). Cell aging in vitro: A historical

perspective. Exp.Gerontol., 22, 231-248.
Witten, M. (1985). A return to time, cells, systems, and aging: III. Gompertzian models of biological aging and some possible roles for critical elements. Mech.Ageing Devel., 32, 141-177.
Witten, M. (1986). A return to time, cells, systems, and aging: IV. Further thoughts on Gompertzian survival dynamics - the neonatal years. Mech.Ageing Devel., 33, 177-190.
Woodbury, M.A. and K.J.Manton (1983). A theoretical model of the physiological dynamics of circulatory disease in human populations. Hum.Biol., 55, 417-441.
Woodhead, A.D. and K.H.Thompson (eds.) (1987). Evolution of longevity in anaimals: A comparative approach. Plenum Press, NY.
Woodruffe, G.E. (1951). A life-history study of the brown house moth *Hofmanophila pseudosprella* (Saint.) (Lep., Oecophoridae). Bull.Res., 41, 529-553.
World Health Statistics Annual. 1968-1987. World Health Organization, Geneva.
Wright, A.W., G.H.Klinck and J.M.Wolfe (1940). The pathology and pathogenesis of mammary tumors occuring spontaneously in the Albany strain of rats. Amer.J.Pathol., 16, 817-834.
Yazan, Yu.P. (1961). Biological characteristics and methods of economic exploitation of a population of migratory elks in the Pechora taiga. Trudy Pechoro-Ilychskogo Zapov., 9, 114-201. (In Russian).
Young, F.E., S.L.Nightingale and R.A.Temple (1988). The preliminary report of the findings of the aspirin component of the ongoing Physician's Health Study. J.Am.Med.Assoc., 255, 3158-3160.
Young, O.P. (1985). Survival of a carrion beetle, *Necrodes surinamensis* (Coleoptera: Silphidae), on a diet of dead fall armyworm (Lepidoptera: Noctuidae) larvae. J.Entomol.Sci., 20, 359-366.
Yu, B.P., S.Langaniere and J.-W. Kim (1989). Influence of life-prolonging food restriction on membrane lipoperoxidation and antioxidant status. In M.G.Simic, K.A.Taylor, J.F.Ward and C.von Sonntag (eds.). Oxygen radicals in biology and medicine. Plenum Press, NY etc.
Yuhas, J.M. and N.K.Clapp (1972). Incidence of leukemia and nonlymphatic tumors in mice with glomerulosclerosis and allied disorders. J.Natl.Cancer Inst., 48, 367-373.
Yčas, M. (1969). The biological code. North-Holland Publ.Co., Amsterdam, London.
Zammuto, R.M. (1986). Life histories of birds: clutch size. longevity, and body mass among North American game

birds. Can.J.Zool., 64, 2739-2749.

Zammuto, R.M. (1987). Life histories of mammals: analyses among and within *Spermophilus columbianus* life tables. Ecology, 68, 1351-1363.

Zammuto, R.M. and P.W.Sherman (1986). A comparison of time-specific and cohort-specific life tables for Belding's ground squirrels, *Spermophilus beldingi*. Can.J.Zool., 64, 602-605.

Zepelin, H. and A.Rechtschaffen (1974). Mammalian sleep, longevity and energy metabolism. Brain, Behav. and Evol., 10, 425-470.

Zhivotovsky, L.A. (1984). Integratsiya poligennykh sistem v populyatsiyakh [Integration of polygenic systems in populations]. Nauka, Moscow. (In Russian).

Zhirmunsky, A.V. and V.I.Kuz'min (1980). Critical periods in human ontogenesis and maximum life span. Dokl.Akad.Nauk. SSSR, 244, 251-253. (In Russian).

Zubov, A.A. and V.I.Kozlov (1982). A composite approach to studying the phenomenon of long life. In Fenomen dolgozhitel'stva [The phenomenon of long life]. Nauka, Moscow, 13-24. (In Russian).

## SUBJECT INDEX